EXPLORING THE HIGHEST SIERRA

EXPLORING THE

HIGHEST SIERRA

JAMES G. MOORE

STANFORD UNIVERSITY PRESS
STANFORD, CALIFORNIA

Stanford University Press

Stanford, California

© 2000 by the Board of Trustees of the

Leland Stanford Junior University

Printed in the United States of America

CIP data appear at the end of the book

FRONTISPIECE: It was precisely this view that awed William Brewer and Charles Hoffmann in the summer of 1864, when they made the first ascent of Mount Brewer and gazed eastward. They had assumed that the mountain was on the Sierra Crest and were dismayed to discover that the main crest in fact lay an additional eight miles farther east. They named pyramidal Mount Tyndall (far left) and helmet-shaped Mount Whitney (right skyline), which they measured as the highest peak in the range. Clarence King—an unpaid field assistant—felt challenged by this unknown terrain and begged Brewer for permission to climb Mount Whitney. Five days later he and the drover Dick Cotter returned; they had climbed Mount Tyndall with difficulty but judged Mount Whitney too far to attempt.

FOR THOSE WHO MAPPED THE WAY SO OTHERS COULD FOLLOW,

AND FOR KAREN, WHO HELPED MORE THAN SHE WILL EVER KNOW.

Contents

Preface ix

1 INTRODUCTION 1

2 EXPLORATION 21

Probing the High Sierra: Fremont and Carson 23   The Pacific Railroad Surveys: Blake and Williamson 30   Breaking New Ground: Goddard and the Trail Blazers 34   The California Geological Survey: Whitney, Brewer, King, and Hoffmann 39   Exalting the Wilderness: John Muir 68   Creating and Enlarging the National Parks 88

3 MAPPING 97

The Territorial Surveys: King and Wheeler 98   The Scaling of Mount Whitney: Errors Compounded 118   Measuring the Ground: The General Land Office 122   Mapping the High Trails: LeConte, Solomons, and the Sierra Club 128   Detailed Topographic Mapping: The U.S. Geological Survey 142   Peaks, Valleys, and Rivers: The Basis of Place Names 161   The Structures of the Land: Geologic Mapping 165

4 TIME, MINERALS, ROCKS, AND PLATES 171

The Geologic Time Scale 171   Dating the Rocks: Radiometric Methods 175   Building Blocks of Rocks: The Common Minerals 176   Solidified Magma: The Igneous Rocks 179   The Global View: Plate Tectonics and the Rise of the Sierra 185

5 METAMORPHIC ROCKS 189

The Four Terranes of Metamorphic Rock 192   Rare Fossils: Key to the Age of Metamorphic Rocks 199   Marble: The Birthplace of Caves 202   The Origin and Assembly of the Metamorphic Terranes 206

6 GRANITIC ROCKS 209

Plutons and the Sierra Nevada Batholith 209   Making Room for the Granite 219   Structures of Granitic Rocks 221   Field Criteria for the Relative Ages of Plutons 232   The Age of the Sierra Nevada Batholith 234   The Depth and Thickness of the Batholith 237   Analyzing the Granitic Rocks 239   Changes in Composition Across the Batholith 249   Mafic Plutonic Rocks 254   The Independence Dike Swarm 256   Plate Tectonics and the Batholith 259

7 MINERAL DEPOSITS 263

The Kearsarge District 265   The Mineral King Bust 267   Copper and Molybdenum Deposits 269   The Tungsten Mines 270

8 CENOZOIC VOLCANIC ROCKS 273

The Initial Volcanic Phase 275   The Second Volcanic Phase 277   The Third Volcanic Phase 277

9 GLACIERS AND GLACIATION 285

The Extent of Glaciation in the High Country 289   Glacial Processes and Their Products 299   The Causes and Timing of Glaciation 310

10 LANDSLIDES 315

11 GEOLOGIC STRUCTURES 319

The Crustal Structure of the Sierra 319   Faults in and near the Sierra Block 322   The Nature of Joints 334   Uplift and Erosion 343

12 AFTERWORD 349

APPENDIX: GEOLOGIC ROAD AND TRAIL GUIDES 357

Kings Canyon Highway, State Route 180 360   State Route 198 and the Generals Highway 370   The Mineral King Road 377   The John Muir Trail 380   The High Sierra Trail 388

Glossary 395

References Cited 409

Index 421

Preface

I first ventured into the highest Sierra in the summer of 1947, with my brother George, who shares my interest in the outdoors and would also become a geologist. We took a Greyhound bus to the town of Lone Pine, in the shadow of the eastern Sierra, and set out backpacking up the Whitney Portal Road. Our plan was to climb Mount Whitney and from there cross the Sierra and descend into Kings Canyon. We were not well prepared and had no idea of the difficulty of the trip, but with the brashness of youth we welcomed the challenge. I was 17, and George, 19.

Our packs were laden with a mix of Boy Scout and army surplus equipment, and a good deal less food than we would want. Modern lightweight gear and dehydrated backpack rations were not yet to be had, and we carried our gear roped onto packboards we had made from segments of old wooden bicycle wheels, using burlap straps cut from potato sacks. Our army surplus pup tent was shaped from two ponchos snapped together and suspended on cord we would tie between two trees or piles of rock. The cooking pot was a large tin can with a baling wire handle, and at those high elevations the cooking was not always what we had expected. I remember waiting patiently for dinner one night while the split peas boiled—but never got soft. We gave up, built a big fire, and had them for breakfast the next morning.

Nearly a week later, tired, hungry, but exuberant, we emerged from the highest Sierra at Cedar Grove, 60 miles from our starting point, with a profound appreciation for the immensity of the range, and with a yearning to return. Now, some 50 years later—usually when some adversity confronts me—I envision George slowly but resolutely marching up the trail in wind-driven sleet toward the crest of 13,000-foot Foresters Pass. His pack-laden silhouette is hunched into the wind and the army poncho is whipping and flapping uselessly in the gale-force wind of a summer storm.

Partly as a result of that first trip, I became inclined toward a career in natural science, and while still an undergraduate at Stanford University I came under the influence of geology professor Aaron Waters, a devoted teacher who would shape my life and career. He had grown up in Washington State, had gone to Yale, and had studied the igneous rocks of Scotland as a Guggenheim Fellow. He was interested in the volcanic geology of the Pacific Northwest, and came to be recognized as a world authority on the subject. I remember his richly informative lectures, given from memory, lectures in which he would survey a subject and circumscribe its limits of knowledge. In defining that which was unknown, he awakened us students to the possibility that we, too, might advance the frontiers of knowledge. The unrelenting challenge that he posed was a force that excited my interest.

For three summers at Stanford I worked as a field assistant with the U.S. Geological Survey, under the stimulating supervision of Don White and George Thompson. We mapped the geology of a region that included Mount Rose in the Sierra Nevada, east of Lake Tahoe, as well as Steamboat Springs and the Comstock Lode, in the Nevada desert to the east. During these summers I learned to locate myself on the map, identify different rock types, measure geologic structures, and compile all these data into a geologic map.

The geology of the Mount Rose area became the subject of my thesis for a Master's Degree at the University of Washington. There I worked under the tutelage of Professor Peter Misch, who evinced great enthusiasm for mountains, geology, and his students. He had grown up in Germany and studied under V. M. Goldschmidt at the University of Göttingen. In 1934 he joined the German Nanga Parbat expedition to the Himalayas as geologist, but when he returned to Germany, he found that university positions were closed to him because of his Jewish ancestry. Eventually, he took a job at Sun Yat-sen University, Canton, China. When the Japanese invaded China, he moved on to Kunming, far to the southwest, and spent the war years there, teaching and mapping geology in the mountains. With World War II over, he accepted a post in Seattle as professor of geology at the University of Washington, and began a geologic mapping program in the high mountains of the northern Cascades. Professor Misch was always available for consultation with his graduate students, and he excelled in leading geologic field trips.

I then transferred to graduate school at Johns Hopkins University and rejoined my former professor, Aaron Waters, who had left Stanford for Hopkins. The Hopkins experience was enriched by Ernst Cloos, professor of structural geology, who, in the 1930s, studied the granitic and metamorphic rocks of the Sierra Nevada. The area I chose for field study leading to a Ph.D. degree was in the highest Sierra, north of Mount Whitney, reaching from University Peak to the Palisades on the main Sierra Crest. Preliminary topographic maps of the Mount Pinchot Quadrangle at the enlarged 15-minute scale, which had just been released, provided a new base ideally suited for geologic mapping. I spent several years in field and laboratory study of this area, all of which is within the region covered by this book. Professor Waters's style had not changed; he eventually taught and inspired so many of us that for several decades the field of volcanic geology in this country (and the world) came to be dominated by his students. I will be eternally grateful to him for one favor in particular. Before my oral examinations for the Ph.D. degree at Johns Hopkins, he asked me what question I would like him to ask me first. I suggested a question to him, prepared the material carefully, and was able to answer well when, indeed, he asked it first off. With the pattern established, and my anxiety laid to rest, the lengthy grilling from him and the other committee members went fairly smoothly and I passed the examination and later was awarded the degree.

Once I had finished at Hopkins, student deferments from the draft were no longer forthcoming, and induction into military service was inevitable. Following my discharge from the U.S. Army two years later, I applied to, and was hired by, the U.S. Geological Survey. My first chief assignment was to work with Paul Bateman in a program of his design directed toward determining the geologic framework of the central Sierra Nevada and its relation to mineral deposits, particularly tungsten. Bateman had begun his work on mineral deposits in the Bishop District in the central Sierra Nevada in 1943, and his work there had led to his doctoral thesis from the University of California at Los Angeles. He became the architect of a strategy to map the geology of a strip 60 miles wide and 100 miles long across the central Sierra Nevada, from the foothills on the west to the White-Inyo Mountains east of the range. The recent completion of a new series of 15-minute topographic quadrangles (one mile to the inch) permitted much more detailed mapping than had previously been possible. Bateman collected closely spaced samples of granitic rocks and from their analyses he could detect small changes in composition across the region, changes that were undetectable by field observation.

The final results of this comprehensive work were published in 1992 when Paul was 82.

My first major assignment was to complete the mapping of the Mount Pinchot Quadrangle, which included the area of my dissertation. As this work evolved, it was decided that I would head up work on a second east-trending strip across the Sierra, bounded on the north by Paul's mapping. This strip included three tiers of 15-minute quadrangles south of the 38th parallel. Paul and I collaborated on the mapping of the Mount Goddard Quadrangle, which fell in his area, since part of it included remote regions most accessible from my center of operations to the south.

Cliff Hopson, a close friend, had been a fellow student and member of the Stanford Alpine Club. As I had, he went on to the University of Washington and Johns Hopkins after Stanford. Cliff worked with me during the early years of mapping in the highest Sierra Nevada for the Survey. The field days were not long enough for Cliff; generally he would become so carried away with his examination of that last outcrop of the day that he would return late and we would eat dinner in the dark.

My work on the Sierra geologic mapping program was interrupted during a two-and-one-half-year assignment to the Hawaiian Volcano Observatory, where I served as scientist in charge. Other administrative duties followed, but there was time for field studies of 18 ongoing volcanic eruptions, not only in Hawaii, but also in Iceland, the Philippines, Italy, Japan, and Washington State. The Hawaiian work had led into investigations of the submarine aspects of volcanic activity, and my participation in several dozen oceanographic cruises. But notwithstanding these interruptions, work on the Sierra Nevada mapping program continued.

The Sierra work was given a fresh stimulus when Tom Sisson, a recent Stanford graduate, joined the project as field assistant in the summer of 1978. Tom, a lean and avid mountaineer, quickly proved his skill in traveling quickly over rugged terrain, and in mapping the geology accurately along the way. Tom's work in the Sierra was interrupted by his return to graduate studies. He later received his Master's Degree from the University of California at Santa Barbara under Professor Cliff Hopson and went on to Massachusetts Institute of Technology, where he completed his graduate work. For his Ph.D. dissertation he unraveled the geologic history of an area of dark plutonic rocks in the Mount Pinchot Quadrangle by employing detailed mapping and laboratory study.

Several other geologists were associated with the geologic mapping of the highest Sierra. They included Dwight Crowder, Frank Dodge, Ed DuBray, Jon Fink, Fraser Goff, Jack Lockwood, Warren Nokleberg, and

Clyde Wahrhaftig. The Sierra work led to the publication of nine 15-minute geologic quadrangle maps at a scale of 1 mile to the inch. We also did reconnaissance mapping in parts of five other quadrangles to fill out the general map of Sequoia and Kings Canyon National Parks and vicinity. This map covers the range from the Central Valley on the west, over the Sierra Crest, down the eastern escarpment, and into Owens Valley. The mapping of each quadrangle, which includes about 240 square miles, required foot traverses along most ridges and streams and much of the terrain between. That we came upon the previously undiscovered remains of three downed airplanes, found during our field work, indicates the remoteness of some of the country covered. Back-country camps were supplied by pack train, helicopter, and backpacking. In all, I spent more than six cumulative years in the field, two and one-half years of it camping out and sleeping on the ground. The mapping of the deepest part of the Kings River Canyon required 85 helicopter flights.

The daily problems that the work posed have faded in light of the wilderness experiences that we enjoyed. The beauty of this high rocky land, the excitement of fitting together the pieces of the geologic puzzle, the new vistas when attaining the crest of a ridge or peak, and the contentment we felt when relaxing and inking maps before the campfire after a hard day—all these made this life of doing science in the mountains a deeply memorable experience.

During the course of my career, and during the research and writing of this book, I have benefited from my affiliation with the U.S. Geological Survey, and with the scientific expertise that is the backbone of the agency. Help and counsel have been freely given by many scientific colleagues at the survey, including Paul Bateman, plutonic geology; John Galloway, archaeology; John Tinsley, speleology; George Moore, plate tectonics; Malcolm Clark and the late Clyde Wahrhaftig, Pleistocene geology; Warren Nokleberg, geologic structure; William Johnson, topographic mapping; and William Glen, book organization and writing style. Those helping with computer drafting include Michael Diggles, David Lewis, Susan Mayfield, George Moore, Steven Scott, Thomas Sisson, Heidi Stauffer, and Danielle Turpin.

Officials of Sequoia and Kings Canyon National Parks have been unfailingly helpful in supporting the field work on which much of this book is based. David Graber, senior scientist at the parks, remains a constant source of information and help, and Bill Tweed provided valuable comments on an early version of the manuscript.

Several libraries, through the efforts of their most helpful personnel, have loaned printed material and located photographs and other illustrative material. They include the Sequoia and Kings Canyon National Park Museum and Photograph Archives, Melanie Ruesch, photograph librarian; Sierra Club Library, Ellen Byrne, librarian; U.S. Geological Survey Menlo Park Library, Nancy Blair and Anna Tellez, librarians; U.S. Geological Survey Denver Library, Joseph McGregor, photograph curator; Stanford University Library, John Rawlings, reference librarian; Tulare County Library; U.S. Museum of Natural History, Smithsonian Institution; Bancroft Library, University of California; the John Muir National Historic Site; and Holt-Atherton Department of Special Collections, University of the Pacific Libraries. Of particular value has been access to two personal libraries: the Warren Yeend library of early California history, and the Nicholas Clinch mountaineering library.

In preparing the manuscript, every effort has been made to eliminate technical terms and jargon, so that the reader does not bog down. Some unfamiliar words are of course necessary. For example, granite, the most abundant rock in the Sierra Nevada is composed of quartz and two feldspars: *orthoclase* and *plagioclase*, and needs to be distinguished from two related rock types that differ slightly: *granodiorite* and *gabbro*. Already, then, we require four unfamiliar terms. But virtually all words that are not in common usage are included in the Glossary, so that the reader need not go to another source to learn about them. Similarly, the book makes use almost entirely of the English system of measurement, though I envision furrowed brows amongst my scientist colleagues for pursuing this practice. In the United States, the use of miles, feet, and pounds remains the overwhelmingly dominant measurement system of the land.

Some time ago, it became U.S. Geological Survey policy to eliminate apostrophes in place names on maps (for example, Bubbs Creek, not Bubb's Creek); because the book reproduces salient portions of U.S.G.S. maps, and because the text so often refers to places shown on these maps, I have adhered to the policy in text discussions as well.

Apart from references to the big trees of Sequoia National Park, I have made no attempt to describe or discuss the wildlife of the Sierra—another whole subject, well treated in field guides and other works.

Diagrams and photographs have been credited to their original source. Those illustrations not credited were prepared or photographed by the author.

The manuscript has been substantially improved by the reviews of colleagues and associates who have read all or part of it. I thank William Glen, George Moore, Kurt Servos, Tom Sisson, and William Tweed for their readings of the manuscript and for their many constructive critical comments. The exhaustive editorial comments of William Carver, in retirement from Stanford University Press, have certainly improved the manuscript, and for his work I am grateful.

J.G.M.

EXPLORING THE HIGHEST SIERRA

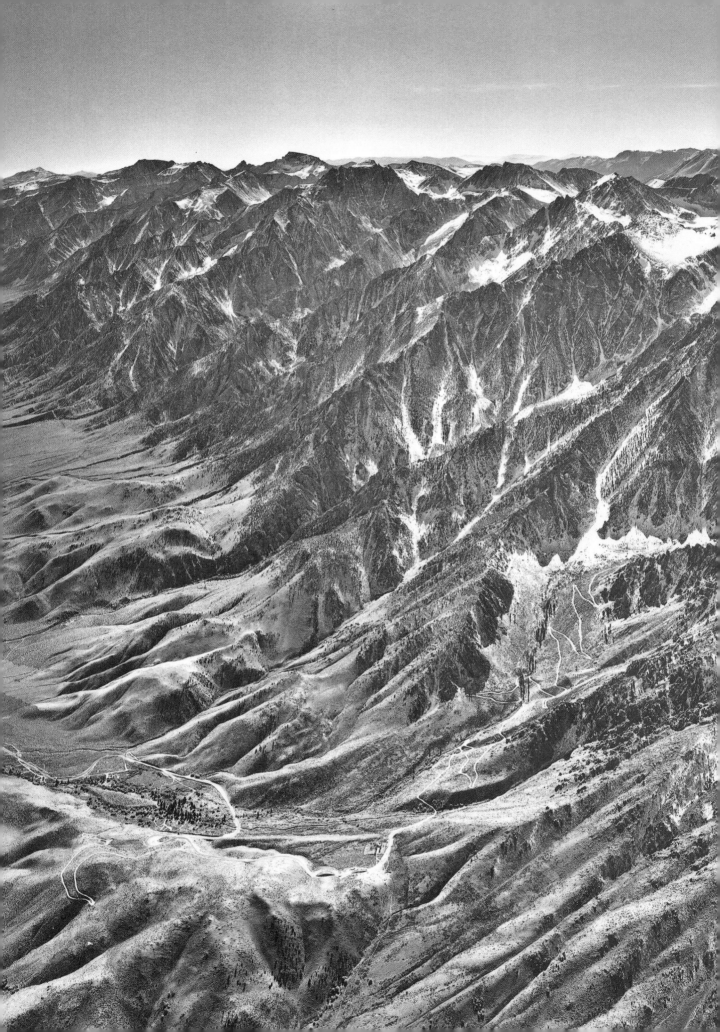

> And after ten years spent in the heart of it, . . . seeing the sunbursts of morning among the icy peaks, the noonday radiance on the trees and rocks and snow, the flush of the alpenglow, and a thousand dashing waterfalls with their marvelous abundance of irised spray, it still seems to be above all others the Range of Light.
>
> —JOHN MUIR (1838–1914) ON THE SIERRA NEVADA

INTRODUCTION

Outside of Alaska, the Sierra Nevada is the highest and most continuous mountain range in the United States. It has always been a major barrier to travel, and it remained unexplored until late in the years of westward expansion and settlement. The highest peak in the range was not identified until the time of the American Civil War, and it was not climbed until nearly a decade later, in 1873. The most forbidding part of these mountains, the southern Sierra in the region of the Sequoia and Kings Canyon National Parks, was especially avoided, because of its extreme height and rugged terrain. The towering east face of that part of the range, extending more than 100 miles in an unbroken wall from Yosemite to Mount Whitney (Fig. 1.1), was humbling to early travelers. To gain access to the inhabited parts of California they had no choice but to divert their route to the north or the south.

1.1. Oblique aerial view of the eastern escarpment of the Sierra Nevada, extending 25 miles from Independence Creek (along road in foreground) to south of Lone Pine. Mount Whitney (14,491 feet) is the high peak on the center skyline. The Independence Fault, a branch of the Sierra Nevada Frontal Fault, follows prominent depressions in five ridge crests near the base of the escarpment. U.S. Geological Survey oblique aerial photograph taken in 1956.

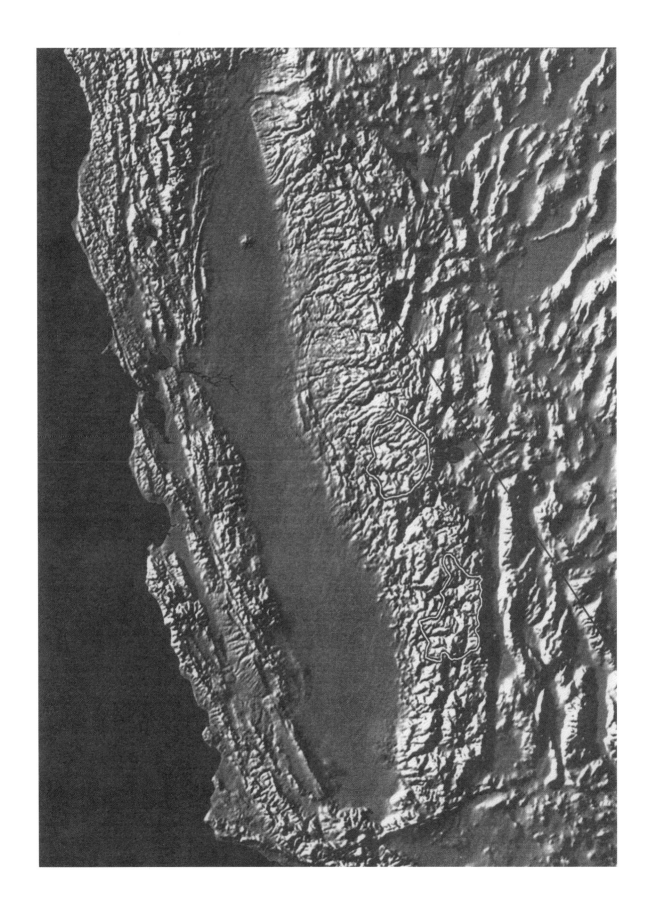

INTRODUCTION

The Sierra Nevada extends for much of the length of the state of California (Fig. 1.2). It is 450 miles long, reaching from the Mojave Desert on the south to the vicinity of Mount Lassen on the north, and it is 40 to 70 miles wide. Flanking it for most of its length, on the west side, is the great Central Valley of California. On the east it is bordered by the vast Basin and Range Province, composed of dozens of small north-trending mountain ranges separated by fault-bounded valleys, stretching across Nevada well into Utah. Close by, on the east, the southern part of the Sierra is flanked for about 100 miles by a particularly long fault-bounded basin, Owens Valley.

This book is concerned with the highest part of the Sierra Nevada, the area lying around and within Sequoia and Kings Canyon National Parks, near the southern part of the range (Fig. 1.3). Included in this area, adjacent to the parks on the east, is the grand eastern escarpment, towering two miles above the floor of Owens Valley. Here the crest of the range includes all thirteen of the Sierra peaks that exceed 14,000 feet in height (Fig. 1.4). Elsewhere in California, only Mount Shasta, a volcanic peak isolated near the northern end of the Central Valley, and White Mountain Peak, in the first desert range east of the Sierra, exceed 14,000 feet.

No roads cross this segment of the Sierra Nevada, and all of the passes—the lowest parts of the crest—exceed 11,000 feet in elevation. In normal years these passes are free of snow and passable on foot or horse for only four or five months, from June or July to October. For those hardy souls who first attempted to explore the range, finding the few routes that led across the crest was a prodigious feat, and mapping them was crucial for all who followed.

But exploration and mapping, whether in mountain ranges or archipelagoes, have always been inherently linked. The paramount records by which explorers demonstrate where they have gone are the maps they make. Without them, one cannot define for the next explorers the limits between the known and the unknown. With them, both exploring and mapping can shift their focus from reconnaissance to systematic study—so as to more precisely locate the mountains and the rivercourses, and to fill in the rest of the details.

The first exploration and mapping efforts in the Sierra were based on a single track or traverse across the unknown country, the terra incognita, and encompassed only that terrain that could be seen from the track. Soon came more detailed work—pursuing other traverses and linking them with the first—so as to fill in the blank parts of the map. The early maps were generally based on dead reckoning, determining locations sim-

1.2. Shaded-relief map of parts of California and Nevada, showing the west-tilted block of the Sierra Nevada, which separates the flat Central Valley on the west from the smaller, north-trending ranges of the Basin and Range Province on the east. The outlines of Yosemite National Park (center) and Sequoia and Kings Canyon National Parks (to the south) are shown in white. The steep unbroken eastern escarpment of the southern half of the range discouraged cross-range travel. U.S. Geological Survey image compiled by G. P. Thelin and R. J. Pike (1991).

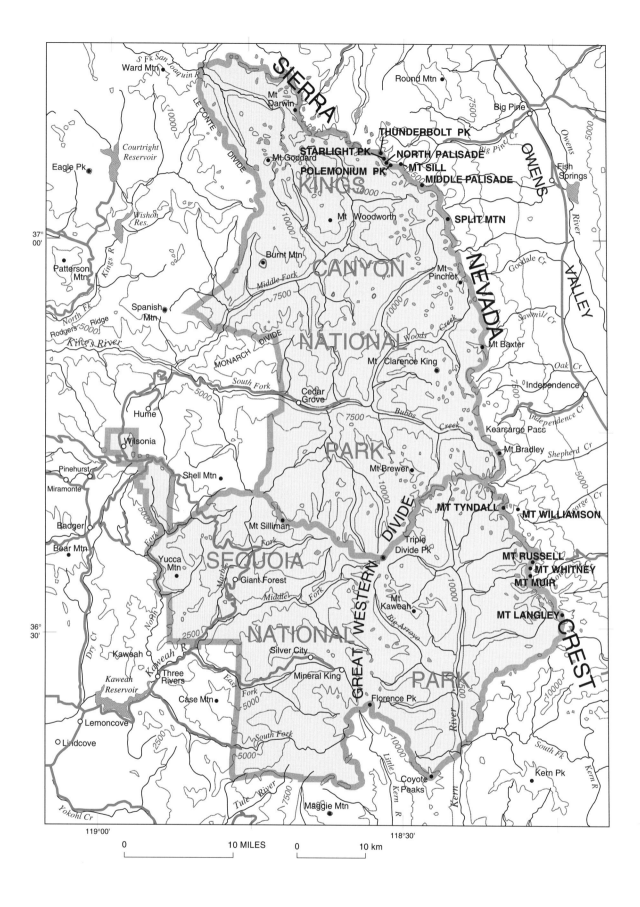

ply by estimating the distance and direction traveled from a given place. Next came the use of simple instruments to establish within the mapped area the approximate position of the various points relative to one another. Finally, the explorer-surveyors employed astronomical transits and chronometers to lend a measure of precision to the previous determinations and to place the mapped area in its proper position on the globe—within the worldwide grid of parallels of latitude and converging meridians of longitude.

Geology, which records the physical aspects of the land and defines earth materials that may be of use to people, was linked with exploration and mapping almost from the start. Distinguishing the major rock formations was essential in describing the terrain and identifying its landmarks. The earliest maps defined sandy deserts, lava flows, and dark- versus light-colored mountains, and many of the earliest expeditions were commissioned to examine the terrain for other geologic features, such as soil characteristics, water quantity and quality, avenues for transportation routes, and mineral deposits.

It was not until the middle 1860s that serious exploration of the highest Sierra began. In 1864 the nascent California Geological Survey mounted an expedition, led by William Brewer, charged with mapping this terra incognita and thus filling in one of the largest blank spaces on the map of California. It sought to determine the nature of the geologic structure of the southern Sierra, with an eye toward discovering whether the mineral deposits of the central and northern Sierra Nevada, which had spawned the gold rush, continued here. The work of the Survey thus laid the groundwork for later topographic mapping and geologic study.

1.3. *(Opposite)* Map of Sequoia and Kings Canyon National Parks and vicinity. All of the Sierra Nevada's thirteen mountains greater than 14,000 feet in height occur on this segment of the range crest; they are indicated by capital letters. Contour interval 2,500 feet.

1.4. Altitudes of the peaks and passes along 170 miles of the Sierra Crest from Olancha Peak to Sonora Pass. The passes are indicated by circles. The 69-mile scale represents the north-south span of the parks area, as shown in Fig. 1.3.

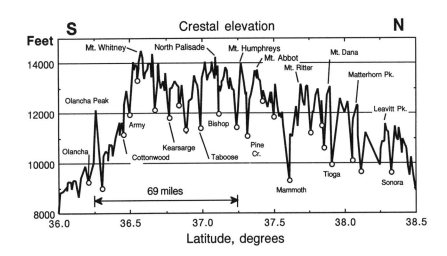

Other institutions and individuals, public and private, built on the early work of the California Geological Survey. The Geographical Surveys West of the 100th Meridian, conducted by the U.S. Army in 1871 and 1875 under the leadership of Lieutenant George Wheeler, investigated part of the east escarpment of the range and made the first detailed surveys of the height of Mount Whitney, tallest of all. The visionary explorer John Muir, a founder of the Sierra Club, made several excursions of discovery into the region in the period 1873–77. And in fact, it was members of the Sierra Club, especially Joseph N. LeConte, who spearheaded the next cycle of exploration and mapping of the mountains, in the period 1893–99.

The first systematic mapping of the mountain topography was the work of the U.S. Geological Survey, in the years 1902–11. In the decades since, scientists have walked the ridges and canyons, made countless observations and measurements, and sampled the rocks by the thousands. From these studies, and from the matrix of geophysical theory, they have pieced together the origin of the rocks and structures and the evolution of the present landscape. This book tells how the parks region was explored and mapped, and recounts what is now known about the geologic processes that formed the rocks and carved them into the fine landscapes that we see today.

The geologic history of the region falls essentially into three periods (Norris and Webb, 1990). An initial period, from about 600 to 100 million years ago, perhaps beginning even earlier, saw marine sediments and volcanic deposits laid down near the old continental margin. These formations were repeatedly deformed as giant slabs of the Earth's crust—tectonic plates—impinged against the North American continental mass. During a second, overlapping period, from about 220 to 80 million years ago, the region was invaded from below by massive amounts of granitic melt (*magma*) that then solidified, mostly below the surface. Some of the magma broke through to the surface and fed volcanoes. In the third period, from roughly 80 million years ago to the present, repeated uplift of the land and constant erosion by both water and ice, stripped off the roof rock, laid bare the intrusive granitic rock, and sculpted the peaks and canyons we see today.

The sediments that were originally laid down on the site of the Sierra Nevada, so long ago, exist now only as remnants, and the legions of fossilized traces of plants and animals that had been incorporated into these sediments are now found only rarely. The deformation and baking of those early rock units, during the upheavals of the succeeding eons, destroyed most of these remnants and altered the remainder. But from the few fossils and sedimentary features available to us, we can piece together the ancient environments where these sediments were deposited, at the continental margin. Included are shallow-water marine settings, nearshore coral reefs that shed limy sediments, deepwater marine environments, and volcanoes both on land and on the sea floor. The oldest well-dated sediments remaining in the highest Sierra are from the Ordovician Period (450 million years ago); the youngest, from the Cretaceous Period (about 100 million years ago), shortly before the onset of the last main phase of invasion of granitic rocks.

The intrusion of vast quantities of molten granitic rock beneath the surface began in the Triassic Period (220 million years ago) and culminated in the Late Cretaceous (about 80 million years ago). The granitic masses seen today represent the eroded chambers into which this molten material rose and solidified. The rising granite melt invaded the original sedimentary rocks, which had already been folded and compressed as a result of previous Earth movements, and the forceful intrusion and great heat of the rising melt further deformed and metamorphosed the preexisting rocks.

During the intrusion of the granitic melts, some leaked or forced their way to the surface and fed numerous volcanoes, and these vents laid down extensive deposits of volcanic rock, partly as lava flows, but also as hot fragmental flows of blocks and ash that swept down the volcanoes' slopes. Much of this material was later invaded by still more granitic melts, which in turn baked, recrystallized, and deformed the volcanic material.

Following the emplacement of the granite, which replaced most of the earlier rock, the region was uplifted by regional tectonic forces and subjected to a prolonged period of erosion, with the result that most of the metamorphic roof rocks were carried away, and much of the subterranean granite became exposed at the surface.

Late in its history, the Sierra Nevada region was further uplifted along a series of giant fractures—faults—along its east side. Thus was created a giant tilted block, high and abrupt on its east side and sloping relatively gently across its broad western expanse (Fig. 1.5). Remarkably, few faults or regional fractures cut the interior of the block. Toward the south the range narrows, and at the latitude of Sequoia National Park its west flank becomes somewhat steeper (Fig. 1.2). Some suggest that this results from the action of a lesser fault system that has cut the west margin of the block and allowed uplift to raise that flank, too, but to a lesser extent.

Today, the great range serves as a barrier to moisture-laden clouds moving east from the Pacific Ocean. Abundant rain and heavy snow fall on the western slope, but the range shields the desert region to the east from precipitation. On the west flank the runoff generated by the rain and melting snow has been channeled into major river systems that drain to the Central Valley. Several of these rivers have carved gigantic, steep-walled canyons, and during the ice ages, glaciers, fed from a broad icecap that mantled the crest of the range, also crawled down these western gorges, grinding the bedrock and producing giant Yosemite-like glacial valleys (Matthes, 1960).

Within the area of the highest Sierra, the Kings and Kaweah Rivers are the master streams; together, they and their tributaries drain most of the west slope (Fig. 1.3). A part of the northernmost sector is drained by the San Joaquin River, and the Tule River drains the southernmost sector. Farther east, the south-flowing Kern River and its glaciers have carved a major canyon, a huge void bounded on the east by the Sierra Nevada Crest, on the west by the Great Western Divide, a range almost as high as the crest. The steep east face of the crest is drained by many small streams, all flowing into the Owens River. The Owens River drains south down the axis of Owens Valley, to empty into Owens Lake, a desert basin with no outlet.

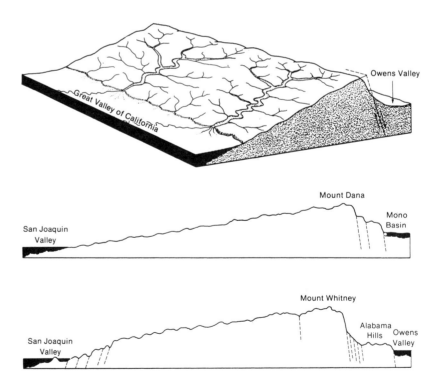

1.5. *Top.* The Sierra Nevada is a giant west-tilted fault block, broken and uplifted relative to Owens Valley, east of its steep east flank. Moisture-laden Pacific air dumps abundant moisture as it rises up the west slope, producing the river flow that has carved giant canyons (Matthes, 1960). *Middle.* Diagrammatic profile across the Sierra Nevada near Yosemite, showing step faults that have uplifted the range above Owens Valley. *Bottom.* Profile to the south across Sequoia National Park, showing the similar west-tilted fault block with major step faults on the east side and faults of lesser displacement on the west side, the latter perhaps accounting for the narrower aspect of the range to the south (compare with Fig. 1.2).

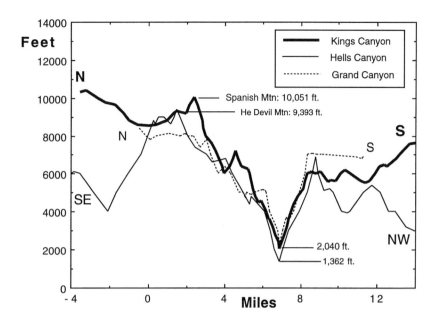

1.6. Comparison of profiles of the Kings River Canyon, Hells Canyon of the Snake River, on the Idaho-Oregon border, and the Grand Canyon of the Colorado River, in Arizona, at their deepest parts. The southeast wall of Hells Canyon is 8,031 feet high, as compared with the 8,011 feet of the north wall of Kings Canyon. Hells Canyon is thus 20 feet deeper, but not as steep. Based on U.S. Geological Survey 15-minute topographic quadrangle maps.

The canyon of the Kings River not only is the largest canyon in the range, but is a close contender for the title of deepest canyon on the continent. Its depth is 8,011 feet, as measured from the summit of Spanish Mountain (10,051 feet) on the north wall of the canyon to river level at 2,040 feet, at a point 4.5 miles to the south near the confluence of the South and Middle Forks. The only other similarly deep canyon in the country is Hells Canyon on the Snake River, at the Oregon-Idaho border. Hells Canyon totals 8,031 feet deep, as measured from He Devil Mountain (9,393 feet) to river level at 1,362 feet, 5.4 miles distant (Fig. 1.6). The Grand Canyon of the Colorado River, though inestimably grand, measures just 5,500 feet deep from its highest side.

The great depth of the canyons of the southern Sierra contrasts dramatically with the height of the peaks. Thirteen 14,000-foot mountains occur along a 43-mile span of the Sierra Nevada Crest between Thunderbolt Peak at the north and Mount Langley at the south (Fig. 1.3). All occur along the crest in the interval between, and west of, the towns of Big Pine and Lone Pine in Owens Valley (Porcella and Burns, 1998). All are carved primarily from granitic rock. The 14,000-foot peaks (Table 1.1) occur in three groups from north to south: the Palisades, Williamson, and Whitney Groups. Mount Whitney, at 14,491 feet, is the highest peak in the range and in the 48 conterminous states (Fig. 1.7). (Mount McKinley, in Alaska, is the tallest mountain, at 20,325 feet, in the United States and in North America.)

1.7. View of the 1,600-foot east wall of Mount Whitney, from the east. This precipice is the headwall of a glacial cirque shaped by vertical joints in the granite mass. The peaks at the left (to the south) of the summit peak are Keeler Needle, Day Needle, and Third Needle. U.S. Geological Survey photograph by Edward Du Bray.

TABLE I.I
Mountains of the Sierra Nevada exceeding 14,000 feet in altitude

Mountain (north to south)	Elevation (feet)	First known ascent
Thunderbolt Peak	14,003	F. P. Farquhar, N. Clyde, R. M. L. Underhill, B. Robinson, L. F. Clark, G. Dawson, J. M. Eichorn (1931)
Starlight Peak	14,200	N. Clyde (1930)
North Palisade	14,242	J. S. Hutchinson, J. K. Moffet, J. N. LeConte (1903)
Polemonium Peak	14,200	Unknown
Mount Sill	14,162	J. S. Hutchinson, J. K. Moffet, R. D. Pike, and J. N. LeConte (1903)
S. Palisade (Split Mtn.)	14,058	Mr. and Mrs. J. N. LeConte and C. Lindley (1902)
Middle Palisade	14,040	F. P. Farquhar and A. F. Hall (1921)
Mount Tyndall	14,018	C. King and R. Cotter (1864)
Mount Williamson	14,375	W. L. Hunter and C. Milholland (1884)
Mount Russell	14,086	N. Clyde (1926)
Mount Whitney	14,491	J. Lucas, A. H. Johnson, and C. D. Begole (1873)
Mount Muir	14,015	Unknown
Mount Langley	14,042	C. King and P. Pinson (1871)

Several of these highest peaks are moderately easy to climb, although at these altitudes some climbers may suffer shortness of breath and mountain sickness. Mount Whitney has a horse trail to the summit, and Mount Muir is only a short distance off the trail to Mount Whitney. Mount Langley was ascended on muleback in 1873 by W. A. Goodyear. Among the more challenging are Mount Russell and the Middle Palisade, which remained unclimbed until the 1920s. Norman Clyde (1962), who knew all these mountains well, noted that Mount Russell "is a fine, craggy mountain, one that delights the heart of a mountaineer . . . but is foolhardy for any but experienced mountaineers to climb. . . . " For Clyde the Middle Palisade was "essentially a crag-and-chimney climb and is not recommended for novices."

The views from all these peaks are outstanding. Situated as they are on

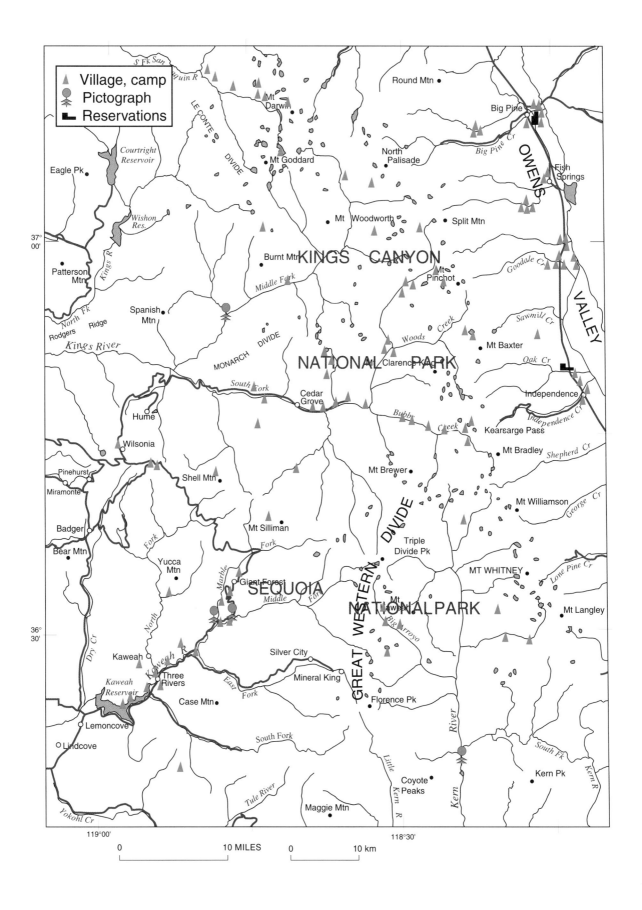

the Sierra Crest, they offer vistas not only of the high peaks and canyons of their own range, but also of the vast fault-bounded trough of Owens Valley, as well as the ranks of desert ranges extending far off to the east in the Great Basin. Walter Starr, Jr. (1934) stated that "the view from the Palisade group is the grandest." He continued:

> The view south over the great culmination of the Sierra Crest looks toward the northern slopes of the peaks and spurs where lie the snowfields and many of the ancient glacial amphitheaters. . . . Below the eastern sheer wall of the Palisades lie their own glaciers. To the west is the high Goddard Divide and the rugged spurs between. To the south, extending to the Kaweah Peaks, the view is indescribable. Mt. Sill has the advantage in position over other Palisade summits, and it can therefore be said to be the peer of all Sierra peaks in the extent and quality of the views it offers. It can be climbed without danger or difficulty.

But of course the early adventurers and surveyors were not the first to enter the highest Sierra. Native American tribes were settled around and within the Sierra Nevada (Fig. 1.8) long before the time the new Americans, those primarily of European descent, began their exploration and settlement of the area. The Indians had occupied Owens Valley, east of the mountains, and the Central Valley, on the west, for thousands of years. The time of arrival of the native peoples to the southern Sierra region is incompletely known because only traces of the early cultures have been preserved. Apparently, the earliest more or less permanent habitations in California date back to about 9,000 years ago (Elsasser, 1972). In the southern Central Valley, archaeological evidence indicates a thriving culture about 4,500 years ago. Likewise, in Owens Valley, east of the mountains, an ancient culture dating back more than 3,500 years has been documented.

The Indian tribes that inhabited the east side of the range were the Paiute, or Shoshone; and those on the west slope were the Monache, or Western Mono. The more numerous Yokuts occupied the Central Valley and the Sierra foothills west of the region settled by the Monache tribal group. The Paiute and Monache tribal groups shared a common language and traded regularly across the high passes. They were familiar with many routes through the mountains and apparently used Kearsarge Pass as the primary route over the range in this region. They visited the higher parts of the range and set up summer camps there to hunt, fish, and gather food.

An extensive trade was conducted between the peoples on the two

1.8. Locations of Indian camps, pictographs, and present-day reservations in the parks area, based on Elsasser (1972), Steward (1933), and the writer's observations. Camp distribution reflects the broad area over which the Indians lived and roamed. The two tiny Indian reservations of today are each less than 1 square mile in size.

sides of the Sierra (Fig. 1.9). Acorns, manzanita berries, salt, buckskin, baskets, shell bead money, and arrow shafts originated from the western Mono. Pine nuts, rock salt, rabbit-skin blankets, mountain sheep skins, moccasins, basket bottles waterproofed with pitch, and volcanic glass (obsidian, for blade chipping) came from the Owens Valley Paiute (Steward, 1933).

Because of the desert heat, the Paiute summer houses were built strictly for shade and consisted of brush or grass bundles supported on upright willow poles. Winter homes were more solidly built. The frame was a cone of poles, each 9 or 10 feet long, set in a shallow pit dug in the ground. Grass, willows, or juniper bark were woven into overlapping mats and lashed to the poles. A few inches of earth covered some houses, and all had a smoke hole in the center at the top. In good weather everyone slept outside, and cooking was also done in the open. When Indians slept outside during cold weather, they would often heat the ground with a fire. The embers would then be swept aside, and clean sand and grass would be spread over the heated sleeping place.

1.9. Indians carrying acorns, the primary food staple on the west slope of the Sierra (Muir, 1887).

The mountain houses used in the Sierra above 6,000 feet were constructed by planting two posts, each 6 to 7 feet tall, in the ground and about 15 feet apart. A horizontal ridgepole was then fixed to the two poles, and sloping poles reached from the ground to the ridgepole. Over this structure was fixed a roof of pine boughs.

For their blankets, the Indians generally used rabbit skins. Some 50 to 70 jackrabbit pelts, each with its fur intact, were cut and fashioned into a long, continuous strip that was then woven into blankets up to 5 by 6 feet in size. Some of the Owens Valley Paiute Indians made simple pottery bowls from clay collected at various valley sites, including Fish Springs. Pots were molded by hand, sun-dried, and baked in the coals of a sagebrush fire. The staple foods of the Paiutes were pine nuts and wild seeds, supplemented with dried insects, birds, and small animals, including rabbits.

The Monache lived in conical huts that were thatched or covered with slabs of bark. Game animals, primarily deer and bear in the forests, were plentiful, and were taken by bow and arrow. Arrowheads were chipped from obsidian collected primarily from the volcanic flows of the Mono region, east of Yosemite. The Paiutes carried unworked obsidian over the range and bartered with the Monache for foodstuffs and skins. Sharp chips can be found at many mountain campsites where points had been made from rough chunks of obsidian carried into the mountains for that purpose. Some knives were made of a hard slate, chipped to a fine edge. Several times I have picked up pieces of slate in the high country, wondering how they arrived in a terrain underlain by granite bedrock, then saw that they had been fashioned into crude knives. Bowls were carved from talc-bearing steatite, a soft metamorphic rock.

The Monache Indians took smaller animals, such as weasels, muskrats, skunks, and squirrels, by trapping, and sometimes smoked them out of their burrows. They used a bola, a sling-like weapon consisting of a round stone lashed to a long cord, to fell birds and squirrels from the treetops. Fish were trapped in streams by means of woven snares and weirs.

For the Monache Indians, on the west slope of the Sierra, acorns were the most important plant food. They collected the acorns in large quantities in the fall and stored them for the winter. They then cracked and peeled them, and ground them into a meal in a conical hole, or *mortar*, in bedrock (Fig. 1.10) by pounding them with an elongate stone, or *pestle*. The resulting meal was leached with fresh water to remove the tannic acid inherent in the acorns, then dried and stored. A nourishing mush was made by boiling the meal in a basket, skillfully woven so as to be water-

1.10. Indian bedrock mortars, a pestle in one of them, on the Middle Fork of the Kaweah River. U.S. National Park Service photograph.

tight. Hot stones, heated in the fire, would be dropped into the water-meal mixture to bring the water to a boil. Hard, fine-grained metamorphic rocks were favored for the heating stones; granite was considered a poor choice. Other important plant foods were pine nuts, manzanita berries, and yucca.

The locations of the bedrock mortars serve as good indicators of the distribution of Indian camps in the area of the parks. The rock of choice was granite; mortars in metamorphic rock are rare. The mortars were hollowed out by grinding and pounding the granite bedrock with an elongate, handheld rock pestle, but a hole of any depth could be attained only after years of pounding of foodstuff materials. Mortars deeper than about 10 inches became inconvenient because of their depth; they were apparently abandoned, and new ones begun. Ten mortars are common at a given site, and at some sites there are as many as 35 holes (Fig. 1.11).

Generally, the mortars were used for grinding acorns. Compilation from my notebooks of descriptions of bedrock mortars in the Giant Forest, Tehipite Dome, and Kaweah 15-Minute Quadrangles shows that although most mortars range from the valleys up to 4,000 feet in altitude (the approximate upper limit of oak trees), they are also rather common between 6,000 and 8,000 feet. These higher-elevation mortars were probably used both to grind acorns the Indians had carried up to the higher summer hunting camps and to grind other foods gathered at the higher

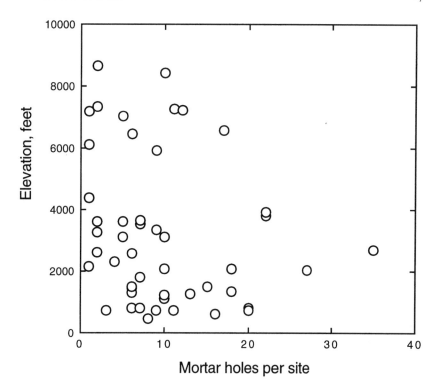

1.11. Elevation and number per site of 464 Indian bedrock mortars in the Giant Forest, Kaweah, and Tehipite Dome 15-Minute Quadrangles. Author's data.

sites, such as manzanita berries. The mortars were also used to grind dried meat and bones.

Isolated localities of Native American rock art can be found in the parks area (Clewlow, 1978). Within the Sierra, the known examples are all *pictographs*, or designs painted on rocks. The artists there made use of black and red paints based on charcoal and iron oxide pigments. The sites that have been studied were apparently painted in the years between 500 and 1200 A.D., to judge from the associated archaeological material.

The pictographs, generally abstract drawings, commonly employ curved lines, though some human stick figures are represented, as at Hospital Rock, on the Kaweah River. In the remote Tehipite Valley on the Middle Fork of the Kings River, pictographs in talus caves depict stylized amoeboid shapes (Fig. 1.12).

Local Indians, when questioned in the nineteenth century, had no knowledge of who painted these designs or for what purpose they were created. Information from other areas, however, suggests that the paintings were made by medicine men (shamans) for ritual use, and to mark the sites of their caches of magic paraphernalia.

Common in the Owens Valley region are *petroglyphs*, or designs abraded or pecked into the rock. The petroglyphs are usually carved on smooth

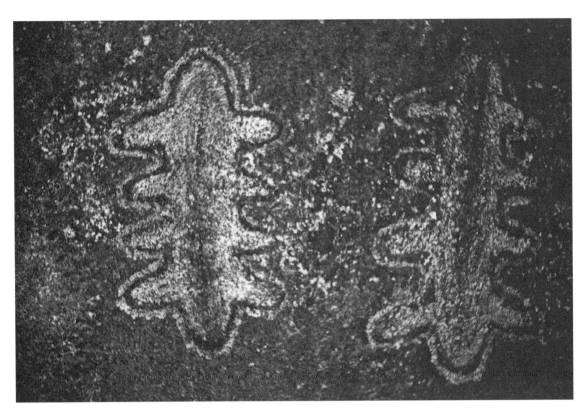

1.12. Pictographs painted on a large granite block in Tehipite Valley, Middle Fork of the Kings River. About natural size.

surfaces of lava outcrops. This rock is well suited for artwork, because the rock surfaces are commonly coated with a layer of dark, durable mineral material called *desert varnish*. Pecking on this surface with a pointed rock flakes off this coating and exposes the lighter-colored rock below. In places the volcanic rock includes a fine, fragmental ash-flow material with a case-hardened outer surface that can be readily pecked through to produce the designs.

As late as 1830, large numbers of Indians were reported to be living in the Central Valley of California, but apparently shortly thereafter epidemics swept through the region, decimating entire populations. Trappers in 1833 reported deserted villages and unburied dead, and in subsequent years markedly fewer natives were encountered.

Following the discovery of gold in the northern Sierra foothills, in 1848, prospecting and grazing expanded rapidly within the range. Competition for available land and resources increased apace, and conflict between the remaining Indians and the settlers was inevitable. The tragedy at Woodville in 1850 (Figs. 2.7 and 2.8) is a stark instance, one that occurred near the parks area (Elliot, 1884). A group of settlers led by a man named Wood built an oak log cabin on the south bank of the Kaweah

River, a few miles west of where the river leaves the foothills and 7 miles east of the site of Visalia (Fig. 2.6). Indians considered this an invasion of their territory, and a band attacked Wood, whereupon he took refuge in the cabin and held them off with musket fire until he ran out of powder and shot. He was finally captured by the Indians, tied to a tree, and skinned alive. The other settlers were outraged and staged punitive raids on the Indians.

In 1856 a band of Indians stole a herd of 100 cattle along the Tule River, defied the settlers, and retreated to a stronghold along the North Fork of the Tule near a small rocky hill. This position, now called Battle Mountain, was in rough and rocky country surrounded by nearly impenetrable manzanita and oak brush. A company of 60 volunteers attempted to attack the Indians, but after two days of siege and several unsuccessful assaults, they retired for reinforcements. A larger "army" was then assembled, including a group of 60 miners from Keyesville, on the Kern River, and a group of volunteers from Visalia led by Captain W. G. Poindexter. The resulting force of about 140 men, now under the command of Poindexter, again attacked the stronghold and maintained a siege for two days, but were unsuccessful in dislodging the Indians and returned to Visalia. Finally, an army of 400 men was raised; this group consisted mainly of volunteers but included 25 U.S. Army soldiers from Fort Miller, to the north, led by one Captain Livingston, and half a company of U.S. Cavalry from Fort Tejon, to the south. They opposed a force of about 700 Indians within the stronghold, including women and children. Captain Livingston, now commander of the combined force, did not attack, but rather positioned a small mountain howitzer on a neighboring hill and began bombarding the center of the Indian position. A group of Indians left their stronghold intent on capturing the cannon, but were repelled. The army then entered the stronghold, where they engaged and defeated the Indians. About 100 Indians were killed, and most of the rest fled to the mountains.

In spite of these massed confrontations, which so often ended in tragedy, contacts with the Indians by individuals were often peaceful. Hale Tharp, one of the thousands who had come overland to work the goldfields in 1851, settled in 1856 on a ranch near Three Rivers. He noted that at that time about 2,000 Indians lived along the Kaweah River between Lemon Cove and Hospital Rock. In the summer of 1858, just two years after the conflict at Battle Mountain, he made a difficult 18-mile trip on horseback up the Middle Fork of the Kaweah River, guided by a local Indian, and at Hospital Rock he visited a large camp numbering about 600 resident Indians. (Tharp was one of the first to visit Giant Forest, the largest grove of sequoia trees in the present Sequoia National Park. By chance he met John Muir in Giant Forest some years later, in 1875, during Muir's third visit to the area of the parks.)

In 1855–56 Allexey W. Von Schmidt, a surveyor working for the General Land Office, made the first land surveys east of the central Sierra Nevada. He extended the Mount Diablo Baseline (we will return to that in Chapter 3) into Owens Valley, and laid out the land lines south down the valley. "This valley contains about 1,000 Indians," Von Schmidt wrote, "and they are a fine looking set of men. They live principally on pine nuts, fish, and hares, which are very plenty" (Chalfant, 1933).

Trouble between settlers and Indians east of the Sierra in Owens Valley broke out somewhat later than it had on the west slope, as settlement of this isolated region proceeded. In 1861 cattle rustling by hungry Paiutes provoked attacks by cattlemen, and there were killings on both sides. Despite negotiations and peace treaties, skirmishes continued. In April 1862 about 50 settlers faced several hundred Paiutes in the Battle of Bishop. The settlers suffered three dead, retreated to Big Pine, and called for help from the Army.

The Government responded, sending several hundred troops from Camp Latham, near Santa Monica, under the command of Colonel George Evans, as well as 50 soldiers from Fort Churchill, Nevada. On July 4, 1862, the combined force established Fort Independence, just north of the present town of Independence, and the balance of power tipped toward the settlers. In July of the following year about a thousand Indians were moved out of Owens Valley and relocated to a reservation near Fort Tejon, south of Bakersfield, and Fort Independence was closed.

Sporadic incidents with Indians in 1865 saw the fort restored to active duty, but it closed down for good in 1877. Today, within the 12,000 square miles of the region of the National parks and vicinity where the Indians formerly lived and roamed, only two Indian reservations remain (Fig. 1.8). The Big Pine and Fort Independence Indian Reservations are each less than 1 square mile in area.

The processes of discovery and mapping in the Sierra favored the settlement and exploitation of this remote mountainous corner of California. The tides of human migration that then flooded into the region inevitably led to the shrinking and suffering of the society they displaced. One must sympathize with the Native Americans who saw their culture and society being overwhelmed by these waves of immigration of new Americans.

Man can learn nothing unless he proceeds from the known to the unknown.

—CLAUDE BERNARD (1813–1878)

Exploration is really the essence of the human spirit.

—FRANK BORMAN, 1969

EXPLORATION

Missionaries from Mexico, intent on exploring California's great Central Valley, were the first Europeans to behold the majesty of the southern Sierra Nevada. They were followed by fur trappers from the Rocky Mountains searching for richer hunting grounds and shorter routes to reach them. And even while California was still Mexican territory, the U.S. Government funded Army surveying expeditions into its hinterlands, in response to the ferment of manifest destiny—the dream of a nation spanning the continent from sea to sea. Once California became a state, further Army surveying expeditions explored and mapped likely wagon and railroad routes across its vast expanses, all the while mindful of tangential military purposes. These surveys delimited the extent of the great mountain range, but it was not until the 1860s that the newly organized California Geological Survey made the first reconnaissance expedition specifically into and across the high mountains of the southern Sierra. Through the closing decades of the century, a dedicated naturalist, John Muir, explored extensively in the Sierra, including the high mountains in its southern reaches, and through his eloquent writings made known both the rare qualities of the region and the urgent need for action toward its preservation.

The earliest explorations in the area adjacent to the southern Sierra were undertaken in the late eighteenth century during the overland journeys of the missionaries from Mexico. In 1776 the Franciscan cleric Francisco Garcés, while exploring the San Joaquin Valley, discovered the Kern River (and named it the Rio San Felipe). In 1806 a party led by Lieutenant Gabriel Moraga explored the southern Sierra foothills west of the modern parks area. It happened that on January 6, party members camped along a swift river flowing from the mountains. This day in the church calendar, the day of Epiphany that honors the visit of the Magi, or Wise Men, to the infant Jesus, suggested the name Rio de los Santos Reyes (River of the Holy Kings). This name has survived as Kings River. But, in all their years in California, neither the missionaries nor their armed escorts penetrated into the more elevated parts of the range.

Making his way up the American River east of Sacramento to the vicinity of Lake Tahoe, early in 1827, the fur hunter Jedediah Smith became the first white man known to have crossed the Sierra Nevada (Merriam, 1923). Smith had entered California in 1826 from the southeast by way of the Mojave Desert, thus skirting the southern end of the Sierra. Along the way he had visited the Spanish missions at San Diego and San Pedro on the Pacific Coast. Then, wishing to join his partners near Great Salt Lake, he attempted to cross the southern Sierra (which he called Mount Joseph) from the Central Valley early in the new year, probably in February. In a letter written July 12, 1827, after he had finally reached Utah, he wrote, "On my arrival at a River which I named the Wim-mel-che (named for a Tribe of Indians who reside on it of that name) I found a few Beaver, & Elk, Deer and Antelope in abundance. I here made a small hunt, and attempted to take my party across. . . . I found the snow so deep on Mount Joseph that I could not cross by horses, five of which starved to death. I was compelled therefore to return to the Valley which I had left" (Farquhar, 1965).

Because the Wim-mel-che Indian tribe is known to have lived along the Kings River, we can presume that Smith attempted to cross the range by ascending this river, a route that is impassable with horses in the winter. In the course of this attempt, therefore, he apparently became the first explorer to approach the region of Kings Canyon, and the first—of many—to experience the challenge and hardships of the highest Sierra. After retreating to the San Joaquin Valley and passing several months there, Smith took his party north some 300 miles and did succeed in crossing the northern part of the range, on the American River route.

Six years later, in 1833 and 1834, another group of fur trappers led by Joseph Walker crossed the Sierra Nevada from the east, through the Yosemite region. They then recrossed the range some 150 miles to the south by way of what has since been known as Walker Pass, thus encircling, but not penetrating, the mountain fastness of the parks. These were the first white men to see Yosemite Valley, as well as the giant sequoia trees, and nearly a decade would pass before another determined explorer, John C. Fremont, would probe the upper Kings River region.

Probing the High Sierra: Fremont and Carson

When her family fell on hard times, Ann Beverly Whiting was forced into a marriage of convenience with a man old enough to be her grandfather. When the opportunity presented itself, she ran off with the young Frenchman who had been her French teacher in Richmond, Virginia. The couple soon moved to Savannah, Georgia, and it was there, in 1813, that she bore Jean Charles a son, whom they called John Charles.

At 14 John Charles became a clerk in a law office, and with the help of a patron attended preparatory school and then Charleston College, in South Carolina. Following graduation he served as a country teacher but in spare time took up the study of maps and astronomy and taught himself the rudiments of surveying. He then signed on as instructor of mathematics aboard the United States naval vessel U.S.S. *Natchez*, which cruised to the coasts of South America. He later accepted a position with the U.S. Topographical Corps to survey a possible railroad route from Charleston to Cincinnati, and in 1836 he joined a survey of the Cherokee country, in the region where North Carolina, Tennessee, and Georgia meet.

At that time the U.S. Government had begun to take an active role in the exploration and mapping of the west. In 1838 the U.S. Army Topographical Engineers Corps was established, as part of the War Department, and employed the distinguished French scientist Joseph Nicollet to lead a survey of the broad northern territory between the Missouri and upper Mississippi Rivers, where immigrants were actively taking up land. Fremont applied for and was granted the position of assistant to Nicollet. The field work and map compilation for this program occupied Lieutenant Fremont (Fig. 2.1) through the 1841 season.

That same year, he was married to Jessie Benton, daughter of Senator Thomas H. Benton of Missouri, who at first was reluctant to grant his approval for the marriage. Later, however, he became an ardent supporter of Fremont, and through his influence Fremont was named to conduct several expeditions. The first of these he commanded the following year, taking his survey party by steamer down the Ohio River and up the Mis-

2.1. In the winter of 1845–46 John C. Fremont led the first exploring expedition into the region of the highest Sierra. He divided his party, one group going south along the east side of the Sierra Nevada and making a western crossing of the divide at Walker Pass on the south, the other group crossing the mountains near Lake Tahoe and traveling down the west side of the range to the Lake Fork (the Kings River). Hence, his party made the first surveys of the entire limits of the range. Later, in 1846, he was appointed military governor of California after the American Bear Flaggers rebelled against Mexican rule. U.S. Library of Congress.

souri and Kansas Rivers and then overland to the Rocky Mountains, where they mapped the region around South Pass, over the Rockies. Chief scout on this and several later expeditions was the legendary Kit Carson.

Fremont's second expedition, in 1843, was charged with the task of linking the exploration of the Great Basin of Utah and Nevada with the U.S. Navy's exploration of the Pacific coastal region, newly reconnoitered in 1841 by Commander Charles Wilkes of the U.S. South Seas Surveying and Exploring Expedition. Wilkes's expedition (1838–42), six ships strong, crossed the Pacific three times, exploring and charting from Antarctica to Hawaii and the Pacific Northwest. In the summer of 1841, Wilkes explored Puget Sound, the lower 100 miles of the Columbia River, and the Oregon coast (with a land party), and then sailed south to San Francisco.

Fremont's party traveled to the Columbia River and on to the coast, making a series of astronomical observations to determine exact locations of several key positions, for mapping purposes. The party then headed south with the objective of exploring the Great Basin and the Sierra Nevada. Early in 1844, in western Nevada, lame horses caused Fremont to abandon plans to head east to the Rockies, but instead to cross the Sierra for the more favorable climate of California.

He explored the east base of the Sierra as far south as the present site of Bridgeport, but then decided to lead his party, in midwinter, west over the range at what is now Carson Pass. They proceeded up the East Fork of the Carson River at the end of January, 1844, and reached the range summit, choked with snow, February 14. Traveling through deep snow, often breaking trail in blizzard conditions, they arrived at length at Sutter's Fort, at the present site of Sacramento, March 6. Though all were terribly worn and hungry, Fremont had brought the entire party across the mountains. Only 33 of their original 67 horses and packmules had survived; most of the remainder had been eaten. After a period of rest Fremont headed south down the Central Valley, along the west base of the Sierra Nevada, crossed Tehachapi Pass, traversed the south end of the range, and returned to Missouri. The lesson was clear—a crossing of the Sierra Nevada in winter was difficult and dangerous, if not foolhardy.

On his third—and most fateful—expedition in 1845, the now Captain Fremont traveled west once again, but split his party in western Nevada so as to reconnoiter both sides of the Sierra. The smaller group, led by Fremont with Kit Carson as scout, traveled up the Truckee River and over Donner Pass, and arrived December 8 at Sutter's Fort. The larger party, headed by Theodore Talbot, went south to Walker Lake and then southwest to the Owens River (the two features were named for two party members, Joseph Walker, scout, and Richard Owens), traveled along the east flank of the Sierra, and crossed the range toward its southern limit at Walker Pass, which Walker knew well from his own trip 11 years earlier. The plan was to rendezvous on "the Lake Fork of the Tulare Lake," which to Fremont meant the Kings River, with which he was familiar from his own previous expedition, but to Talbot, as would be learned later, it meant the Kern River. When Fremont arrived late in the year at the supposed meeting place along the Lake Fork River, no party had arrived from the south. For three weeks the Talbot-Walker party dutifully waited on the Kern River as their supplies dwindled.

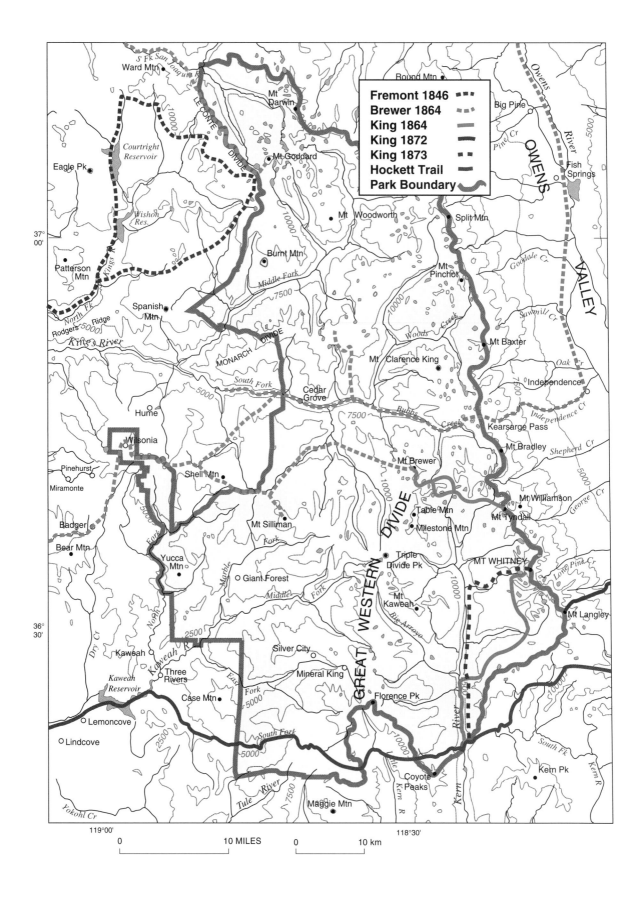

Fremont, meanwhile, drove deeper into the range in what was probably the first exploration into the region of the parks (Figs. 2.2 and 2.3). In December he led his party up the Kings River, far into the mountains. With some difficulty they attained the headwaters of the North Fork of the Kings River, perhaps into upper Blackcap Basin (Farquhar, 1965). "We forced our way up among the head springs of the river and finally stood upon the flat ridge of naked granite which made the division of the waters and was 11,000 feet above the sea," wrote Fremont in his memoirs, written in later life (Fremont, 1887). "Lying immediately below, perhaps 1,000 feet, at the foot of a precipitous descent was a small lake which I judged to be one of the sources of the San Joaquin." Fremont realized that the country was much too rough to permit them further penetration into the range in winter, and the party turned, making a broad loop apparently in the upper drainage of the North Fork of the Kings River (Fig. 2.2). This reckless journey into the High Sierra in the winter of 1845–46 almost ended in disaster when a heavy snowstorm overtook the travelers and nearly blocked their escape from the mountains.

The two parties finally reunited at San Jose, at the southern extremity of San Francisco Bay, in mid-February 1846. It was during that summer that a group of restive American settlers, the Bear Flaggers, took over the Mexican garrison in Sonoma, north of San Francisco, and raised the flag of the California Republic on June 15. Fremont, who commanded the only U.S. Army unit then deployed in California, lent his support to the group. On July 4 he addressed the people at the plaza of Sonoma and declared martial law. On July 7, the American flag was raised at Monterey by Commodore John D. Sloat of the U.S. Navy, and California became a part of the United States. Fremont was named military governor of California by Commodore Robert F. Stockton, in command of U.S. naval forces in San Francisco. General Stephen W. Kearny, arriving overland through Arizona, then took command of U.S. Army forces in California. Following the conquest of California, Fremont was court-martialed on grounds of his supposed insubordination to Kearny.

2.2. Early trails of discovery in the highest Sierra Nevada, 1846 to 1873, opened by John C. Fremont, William H. Brewer, and Clarence King. Just south of the region of the highest Sierra is the Hockett Trail, a trail built to provide access to Fort Independence and to silver mines east of Lone Pine.

A map of Fremont's explorations, dated 1848, was published in 1850, by order of the Senate of the United States (Fig. 2.3). Because of the extraordinary events following the expedition, which included his court-martial for mutiny, a presidential pardon, and his resignation from the Army, Fremont himself was unable to complete the report of his third expedition. The map was in fact prepared from available data by Charles Preuss, a German topographer who accompanied Fremont on his first, second, and fourth expeditions, but was not present on the third. Preuss's

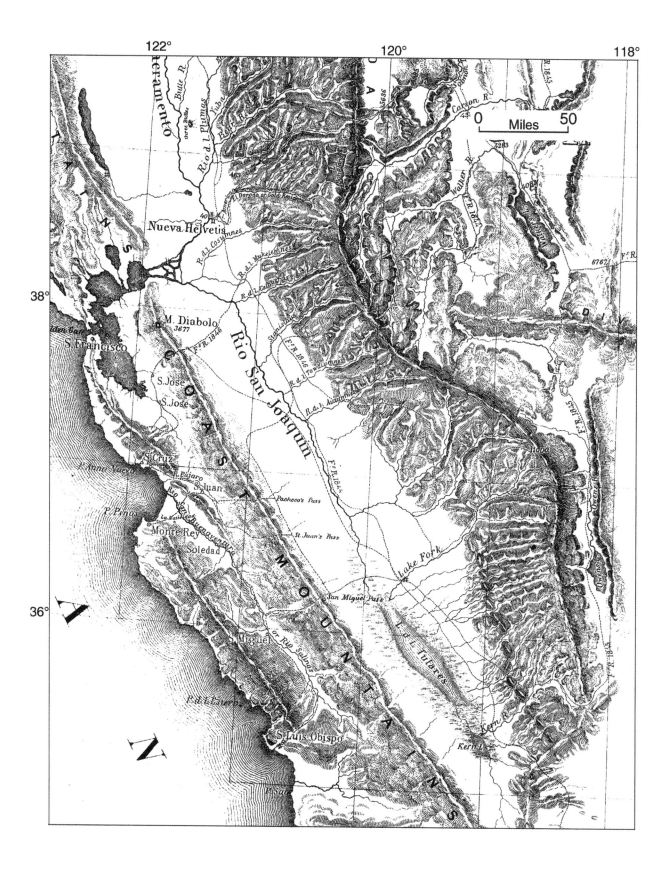

use of field sketches and his skill in representing topography on a map by the use of fine line-work—hachures—was unsurpassed (Fig. 2.3). This map (scale 1:3,000,000; 45 miles to the inch) is by far the best representation of the Sierra Nevada that had yet been made, in part because of Preuss's skill, but also because Fremont's group had encircled the central and southern parts of the range. The map, however, shows only two named rivers in the entire southern half of the range, the Lake Fork (now the Kings River) and the Kern River, whereas in the northern part of the range more than a dozen rivers are labeled. (The presence of named features is often an indication of the level of detail attained in mapping.) The several good passes in the north, and the relatively heavy travel across them, in any event provided the impetus for more extensive mapping in that region.

Fremont had expected future assignments on government surveys, particularly for those concerned with railroad routes, but his strong antislavery stance, as well as the court-martial, eroded his political support, and he was forced to raise private funds for future mapping expeditions. His fourth expedition, in the winter of 1848–49, in search of a transcontinental railroad route in the San Juan Mountains of the southern Rockies, was a disaster. The party suffered terribly from one of the most severe winters in memory, and ten men, about a third of the survey team, died of starvation and cold. A similar fate befell the fifth expedition, which he had mounted in the winter of 1853–54. This one managed to cross the San Juan Mountains and the Colorado Plateau to reach the Mormon settlement of Parowan in what is now southwestern Utah. One man froze to death and many were frostbitten.

In 1856 Fremont entered the presidential campaign as the nominee of the Republican Party, on the strength of his great popularity, but was defeated by James Buchanan. He was appointed Brigadier General in command of the Department of the West in 1861, but his service during the Civil War lasted just 100 days. Following Fremont's explorations in California, nearly a decade would pass before action would be taken on the pressing need for government surveys of transportation routes.

In later life General John C. Fremont would serve as the appointed governor of Arizona Territory, vigorously promoting the westward expansion of the railroads. His greatest legacy, however, remains his work as an explorer and pathfinder. An epitaph reads "From the ashes of his campfires have sprung cities." I particularly salute his early penetrations, during winter, into the completely unknown territory of the highest Sierra.

2.3. Part of Fremont's 1850 map, showing the route of his winter journey (1845–46) up the Lake Fork (the Kings River), which attempted, unsuccessfully, to make a crossing of the range. They attained an elevation of 11,000 feet (as labeled), but turned back in the face of blizzard conditions. The Talbot party, which traveled south down the east side of the Sierra, was to rendezvous with the Fremont party on the Lake Fork, but Talbot mistook the Kern for the Lake Fork and waited three weeks there. The two parties finally met later at San Jose. Note that no habitations are shown in the San Joaquin River Valley. This is the first map to use the name Golden Gate (Taylor, 1850).

The Pacific Railroad Surveys: Blake and Williamson

In 1853–54 Congress had authorized the War Department, headed by Jefferson Davis, Secretary of War, to undertake a survey of railroad routes from the Mississippi River to the Pacific Ocean. Lieutenant Robert S. Williamson of the Corps of Topographical Engineers was assigned the task of seeking out such routes in California. The sum of $30,000 was set aside for the survey party, which was to be accompanied by a mounted escort of three officers and 25 privates. Five teamsters and eight others, who were to serve as field-men and cooks, were also hired.

Rounding out Williamson's survey group were four specialists: a physician-naturalist, a civil engineer, an assistant civil engineer and artist, and a geologist-mineralogist. The geologist was William Blake (Fig. 2.4), a last-

2.4. William P. Blake, geologist-mineralogist on the 1853 Pacific Railroad Survey headed by Lieutenant Williamson, which surveyed the western foothills of the southern Sierra Nevada. His report contains some of the first scientific descriptions of this region, and his 1855 map (Fig. 2.7) is the first geologic map to include a large part of California. Blake coveted the directorship of the Geological Survey of California, but when Josiah Whitney was appointed to the position, he accepted, in 1864, a position as professor of geology at the College of California at Berkeley, which later became the University of California (Raymond, 1911).

EXPLORATION

2.5. U.S. Army Engineers party surveying for railroad routes in the summer of 1853. This expedition, led by Lieutenant Robert Williamson, traversed south down the west side of the Sierra Nevada and mapped the position of the major rivers, but failed to find anything resembling a route over the highest Sierra. From a field sketch by Charles Koppel (Blake, 1858).

minute addition to the team whom Williamson had located by requesting aid from a friend at the Smithsonian Institution. Blake was born in New York City and attended Yale College. He received his Ph.B. degree in 1852 after a two-year course of study in chemistry and natural history in the department that later (in 1861) would be called the Sheffield Scientific School. Blake continued his work with the War Department Pacific Railroad Surveys until 1856, and his writings were among the first scientific accounts of the areas surveyed. He published the results of the 1855 exploration in a comprehensive report in 1857, and later, in 1864, he was appointed professor at the College of California at Berkeley, which would become the University of California.

The Williamson party set out from the military post at Benicia, traveled by way of Livermore Pass to the San Joaquin Valley, and arrived in the Sierra foothills near the Tuolumne River (Fig. 2.5). Continuing south, retracing Fremont's route, they passed the Merced and Chowchilla Rivers and stopped at Fort Miller (Fig. 2.8) on the San Joaquin River for several days of reconnaissance surveying. It was here that they reconnoitered the remnants of lava flows that occupy the San Joaquin River valley in the Sierra foothills. Then, proceeding south along the west base of the range, they passed the region of the highest Sierra near the present site of Visalia and noted the rugged mountains to the east (Fig. 2.6). "The Sierra Nevada . . . presents a magnificent spectacle from this place," Blake (1858)

2.6. *(Above)* View east toward the snow-covered Sierra Nevada from the Four Creeks region, near the site of Visalia in the Central Valley. This lithograph was produced from field sketches (probably by Charles Koppel) drawn during the 1853 Pacific Railroad Survey (Blake, 1858).

2.7. *(Opposite)* Part of a geologic map prepared by William P. Blake, geologist with the 1853 Pacific Railroad Survey in California headed by Lieutenant Robert Williamson. The map, published in 1856, is the first regional geologic map of most of the state of California. Note that the only habitations in the San Joaquin Valley south of Stockton are Fort Miller and Woodville. Nine different geologic units were distinguished by watercolors that were individually hand-painted on all published maps. All of the parks region is depicted as granitic and metamorphic rock. Special units mapped are metamorphic slates (M), basaltic lava (L), and Tertiary strata (cross pattern) (Blake, 1858).

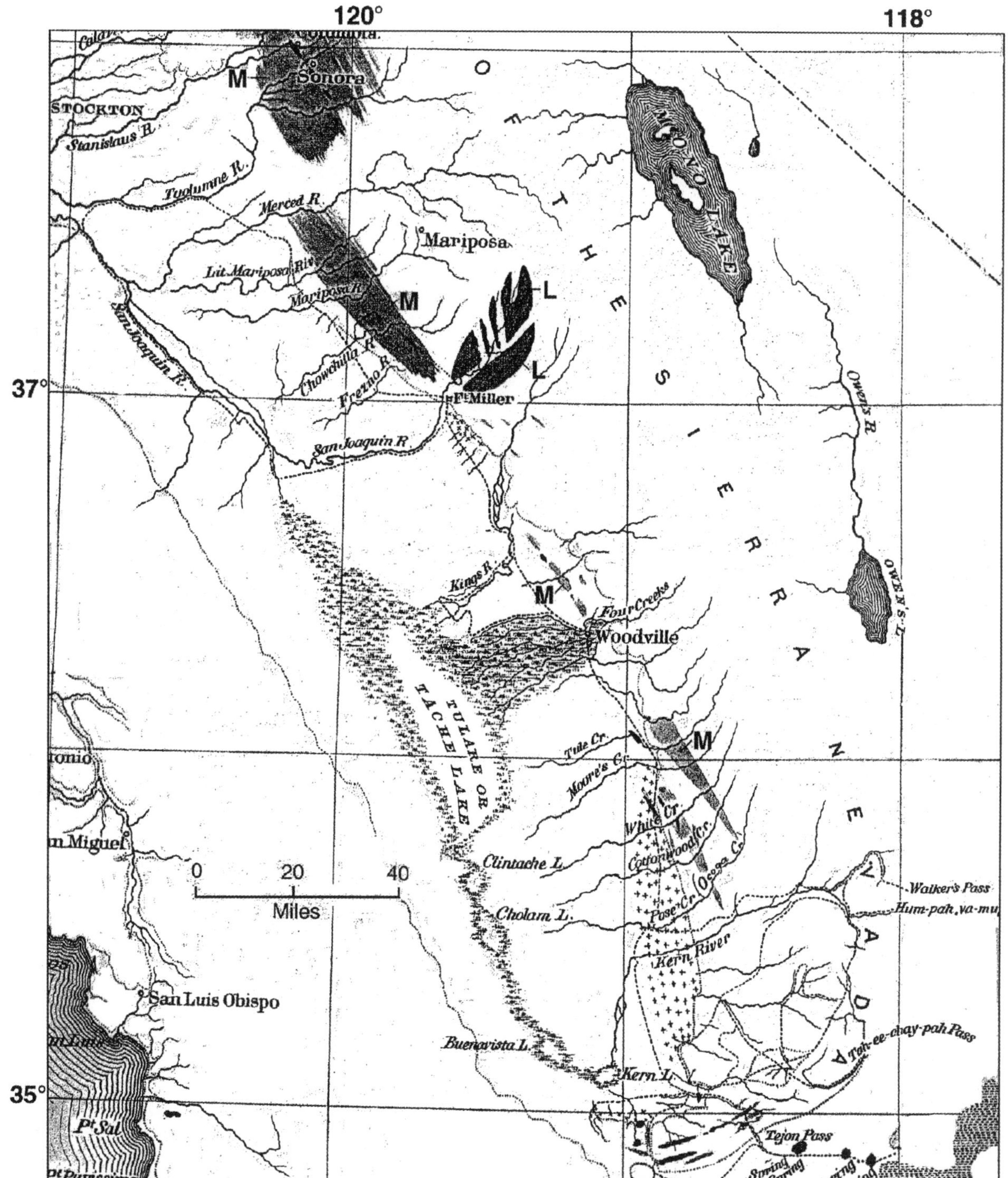
120°
118°
37°
35°
Columbia
Sonora
M
STOCKTON
Stanislaus R.
Tuolumne R.
Merced R.
Mariposa
Lit. Mariposa Riv.
Mariposa R.
M
L
L
Chowchilla R.
Frezno R.
San Joaquin R.
Ft. Miller
San Joaquin R.
O F T H E S I E R R A N E V A D A
MONO LAKE
Owen's R.
OWEN'S L.
Kings R.
M
Four Creeks
Woodville
TULARE OR TACHE LAKE
Tule Cr.
M
Moore's Cr.
White Cr.
Cottonwood Cr.
Clintache L.
Pose Cr.
Cholam L.
Kern River
Walker's Pass
Hum-pah-va-mu
0
20
40
Miles
San Miguel
San Luis Obispo
Buenavista L.
Teh-ee-chay-pah Pass
Kern L.
Tejon Pass
Spring
Spring
Spring
Spring
Pt. Sal
Pt. Purissima

wrote, looking east from near the Kings River. "The chain appears to reach a great altitude, and to rise abruptly from the surrounding subordinate ridges. . . . Snow was resting on the summits in broad, white fields that glistened under the rays of an unclouded sun, and by its rapid melting kept the rivers well supplied with water." The party continued south past the Kaweah, Tule, and Kern Rivers and climbed over the southern tail of the Sierra Nevada at Tejon Pass. From their encampment near the pass they explored the southern end of the range, and realized that for the 250-mile span of the range extending south from Donner Pass the most usable crossing over the Sierra was Tehachapi Pass (Williamson used the Indian word Tah-ee-chay-pah).

Blake published the results of the survey in a comprehensive report (Blake, 1858). Included was a geologic map (dated 1855) that covered about half of the state of California and included nine separate geologic units, separately designated by hand-painted watercolors (Fig. 2.7). Within the area of the parks (which the expedition had not visited), the map colors indicate only granitic and metamorphic rocks. Fourteen named streams are shown on the west slope of the range south of the Stanislaus River, in contrast to the four shown on the Fremont map published seven years earlier. A somewhat distorted Mono Lake appears east of the Sierra Nevada on the Blake map. This large lake is absent from Fremont's map, since his routes had bypassed it and it was never seen. The only habitations shown in the Central Valley south of Stockton are Fort Miller and Woodville; Visalia and Fort Tejon are not shown. Blake's was the first regional geologic map in California, and was a valuable resource for the compilation of later geologic maps. His map, however, does not attempt to depict topography, and is clearly not as advanced as the hachured topographic map that would be published in 1857 by George Goddard.

Breaking New Ground: Goddard and the Trail Blazers

In 1857 the state of knowledge of Sierra Nevada geography was outlined in the most detailed map of California to date, a map prepared by civil

2.8. Part of the State of California map compiled by George H. Goddard in 1857. This map was a primary source for the Brewer party, but they discovered in 1864 that the Kern River did not have its headwaters near Walker Pass, but turned north for some 70 miles, north of the latitude of Owens Lake. Settler Wood was skinned alive in 1850 by Indians at the site of Woodville (east of Visalia, indicated as "Visaija" on the map).

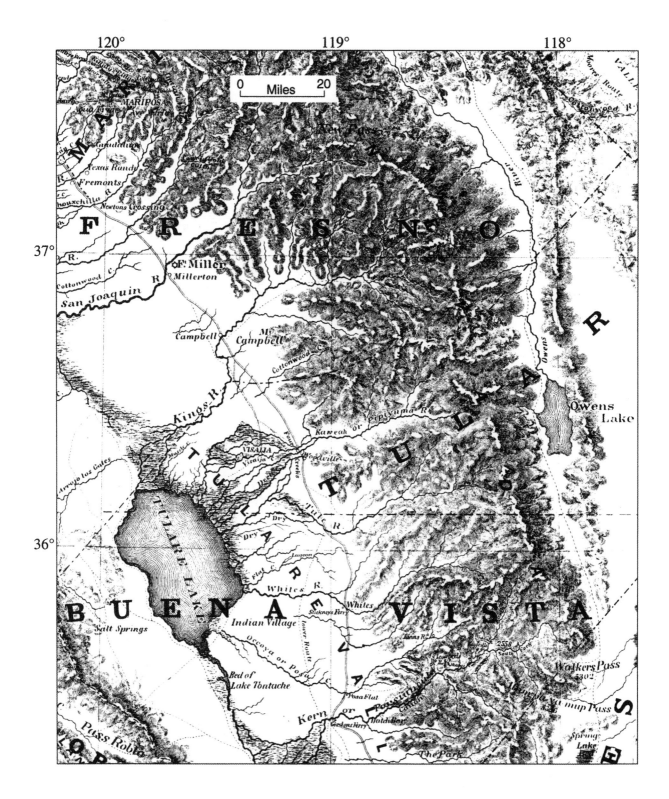

engineer, surveyor, architect, and artist George H. Goddard (Fig. 2.8). Goddard was born in Bristol, England, and was educated at Oxford. He, like many others, was drawn to California by the prospect of making his fortune in the gold rush. In 1850, at the age of 33, he sailed to California by way of Cape Horn. The journey from Plymouth to San Francisco, with an eight-day stopover in the Falkland Islands before rounding the Horn, took 180 days. His wife and two children remained in Europe until 1858, when the family was finally reunited. Within a year after their arrival in California a third child was born.

Goddard went to the goldfields and became a partner in a drugstore in Columbia, but by 1853 he had begun a career as a surveyor. He also maintained a professional interest in architecture, and began a serious interest in landscape painting. His first expedition was a survey of a possible railroad route over the Sierra at Sonora Pass, a route already traversed by a wagon road. This work led Goddard to other mapping excursions around the state. All the while he took copious notes and measurements that would help him later in compiling the California map.

In 1855, working for the California Surveyor General, he made a reconnaissance survey of the route over Carson Pass, and also surveyed the boundary between California and Utah Territory (in what is now western Nevada). For this work Goddard carried altitude and azimuth transits and chronometers, so that he could accurately determine by astronomical observations the latitude and longitude of important survey stations. When location and circumstance permitted, he checked the chronometers against time signals received on newly erected telegraph lines.

Goddard employed mercury and aneroid barometers for the determination of altitude. In his report of September 1, 1855, he wrote, "During the day the aneroid appeared more sensible to small changes in the atmosphere than the mercurial barometer. By half-past nine in the evening both instruments stood precisely the same, at 22.71 inches. Unfortunately, this was the last observation I had with the mercurial barometer, as it was broken soon after. This was a cause of great regret to me, since I had no means left of checking the aneroid" (Goddard, 1855). When camped, Goddard also routinely measured the temperature of boiling water. This temperature is dependent on atmospheric pressure, and hence is a measure of altitude and a check on the barometers. The temperature of pure boiling water under standard conditions at 5,000 feet altitude is 202.8°F (94.9°C), and at 10,000 feet altitude is 193.6°F (89.8°C).

The Goddard map of the state of California, published in 1857 by order of the Surveyor General of California (Fig. 2.8), is based partly on the

Fremont map of 1848 (Fig. 2.3), though many details have been added and other improvements have been made. The map (Goddard, 1857) shows no trails crossing the Sierra in the 200-mile span of the range crest from Sonora Pass, north of Yosemite, to Walker Pass, 50 miles south of Owens Lake. No habitation is shown in Owens Valley. The U.S. Army's Fort Independence is not shown; it was established five years later, on July 4, 1862, as a result of skirmishes between settlers and Indians. The main settlement in the Central Valley of California adjacent to the parks area was Visalia (spelled "Visaija" on the map). Fresno, Bakersfield, and Merced had not yet been settled. The map shows Tulare Lake to be about 40 miles long, whereas on the Fremont map (Fig. 2.3) it is 60 miles long. This seeming discrepancy reflects the extreme shallowness of the lake and its great expansion over flat ground during wet seasons. In early days, prior to the construction of dams, levees, and irrigation canals, much of the Central Valley was flooded during wet winters, and travel was at best uncertain.

Compared with the Fremont map, the Goddard map shows more place names in the region of the two southern parks. That part of the San Joaquin River having its headwaters in the Sierra is named, and the name Kings River is used (as on Blake's map, Fig. 2.7) in place of Fremont's "Lake Fork." Some of the main river names now in use are coupled with other names, such as Kaweah *or* Pipiyuma River, and Kern *or* Porsiuncula River. Goddard's map does not depict the giant south-trending canyon of the Kern River in the heart of the southern Sierra. The range west of the canyon, the Great Western Divide, and that east of the canyon, the main Sierra Crest, form a formidable double crest to the range that was not yet known.

Records of a few incursions into the mountains are available for the period after the publication of the Goddard map and before the 1864 discoveries of the Brewer party. In 1858 J. H. Johnson and five others from Tulare County, in the San Joaquin Valley, were guided by an Indian to the main Sierra Crest at the headwaters of the Kings River. From there they crossed what would soon be called Kearsarge Pass and descended to Owens Valley (Chalfant, 1933). This was apparently the first party of Anglo-Americans to have crossed the highest part of the Sierra Crest and to have seen the main Kings Canyon. For the journey they made, no route, other than one passing through or near the canyon, is feasible (Farquhar, 1965).

John Muir (1968) wrote in 1894 that "Between the Sonora Pass and the southern extremity of the Alps, a distance of nearly 160 miles, there are only five passes through which trails conduct [one] from one side of the range to the other." But four of these passes are to the north of the parks region, and Kearsarge Pass is the only one then known to Muir that crosses the crest of the range within the parks area. This pass, at 11,800 feet, connects two trails, one from the west leading from the Central Valley up to the headwaters of the South Fork of the Kings River, the other from the east, from near the present town of Independence, leading up to the headwaters of Independence Creek.

In 1861, John Jordan established a trail across the range south of the highest Sierra region. Jordan's trail provided a shorter and quicker route from Visalia, in the Central Valley, to the silver mines that had been opened in the Coso Range near Owens Lake, east of the Sierra. The Jordan trail headed east up Yokohl Creek (in the extreme southwestern corner of the map of Fig. 2.2) to Hossack Meadow and Kern Flat on the Kern River. It crossed the main Sierra summit via Jordan Hot Springs, Monache Meadow, and Olancha Pass, and descended the Walker Creek drainage to Owens Lake. Jordan was drowned the following year while crossing the Kern River on a raft.

The Jordan toll trail was soon paralleled by a second trans-Sierra trail, somewhat to the north, that traversed the southern part of the present Sequoia National Park. The Hockett Trail (Fig. 2.2), built by a cattleman, served as a more direct route between Visalia and the Army post at Fort Independence in Owens Valley. It also accommodated the delivery of supplies to the Cerro Gordo mines near Owens Lake, east of the Sierra. Cerro Gordo had become an important source of silver, and the Union forces did not want the mine output to fall into Confederate hands. The three men who were authorized to build the toll trail, with the help of soldiers from Fort Independence, were Bill Cowden, John Hockett, and Lyman Martin. Construction began in 1862. The Hockett Trail led up the South Fork of the Kaweah River, through what is now called Hockett Meadow, and into the basin of the Little Kern River near Quinn Peak. From there it descended into the Kern Canyon, up Golden Trout Creek (then called Whitney Creek or Volcano Creek), through Tunnel and Mulkey Meadows, and over the main Sierra Crest through a pass some 3 miles southeast of Cottonwood Pass. From there the trail descended to Horseshoe Meadow, Cottonwood Creek, and the town of Lone Pine, in Owens Valley. The trail was generally blocked by heavy snow from November through May, but remained for many years the most heavily traveled trail over this part of the southern Sierra.

The leaders of the trail crew hired John O'Farrell as a game hunter to supply meat for the men working on the trail. O'Farrell was also a prospector and loved to roam into the high mountains, and it was on one of these treks that he discovered one of the earliest, and most celebrated, mineral deposits found in this region—Mineral King.

EXPLORATION

The California Geological Survey: Whitney, Brewer, King, and Hoffmann

The first reconnaissance exploration of the high country in the region of the present Sequoia and Kings Canyon Parks was conducted by the California Geological Survey. The Survey, created by act of the state legislature in 1860, was established "to make an accurate and complete geological survey of the state." It was headed by Josiah Dwight Whitney (Fig. 2.9).

Whitney, son of a wealthy New England banker, attended Yale College and studied chemistry, mineralogy, and astronomy under Professor Benjamin Silliman, Sr. He graduated in 1839, and then went on for five years' further instruction in mineral science in Europe. After returning to the United States he worked with state geological surveys in New Hampshire, Iowa, and Wisconsin before being appointed, at the age of 41, as director of the California Survey.

When gold was discovered in California, 12 years prior to the establishment of the Survey, Whitney got the notion that a geological survey of the state should be undertaken, and that he was ideally suited to lead it (Block,

2.9. Josiah Dwight Whitney at about the age of 30. Whitney became the first Director of the California Geological Survey in 1860 and led it for over a decade in the face of daunting opposition to its funding from a hostile legislature. His fault-collapse model for the origin of Yosemite Valley distinctly differed from John Muir's model of a glacial origin (Brewster, 1909).

1982). He marshaled influential friends in California to campaign for such a program. These included S. Osgood Putnam, his brother-in-law and secretary of the California Steam Navigation Company, Judge Stephen J. Field of the California Supreme Court and brother of a close friend, and John Conness, a state legislator who would become a U.S. Senator.

When negotiations finally moved toward establishment of the California Geological Survey, Whitney gained endorsements from many eminent men in the eastern scientific establishment. Other qualified scientists had applied for the directorship of the Survey, and the most persistent was William B. Blake, the geologist on Lieutenant Williamson's Pacific Railroad Survey. It was Blake who had made a geological survey of the Mother Lode gold country in 1854. But in the end, it was Whitney who was appointed director of the fledgling California Geological Survey in 1860 (Fig. 2.10), at a salary of $6,000 per year.

The first assistant Whitney hired was William H. Brewer (Fig. 2.11), a professor of chemistry at Washington College, Pennsylvania, who was a graduate of the scientific school at Yale and had studied botany in Europe for two years. Brewer was appointed to head the Survey's Botanical Department. Whitney also hired William Ashburner, a student of Professor Louis Agassiz at Harvard, as his first assistant in geology. Shortly after arriving in California, Whitney accompanied Colonel John C. Fremont to his Mariposa estate to examine promising gold deposits. After a few months in California, Whitney wrote his brother, "I have found out that the state of California is a prodigiously large one. Not that I did not know it before; but now I have a realizing sense of it. It is as big as Great Britain, Ireland, Belgium, Hanover and Bavaria put together! If I had a complete map of the state, a corps twice as large as I now have, and worked as fast—on the geology only—as the English government surveyors do, I should finish in just 150 years."

In 1861, in San Francisco, Whitney met a young German engineer, Charles F. Hoffmann (Fig. 2.12), who had participated in surveying the route of a wagon road across the Rocky Mountains to California. He had considerable artistic talent, and was considered by Whitney to be the ideal man to make the triangulation surveys and construct the maps that were so necessary for the geological surveys.

Whitney, accompanied by Brewer and Hoffmann, examined Yosemite Valley and the adjacent High Sierra in the summer of 1863. They climbed a prominent peak on the crest and named it Mount Dana, honoring James Dwight Dana, professor at Yale and vice president of the National Academy of Sciences. Describing the naming procedure employed by the Survey, he states, (Whitney, 1874), "The principles we have

2.10. The insignia of the California Geological Survey, showing changes through the years from Whitney (1865) (*top left*), Whitney (1874) (*bottom*), and Gray (1880) (*top right*). Many features of the seal are reminiscent of the photograph of the 1864 field men (Fig. 2.11), and suggest that this photograph served as a guide for the artist who created the seal. The hats and clothing of the men in the seal resemble those in the photograph. The man on the left has a neckerchief similar to that worn by Clarence King, and a sheath knife similar to that worn by Brewer. An instrument case with rounded sides beneath the tripod is similar in shape to that beneath the chair in the photograph. The mountain in the early seals may be Cathedral Peak in Yosemite National Park (*top left*). Later, however, the mountain image was changed, and is clearly Mount Whitney as observed from Lone Pine to the east (*bottom*). Still later, the seal was changed to another view of Mount Whitney as viewed from the south (*top right*). The Latin motto on the seal, *Altiora Petimus* (we seek higher things) also appears on the title page of Clarence King's *Mountaineering* (1935 [1872]). A somewhat altered version, *Altiora Peto* (I seek higher things), appeared in 1892 as the motto of the Sierra Club on its official seal (Fig. 3.14). It is interesting that John Muir, first president of the Sierra Club, who had serious professional conflicts with Whitney, would have adopted for the Sierra Club so close a version of the motto of the Geological Survey of California.

followed in the Geological Survey, in giving names to prominent natural objects, and especially mountains, which had previously been unnamed, are simple, and such as must commend themselves to all reasonable people. We have selected for this purpose the names of explorers, surveyors, geographers, geologists, and engineers, and especially of such as have worked or lived in the region in which the point to be named was situated. . . . We have, in a few cases, selected, 'honoris causa,' the names of very eminent geographers, geologists, or physicists, who have labored successfully in general science, and whose results have thus become the property of the world."

2.11. The 1864 field party of the California Geological Survey (Charles Hoffmann absent). From left to right: James T. Gardiner, Richard D. Cotter, William H. Brewer, and Clarence King. Gardiner is holding a sextant used for mapping and astronomical location; Cotter is armed with musket, dagger, and pistol; both Brewer and King have mercury barometers, used to measure elevations, slung on their shoulders; and King holds a geologist's hammer. U.S. Geological Survey Library.

2.12. The staff of the California Geological Survey in about 1870. From left to right: William H. Pettee, Josiah D. Whitney, Alfred Craven, Watson A. Goodyear, Carl Rabe, and Charles F. Hoffmann. All but Pettee played a role in the Mount Whitney saga. Hoffmann (topographer) was on the 1864 Brewer expedition that identified and named Mount Whitney, honoring Whitney (director of the survey). Craven (assistant topographer), Goodyear (geologist), and Hoffmann were on the mapping expedition of 1870 to Owens Valley that incorrectly identified and surveyed Sheep Mountain (now Mount Langley) as Mount Whitney. In July 1873, Goodyear rode a mule to the summit of Sheep Mountain and reported that the true Mount Whitney was about 5 miles to the north. Rabe (cook and topographic assistant) was in the 1873 party that made the third ascent of the true Mount Whitney, and made the first barometric measurements of its height. Bancroft Library.

Whitney was most impressed by the great height of this part of the range, and wrote to his brother (Block, 1982), "The region west and southwest of Mono Lake we found to be the highest part of the State and wonderfully grand—I cannot imagine that there is anything in the United States to compare with what we saw from the summit of Mt. Dana, over 13,400 feet high, and probably the highest point in the State except Mt. Shasta." He went on to remark, "Our most brilliant discovery was of a vast region once occupied by glaciers and most beautifully polished and grooved with all the phenomena of moraines."

In the summer of 1864 a California Survey party headed by William H. Brewer entered the mountains east of Visalia (Fig. 2.11). The others in the party (with their official titles) were Charles F. Hoffmann (Principal Topographical Assistant), James T. Gardner (Volunteer Assistant in the Topographical Field-work), Clarence King (Volunteer Assistant in the Geological Field-work), and Richard D. Cotter (Packer). The volunteers were unpaid. (Gardner spelled his name without the "i" when with the California Survey, but used the spelling "Gardiner" in later life. On modern maps the name is Mount Gardiner.)

Clarence King was born in Newport, Rhode Island, in 1842. His father was a merchant who traveled frequently to China and died there of fever when Clarence was six. In 1855 the family moved to Hartford, Connecticut, and here Clarence met Jim Gardiner, a boy his age who would remain a lifelong friend. King entered the Yale Scientific School at New Haven in 1860, and at his urging Gardiner entered the school two years later. But King was eager to try his luck in the West, and perhaps also was apprehensive of Army service, since Civil War conscription was under way. In the spring of 1863, before Gardiner had taken his examinations, King enticed his friend to go west with him. The two headed off by railroad to St. Joseph, Missouri. Here they joined a wagon train, and it was while crossing the plains near Chimney Rock that they met Dick Cotter, a young drover, who would later accompany King on several expeditions. They crossed the Rocky Mountains at South Pass in June and arrived at the Comstock Lode in western Nevada August 6. By chance, the three met William Brewer on a river steamer between Sacramento and San Francisco. Probably aided by the common Yale background of all but Cotter, they were taken on as unpaid volunteers with the California Geological Survey.

That fall King joined Brewer on a reconnaissance geologic survey of the northern Mother Lode in the Sierra. Twice they climbed Mount Lassen, to take advantage of its towering height for mapping purposes. The small party then traveled north into Oregon, and returned to San Francisco by way of Crescent City and down the north coast of California.

King's next assignment, in the winter, was to make a general survey of the Mariposa Estate, which had been purchased by John Fremont after his third expedition in 1846. The property, now controlled by the Mariposa Mining Company, was considered one of the most important gold-vein regions in the Mother Lode. It was while working here that King saw to the east the vast array of high peaks along the Sierra Nevada Crest. Observations made in the clear winter weather by King and Hoffmann during January 1864 from the summit of Mount Bullion, 5 miles west of Mariposa (Fig. 2.7), revealed an "immensely high mountain mass south . . . somewhere near the Owens Lake." The following summer, expecting that the highest peaks in the range might be in this area, and desiring to fill in the largest blank area on the map of California, the survey mounted an expedition into the region.

King, short but well built and possessed of great physical stamina, was a man of remarkable gifts. He was an eloquent and persuasive talker, a talented writer, and a good scientist. Considering King's youth (he was just 22), we may well conclude that it was the force of his arguments that was decisive in taking the expedition to this sector of the Sierra, which had come to interest him after his observations from Mount Bullion. He was also successful in bringing along his two friends, Gardiner and Cotter, and these three, all unpaid, actually outnumbered the other two members of the expedition, Brewer and Hoffmann.

The decision having been made, the five men traveled to Visalia, passed by Thomas sawmill (at the present site of Sequoia Lake, 2 miles west of Wilsonia; Fig. 1.3), and observed that felling and splitting of the giant sequoia trees of the Grant Grove region was rampant. These majestic trees were being cut primarily to make fenceposts. The party continued through Big Meadows and climbed a peak on the Kings–Kaweah Divide (Fig. 2.13) that they named Mount Silliman (after Yale Professor Benjaman Silliman, Jr.). From this peak they had a magnificent view of the country above timberline, including the Kings–Kaweah Divide, and eastward toward an impressive mountain wall that they took to be the principal crest of the range. (The fact that the main crest of the Sierra was obscured by the intervening Great Western Divide was unknown to them at the time.) They now directed their efforts toward the highest peak on this ridge crest, believing it to afford a good site for observation of the whole range and the desert country to the east, including Owens Lake, a major landmark.

2.13. Mount Silliman seen (*above*) from the north (U.S. Geological Survey photograph by Grove K. Gilbert), and (*below*) from the northwest, after a sketch by Charles Hoffmann (Whitney, 1865). When the 1864 Brewer party climbed this peak, they saw a high mountain crest 10 miles to the east (Fig. 2.16) that they took to be the main Sierra Crest. After climbing Mount Brewer on this crest, they realized that it was only a secondary ridge, the Great Western Divide, and that the main crest was still nearly 10 miles farther east.

EXPLORATION 47

The expedition's mapping strategy was to occupy high peaks or other vantage points and measure the angular relations between these points and other major identifiable landmarks. The transit, sextant, and compass were the instruments used for taking angular measurements on such landmarks as peaks and distinctive parts of ridges, valleys, and lakes. Measurements of vertical angles determined the relative elevations between an occupied peak and other landmarks.

At that time the preferred method for measuring the altitude of occupied points was by readings from a mercury barometer (also called a *cistern barometer*, from the form of its mercury reservoir; Fig. 2.14). This instrument measured atmospheric pressure by the height of a column of mercury that balanced the weight of the atmosphere. Inasmuch as 30 inches of mercury is about equal to atmospheric pressure at sea level (the height of the mercury falling proportionately at higher altitudes), and since an instrument of this sort thus had to be at least that long, these instruments were long, cumbersome, and fragile. The glass barometer was housed in a brass tube and an outer leather or wooden case for protection, but breakage of the barometers in the field was nonetheless a common problem. The height of the mercury at the top of the barometer was measured to a precision of two one-thousandths of an inch by means of a vernier scale. Because the height of the mercury column is affected by temperature as well as pressure, temperature readings were required so that appropriate corrections might be made. In theory, then, one could detect a change in the mercury height by raising the barometer only a few feet, but a moderate weather change could change the mercury height one hundred times that amount.

The much more compact aneroid barometer, a mechanical device, had been used by Goddard in 1855, but did not come into general use until later. In order to account for natural changes in barometric pressure caused by shifting weather conditions, corrections were made (at a later time) by comparing concurrent readings taken from a separate stationary instrument kept at a base station. For the 1864 field campaign, these readings were made and recorded daily by Army personnel at Camp Babbitt, near Visalia.

From Mount Silliman the party descended Silliman Crest near Kettle Peak and followed Sugarloaf Creek to camp near Sugarloaf Rock (Fig. 2.15). The next day they saw a system of terraces of debris left by glaciers on canyon sides at the previous high-ice level. These moraines flanked the canyon of Roaring River (Fig. 9.9). "Reaching the brink of this gorge," King (1935) wrote later,

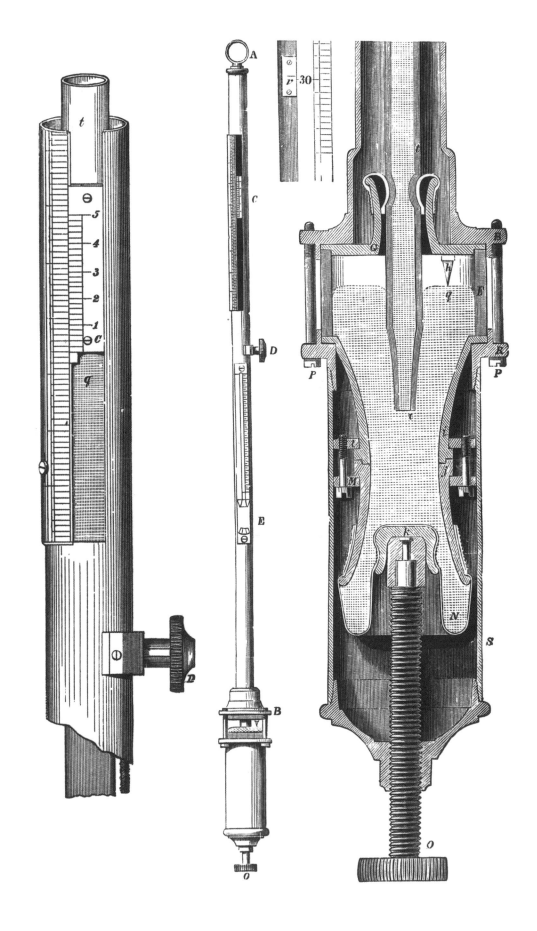

we observed, about half-way down the slope, and standing at equal levels on both flanks, singular embankments—shelves a thousand feet in width—built at a height of fifteen hundred feet above the valley bottom, their smooth, evenly graded summits rising higher and higher to the eastward on the canyon-wall until they joined the snow. They were evidently the lateral moraines of a vast extinct glacier, and that opposite us seemed to offer an easy ride into the heart of the mountains. . . . By a series of long zigzags we succeeded in leading our animals up the flank to the top of the north moraine, and here we found ourselves upon a forest-covered causeway, almost as smooth as a railroad embankment. Its fluted crest enclosed three separate pathways, each a hundred feet wide, divided from each other by roughly laid trains of rocks, showing it evidently to be a compound moraine. As we ascended toward the mountains, the causeway was more and more isolated from the cliff, until the depression between them widened to half a mile, and to at least five hundred feet deep. Throughout nearly a whole day we rode comfortably along at a gentle grade.

At the head of the moraine, the trail became rougher and, finally, when they could take the animals no farther, they made camp at a mountain lake at the base of the pyramidal peak (Fig. 2.16) they had seen from Mount Silliman.

Arising before dawn the next morning, Hoffmann and Brewer climbed the peak with difficulty, and from the summit they commanded a view of a desolate and wild landscape with a vast array of high peaks. Brewer (1930) wrote: "Such a landscape! A hundred peaks in sight over thirteen thousand feet—many very sharp—deep canyons, cliffs in every direc-

2.14. Mercury (or cistern) barometer manufactured by James Green, first described in 1856. Mercury (q) in the lower part of the barometer (figure at the right) is contained in a cistern with a leather bottom (N) inside the brass cylinder (S). The cistern is fabricated with boxwood sides and top (G, E, i, j, k). Boxwood is pervious to air but impervious to mercury, and therefore there need be no opening in the cistern to allow access of air, a feature important in a portable instrument. To calibrate the instrument, a twist of the thumb screw (O) squeezes the bottom of the bag and raises the surface of the mercury to the tip of the ivory point (h) visible on the right side of the mercury surface. When adjusted, the height of the mercury column (which is supported by the local air pressure) can be measured to $2/1000$ inch by moving a vernier scale (C) to the top of the mercury column, as seen within the glass tube (t). The reading of atmospheric pressure, in inches of mercury, can be converted to apparent altitude with appropriate corrections, including temperature as measured by the thermometer (E). Smithsonian Institution (Middleton, 1969).

2.15. Part of the Hoffmann map of 1873, showing the route of the Brewer party across the Sierra in 1864. It was from the summit of Mount Brewer (center) that Brewer and Hoffmann sighted the highest peak in the range and named it Mount Whitney (lower right). Clarence King begged to backpack to it and climb it. He and Richard Cotter returned five days later; they had got only as far as Mount Tyndall on the main Sierra Crest (Hoffmann, 1873).

tion . . . sharp ridges almost inaccessible to man, on which human foot has never trod." Much to their dismay, however, they discovered that this peak, Mount Brewer (13,570 feet), which they named for their party chief, was not on the main Sierra Nevada Crest, but on a lesser ridge, the Great Western Divide. The main Sierra Crest lay a formidable 8 miles farther east across a wild and rugged region at the headwaters of the Kings and Kern rivers. They named several lofty peaks, including Mount Tyndall (for an eminent British geologist) on the main crest to the east, Mount Goddard (for the engineer who had compiled the best California map to date) 30 miles north-northwest, and Mount Whitney off to the southeast. Mount Whitney was of course named for the director of the California Geological Survey. On setting the level in place, they found that Mount Whitney was clearly the highest mountain in sight and probably the highest in the entire range and in the United States.

Brewer and Hoffmann returned to camp that night exhilarated and exhausted. Hoffmann showed the others his field sketches of the main range crest, and Brewer explained why a crossing of the terrain that lay between them and the crest was utterly impossible. Clarence King, however, did not accept this judgment. He was excited about further exploration among the high peaks and begged Brewer for permission to undertake a backpacking trip to explore them, with the goal of climbing Mount Whitney. Brewer thought the plan sheer madness but, anxious to obtain the observations from the main Sierra Crest that would outline this part of the range, finally agreed.

The next day King and Cotter shouldered 35- to 40-pound backpacks with blankets and five days' provisions. Brewer, Gardiner, and Hoffmann, helping with the packs, climbed with them to the saddle south of Mount Brewer. King (1935) described the view from the pass:

> Rising on the other side, cliff above cliff, precipice piled upon precipice, rock over rock, up against sky, towered the most gigantic mountain-wall in America, culminating in a noble pile of Gothic-finished granite and enamel-like snow. . . . I did not wonder that Brewer and Hoffmann pronounced our undertaking impossible; . . . our friends helped us on with our packs in silence, and as we shook hands there was not a dry eye in the party. Before he let go of my hand Professor Brewer asked me for my plan, and I had to own that I had but one, which was to reach the highest peak in the range.

From the pass, Brewer, Gardiner, and Hoffmann turned north to climb again to the summit of Mount Brewer and complete their observations.

2.16. Mount Brewer (13,570 feet) as viewed (*above*, left of center) from the northeast as seen from Glen Pass (U.S. National Park Service photograph), and (*below*) from the southwest in a sketch by Brewer (Whitney, 1865).

King and Cotter traversed south on the crest of the Great Western Divide, looking for a way into the Kern River Drainage, but after climbing over the next eminence south of Mount Brewer, a high peak now called South Guard (13,224 feet), they found that progress along the ridge crest was blocked by a series of jagged pinnacles. Thwarted in that endeavor, they descended by rope near Longley Pass to a lake (probably the lake marked 11,459 on the Mount Whitney 15-Minute Quadrangle map) draining into East Lake on the Kings River side of the Kings–Kern Divide. From there they turned south and climbed into a cirque at the north base of the divide, where they camped above timberline. King recounts cutting shavings off the wooden barometer case to build a tiny fire to " . . . warm water for a cup of miserably tepid tea." It was a cold night; King reported temperatures of 20°F at 9 p.m. and 2°F when they arose in the morning.

Beginning before first light the next morning they climbed the precipitous Kings–Kern Divide by a route that is difficult to reconstruct but may have been near and west of what is now called Mount Jordan (13,344 feet; Alsup, 1987). At one ledge, where their advance was blocked by a 30-foot vertical cliff, King lassoed a spike of rock protruding from the next ledge and climbed the rope hand over hand. After King hauled up the packs, Cotter too climbed the rope, and before long they were at the summit of the divide. The descent down the headwall into the upper Kern River Drainage was a hair-raising adventure. They roped down a series of almost vertical faces, until they finally reached the snow slopes at the base of the cliffs.

A long traverse eastward, high in the headwaters of the Kern River, brought them at dark to a campsite probably on Tyndall Creek at about 11,600 feet. Arising at 3:30 a.m. the next morning, they began climbing the north side of Mount Tyndall (Fig. 2.17), soon realizing that the highest peaks to the south were too distant to attempt. Climbing up steep rock slopes, they were often compelled to cut steps in steep snow and ice. The culminating challenge was a giant column of ice that stood out from the final cliff below the summit. King (1935) recalled,

> We climbed to the base of this spire of ice, and, with the utmost care, began to cut our stairway. The material was an exceedingly compacted snow, passing into clear ice as it neared the rock. We climbed the first half of it with comparative ease; after that it was almost vertical, and so thin that we did not dare to cut the footsteps deep enough to make them absolutely safe. There was a constant dread lest our ladder should break off, and we be thrown either down the snow-slope or into the bottom of the crevasse. At

last, in order to prevent myself from falling over backwards, I was obliged to thrust my hand into the crack between the ice and the wall, and the spire became so narrow that I could do this on both sides; so that the climb was made as upon a tree, cutting mere toe-holes and embracing the whole column of ice in my arms. At last I reached the top, and with the greatest caution, wormed my body over the brink.

King's account of the difficulty of the climb up Mount Tyndall has been criticized by some as exaggerating what is considered today to be a moderate climb. We must remember, however, that he and Cotter were in a totally unknown wilderness never before penetrated by explorers, an alien place that was two long days distant from their base camp, which was itself in a remote place. It is understandable that they would perceive this forbidding country differently than we do today.

On the summit of Mount Tyndall, on the main Sierra Crest, the mercury level in their barometer stood at 17.99 inches, water boiled at 192°F, and they made the altitude at 14,386 feet (the 1956 U.S. Geological Survey map shows it at 14,018 feet). King (1935) stated that

> Upon sweeping the horizon with my level, there appeared two peaks equal in height with us, and two rising even higher. That which looked highest of all was a cleanly cut helmet of granite upon the same ridge with Mount Tyndall, lying about six miles south, and fronting the desert with a bold square bluff which rises to the crest of the peak, where a white fold of snow trims it gracefully. Mount Whitney, as we afterwards called it in honor of our chief, is probably the highest land within the United States (see Fig. 2.18).

In describing the bedrock, King mentions "the large crystals of orthoclase with which the granite is studded being cut down to the common level, their rosy tint making with the white base a beautiful burnished porphyry." From Mount Tyndall they became aware how the summit region of the Sierra, as seen at this latitude, is divided into two giant paral-

2.17. Looking southeast past Mount Tyndall (lower right, 14,018 feet) to Mount Versteeg at the bend in the ridge opposite the most distant lake (Lake Helen of Troy, which lies at 12,515 feet). Mount Tyndall was climbed in 1864 by Clarence King and Richard Cotter, when they realized that Mount Whitney was much too far to attempt. The west flank of Mount Williamson is at the upper left. Rock glaciers, fed from avalanche chutes on the east flank of Mount Tyndall, encroach on the lake basin at the lower left. U.S. Geological Survey aerial photograph by Austin Post.

2.18. View southeast across Mount Whitney (W) to Mount Langley (L, 14,042 feet), showing the similarity in shape of these two mountains. Clarence King climbed Mount Langley in 1871 in cloudy weather believing it was Mount Whitney. The peak in the right background (C) is Cirque Peak (12,900 feet), and Olancha Peak (12,135 feet) stands on the distant skyline above it. U.S. National Park Service photograph by Francois Matthes.

lel ridges, with the Kern River flowing in the tremendous south-directed gorge between them. They also realized that the Kaweah Peaks, a prominent jagged ridge (Fig. 2.30) that could be seen to the east from near Visalia in the Central Valley, were visible to them to the southwest, near the south end of the Great Western Divide. East of the higher, main Sierra Crest they beheld a magnificent view of Owens Valley, Owens Lake, and the desert ranges extending deep into Nevada. To the northeast, east of the Sierra, they perceived a distant high range capped with snow, the White Mountains. From the summit, King named a major peak somewhat east of the range crest, only one mile distant, Mount Williamson, in honor of Colonel Robert Williamson, leader of the Pacific Railroad Surveys in 1853 and 1855.

Overwhelmed by the view, King waxed eloquent. In his book *Mountaineering in the Sierra Nevada* (1935), he described the landscape seen from the summit of Mount Tyndall,

> The whole region, from plain to plain, is built of this dense solid rock [granite], and is sculptured under chisel of cold in shapes of great variety, yet all having a common spirit, which is purely Gothic. . . . Yet, as I sat on Mount Tyndall, the whole mountains shaped themselves like the ruins of cathedrals—sharp roof-ridges, pinnacled and statued; buttresses more spired and ornamented than Milan's; receding doorways with pointed arches carved into blank facades of granite, doors never to be opened, innumerable jutting points with here and there a single cruciform peak, its frozen roof and granite spires so strikingly Gothic I cannot doubt that the Alps furnished the models for early cathedrals of that order.

The granite landscape is, indeed, overwhelming in this region, and King was obsessed by it. The area of his view, including Mounts Brewer, Tyndall, and Whitney, is underlain by a giant sequence of granitic masses, or plutons, called the Mount Whitney Intrusive Sequence; 50 miles long and more than 10 miles wide, it includes the largest individual plutons in the Sierra Nevada. For a considerable distance, the almost monotonous granite landscape is nowhere moderated by darker rocks.

Rocks of these granitic masses are particularly coarse-grained and contain large crystals (phenocrysts) of potassium feldspar set in a finer-grained matrix. The conspicuous phenocrysts reach several inches in size and commonly protrude from the surface, because of differential weathering. This interesting textural variation was not lost on King, who found that the protruding crystals were hard to sleep on but offered good handholds for climbing.

In *Mountaineering* he mentions that when sleeping on a granite slab at the base of the Kings–Kern Divide, "A single thickness of blanket is a better mattress than none, but the larger crystals of orthoclase, protruding plentifully, punched my back and caused me to revolve on a horizontal axis with precision and frequency." In describing the descent of the Brewer wall, he wrote, "When within about eight feet of the next shelf, I twisted myself round upon the face, hanging by two rough blocks of protruding feldspar, and looked vainly for some further hand-hold; but the rock, besides being perfectly smooth, overhung slightly, and my legs dangled in the air." On the final ascent of Mount Tyndall, he states, "We climbed alternately up smooth faces of granite, clinging simply by the cracks and protruding crystals of feldspar, and then hewed steps up fearfully steep slopes of ice, zig-zagging to the right and left to avoid the flying boulders."

With observations from the summit of Mount Tyndall completed, King and Cotter descended the mountain on the west side and circled back to their previous campsite near Tyndall Creek, where they spent the night. The next day they avoided the previous route because of the grave difficulty they would have faced in ascending the wall they had roped down. Instead, they crossed the Kings–Kern Divide to the east, probably at Millys Foot Pass (3 miles east of South Guard and about 1.5 miles east of their previous crossing west of Mount Jordan), and made their fourth and final backpack camp near Lake Reflection. The next day they returned via Longley Pass and the shoulder of Mount Brewer, which they had previously traversed. King writes about their return to the main camp: "Our shouts were instantly answered by the three voices of our friends, who welcomed us to their camp-fire with tremendous hugs." After they had described the results of their travels, Brewer remarked, "King, you have relieved me of a dreadful task. For the last three days I have been composing a letter to your family, but somehow I did not get beyond 'It becomes my painful duty to inform you.'"

Once again consolidated, the full party returned to Big Meadows, but because of a severe toothache, Brewer decided to seek medical help back in Visalia. King took this opportunity to accompany him to Visalia, and from there to make an attempt on Mount Whitney from the southwest.

With an escort of two men, King headed into the mountains on the Hockett Trail, which was still under construction at that time. He crossed the Kaweah-Kern Divide between the South Fork of the Kaweah River and the Little Kern River (which was called the North Fork of the Kern in the Whitney report), then across Coyote Pass to the main Kern River. Because the Hockett Trail had not yet proceeded beyond this point, King and his followers headed up the river for several miles before climbing cross-country to the east, out of the south-trending canyon. The Whitney report (1865) states that in the uplands east of the Kern

> in the midst of every difficulty, Mr. King worked for three days before he could reach the base of the mountain. . . . The highest point reached . . . was, according to the most reliable calculations, 14,740 feet above the sea-level. At the place where this observation was taken, he was, as near as he was able to estimate, between 300 and 400 feet lower than the culminating point of the mountain, which must therefore somewhat exceed 15,000 feet in height. So far as known it is the highest point in the United States, and the elevation attained by Mr. King was greater than any other person has reached, within our territories, or anywhere on the continent north of Popocatepetl.

King was disappointed when this attempt at the mountain fell short, and was no doubt all the more determined to make another attempt

2.19. General view of Kings River Canyon, as viewed from the trail descending into the canyon from the southwest, the general route taken by the Brewer party. Drawn by Charles Robinson (Muir, 1891).

when the time was right. Mount Whitney, however, was finally first ascended nine years later, in 1873, by three residents of Lone Pine, as will be recounted later.

King then retraced his route through Visalia and joined the Brewer party, which was now resupplied with a month's provisions on three pack mules at Big Meadows. Because encounters with hostile Indians were possible in the mountains and the areas to the east, the party was enlarged by the addition of an escort of seven soldiers. The group then struggled on poor tracks northeast toward the brink of Kings Canyon, the great glacial "yosemite" of the South Fork of the Kings River (Figs. 2.19, 2.28). They descended on a "horrible" trail apparently west of the Don Cecil Trail, perhaps down Lightning Creek, to judge from the route on the Hoffmann map of 1873 (Fig. 2.15). Brewer (1930) wrote, "We sank into the canyon of the main South Fork of Kings River, a *tremendous* canyon. We wound down the steep side of the hill, for *over three thousand feet* [italics his], often just as steep as animals could get down."

From the Whitney report (1865) we hear that

> The canyon here is very much like the Yosemite. It is a valley from half a mile to a mile wide at the bottom, about eleven miles long and closed at the lower end by a deep and inaccessible ravine like that below the Yosemite, but deeper and more precipitous. It expands above and branches at its head, and is everywhere surrounded and walled in by grand precipices, broken here and there by side canyons, resembling the Yosemite in its main features. The Kings River canyon rivals and even surpasses the Yosemite in the altitude of its surrounding cliffs, but it has no features so striking as Half Dome, or Tutacanula [El Capitan], nor has it the stupendous waterfalls which make that valley quite unrivaled in beauty.

(Undoubtedly, John Muir read the official Whitney report of the expedition with intense interest. The allusion to Yosemite Valley suggested to him, no doubt, that Kings Canyon may have had a similar origin. The statement that it "rivals and even surpasses the Yosemite" piqued his interest, and after his first visit to the region, in 1873, he would use comparable phrases.)

Members of the Brewer party next attempted to explore side canyons, but in most places the climb out of the canyon was very difficult. A soldier discovered that a rough route on the north side, up Copper Creek, was passable, and they climbed up this drainage to the crest of the Monarch Divide near Granite Pass (10,673 feet), between the canyons of the South and Middle Forks of the Kings River. From the crest, they took angles on Mount Goddard, 25 miles north-northwest on the San Joaquin–Kings Divide, and the Palisades, 15 miles northeast on the main Sierra Nevada Crest. Two nearby peaks, about 10 miles east, were named Mount Gardiner and Mount King (Fig. 2.20) to honor members of the party. Mount King was later changed to Mount Clarence King, and the name Mount Cotter was added to a lesser peak between Mounts Gardiner and Clarence King.

After extended reconnaissance, Hoffmann found that the pack animals could not descend into the Yosemite-like Middle Fork Canyon. The party therefore retraced its steps, descending into the South Fork Canyon, and then headed east toward the main Sierra Crest up the south branch of the river (now Bubbs Creek). This venture might have failed had it not been for the good fortune of their meeting a group of prospectors who had just built a rough trail for their horses from Owens Valley over Kearsarge Pass (Fig. 2.21).

One of the prospectors was Thomas Keough of Independence, who in later years (Colby, 1918) described the trip from Owens Valley:

2.20. Mount King (12,905 feet; name later changed to Mount Clarence King), as viewed (*above*) from the southeast (U.S. National Park Service photograph by William Huffman), and (*below*) from the west (from a sketch by Brewer, as rendered in Whitney, 1865). On the left skyline (above) are Mounts Gardiner and Clarence King. The conical crag visible in the middle ground to the left of the large lake (Rae Lake) is Fin Dome.

2.21. View north from the summit of University Peak toward Kearsarge Pass (11,600 feet), the major pass over the highest Sierra. Dark-topped Mount Gould (13,005 feet) is at the center, and the peaks of the Palisades are visible on the main Sierra Crest on the skyline at right. The arrow points to one of several trail strands that lead left, up toward the pass.

Our first task was to build a trail up Little Pine Creek [now called Independence Creek] on the east cliff of the mountains. I have sometimes heard it said that the trail over what is now called 'Kearsarge Pass' is an old Indian trail. The fact is, however, that our party built this trail in order to get our animals up over the top of the Sierras. It might have been possible for a man to work his way on foot up over this pass, but there was no sign even of a foot-path until we built the trail in the summer of 1864 when we started on this prospecting tour. . . . It was a rough trail we built, but it sufficed for our purposes and we got our animals up over it.

This statement of the lack of a trail is surprising in view of other evidence that suggests that Indians used the pass frequently, albeit without horses.

After crossing the pass, Keough continues, "We went westerly down the South Fork of the King's River until the cañon became impassable. In the cañon we met a number of scientists headed by Professor Brewer. They named Mt. Brewer after him. Prof. Brewer was trying to find a way across the mountains, and we told him how to get into Owen's Valley over the pass by the trail we had just built."

Keough goes on to relate,

> The rest of our party, who left us soon after we climbed up over Little Pine Pass, found a gold mine near the pass on their way home which they called the 'Cliff Mine'. This mine developed into quite a rich ledge, and it was through this discovery that the pass came to be known as 'Kearsarge Pass'. To the south a mining district was named by Rebels the 'Alabama District' after the Confederate cruiser 'Alabama'. Our crowd, however, were all Union men, and when the news came that the Kearsarge had sunk the Alabama, our boys named the district where the Cliff Mine was the 'Kearsarge District' to taunt the Rebels.

By following the prospectors' very rough trail, the Brewer party gained the summit of the range at Kearsarge Pass (11,823 feet) on the main Sierra Crest and descended the steep escarpment down to Owens Valley, near the present town of Independence.

"Camp Independence was located in the valley, and for a year fighting went on, but at last the Indians were conquered—more were starved out than killed," wrote Brewer (1930). "They came in, made treaties, and became peaceful. One chief, however, Joaquin Jim, never gave up. He retreated into the Sierra with a small band, but he has attempted no hostilities since last fall."

After a long trip north up Owens Valley, with daytime temperatures at 102–106°F, the party reentered the mountains at Rock Creek, crossed Mono Pass, and camped on Mono Creek in Vermilion Valley (now flooded by Lake Edison) in the drainage of the South Fork of the San Joaquin River. They ascended to the plateau south of the San Joaquin River and according to one assumption passed through Helms Meadow and east across Dusy Creek.

Brewer and Hoffmann were determined to climb Mount Goddard (Fig. 2.22), the high peak underlain by dark metamorphic rock that they had seen and named from Mount Brewer, and that had been identified again from the crest of Monarch Divide. This point, they knew, would command vistas of a huge region of unmapped territory, and would thus be important for the topographic surveys. They traveled as far as horses could go and camped for the night at 10,000 feet, perhaps in upper Red Mountain Basin. The next morning a party of four—Brewer, Hoffmann, Cotter, and Spratt (a soldier)—headed toward the peak. After crossing six granite ridges, all over 11,000 feet in altitude, Brewer and Hoffmann gave out from sheer exhaustion when they viewed the mountain from a 12,000-foot peak and saw that it was still about 6 miles distant. They

2.22. Looking southeast across the east ridge of Mount Goddard, with the Ionian Basin in the background at top. Note the large cracks in the ice (bergschrunds) at the heads of small glaciers, and the piles of eroded material (moraines) at the glacier toes. U.S. Geological Survey aerial photograph by Austin Post.

were probably on the crest of the LeConte Divide near Hell-for-Sure Pass, viewing the mountain across Goddard Canyon. Too weary to go on, they turned back, walked until long after dark, and camped at 11,000 feet, with little food and no blankets. They fired a stump and huddled around it for a long freezing night.

Cotter and Spratt, however, continued on toward the mountain. They traveled all night, walked continuously for 36 hours, 26 without food, and finally were forced to turn back after getting within 300 feet of the summit. Brewer pointed out the abundant metamorphic rock in this region, primarily mica slate, and Cotter described alternating veins or beds of slate and granite underlying the Goddard summit. This mass of dark metamorphic rock underlying Mount Goddard is 25 miles long, the largest separate mass of metamorphic rock in the central Sierra Nevada.

After this unsuccessful foray, the party returned to Vermilion Valley, where their provisions were replenished by soldiers from Fort Miller. They then traveled to Wawona, near Yosemite, and Brewer joined another group to explore some of the Sierra Crest near Mount Dana. But upon his return he found Hoffmann with a very sore leg, unable to walk, and the party carried him by litter to Mariposa, where carriage and river steamer transported them back to San Francisco.

This 1864 expedition of the Brewer party, despite its extraordinary accomplishments, was of short duration and low budget. Of the five members, both King and Gardiner were unpaid volunteers. In his synopsis of the field work, Whitney observed that "the Sierra Nevada [is] a chain of mountains nearly as extensive as the Alps . . . and, when we consider that the number of Alpine explorers and of the published volumes of their results may be counted by the hundreds, their researches extending over nearly a century, we feel that we need not apologize for the imperfections of our work, believing, as we do, that we have done the best which our time and means have permitted."

Despite the reconnaissance nature of this work, these men had made the first topographic, geologic, and botanical survey of a vast area. They identified, mapped, and named the highest peak then in the nation, and they crossed the most usable pass over a lengthy section of the range crest. They were the first to outline the course of the Kern River, which flows in a major canyon some 75 miles south from the Kings–Kern Divide before turning west to the Central Valley. They defined the positions of the Great Western Divide to the west of the Kern Canyon and the main Sierra Crest to the east of the Canyon. They showed that evidence for recent glacial action is widespread in the higher parts of the range. They

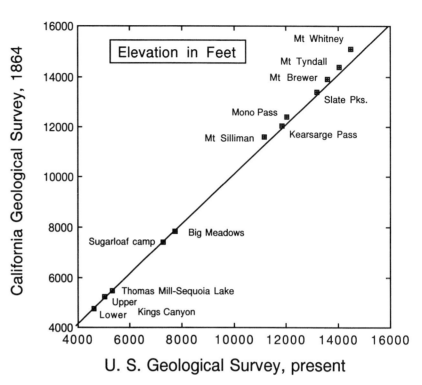

2.23. The elevations of selected points, as measured by barometer by the Brewer party, contrasted with modern U.S. Geological Survey elevations measured by leveling and triangulation. The Brewer party elevations were good, considering the equipment and methods available to them; although their measurements fall close to the line of equal altitudes, curiously many fall above it, indicating that Brewer party altitudes were consistently slightly high.

determined for the first time the altitude of many points on their route, utilizing the mercury barometer, and their data compare favorably with modern determinations (Fig. 2.23).

The expedition demonstrated that granite overwhelmingly predominates in the southern Sierra, whereas in the gold country to the north a broad metamorphic belt flanks the range along the west. In the 26-page section of the Whitney report (1865) on "The Region About the Head of Kern and King's Rivers," the words "granite" or "granitic" are used 49 times. Brewer's background was in botany, and his position on this survey was that of botanist, but he had a keen interest in geology. In a letter written to Professor George Brush of Yale during the expedition, he praised the work of Clarence King, who had been a student of Brush. Brewer remarked that King had been converted from the metamorphic theory of the origin of granite to the intrusive theory. This debate whether granite is formed by the solid recrystallization of preexisting rocks (granitization) or by crystallization from molten material (magmatism) continues today.

Six years after the epic 1864 journey, the California Geological Survey mounted another expedition, this time to explore the vicinity of the Inyo Mountains, Owens Valley, and the east side of the highest Sierra. The party consisted of topographer Charles Hoffmann, assistant topographer Alfred Craven, and geologist Watson A. Goodyear. They crossed the Sierra Nevada via Walker Pass and spent several months east of the Sierra. Hoffmann, who had seen and taken bearings on Mount Whitney from Mount Brewer six years earlier, attempted to identify it from Owens Valley and to refine its position by the measurement of angles from known points in the valley.

"That was the time that the mistake was first made in the identity of Mount Whitney," wrote Goodyear (1888),

> It was a very natural mistake. For 'Sheep Mountain' [the present Mount Langley], which lies some five or six miles southeast of Mount Whitney, and opposite the upper end of Owens Lake, stands several miles nearer to the eastern foot of the range, so that looking up from the valley, ten thousand feet below these summits, the 'Sheep Mountain' *looked* the highest, and at the same time its *shape* bears a close resemblance to that of Mount Whitney itself as seen from the valley near the lake" [italics his].
>
> Mr. Hoffmann, therefore (who was the only one in our party who had ever seen Mount Whitney before), mistook the 'Sheep Mountain' for Mount Whitney, and we continued to call it so all that summer. But after our return to San Francisco, when Mr. Hoffmann came to plot his work, he found a large and unaccountable discrepancy in the location of Mount Whitney between his work of 1864 and 1870, the latter work placing Mount Whitney five or six miles further southeast than the former. He was sure that his work of 1870 was correct; for the bearings taken by him that summer from various points in the valley and in the Inyo Mountains, were numerous and agreed well with each other. But in 1864 he had only two bearings by means of which to locate Mount Whitney: One of these was taken by himself with a transit instrument from the top of Mount Brewer, and on that bearing he felt that he could rely. But the other one was a bearing taken by Mr. King with only a pocket compass, from the top of Mount Tyndall, and this bearing, he concluded, must have been erroneous. And thus the mistake remained undiscovered.

Hoffmann compiled all available data and in 1873 published his map, the most detailed topographical map of central California yet assembled, at a scale of 6 miles to 1 inch (Fig. 2.15). This map was a great advance over those previously available, and details from it were incorporated in later maps for years to come. The map, however, showed Mount Whitney at about the same latitude as Mount Kaweah, rather than in its true position, which is about 4 miles north of that latitude (Fig. 2.2).

Considering the methodology it was able to employ, the accomplishments of the California Geological Survey were indeed pathbreaking, and the several Federal Government surveys that were formed later to investigate the western territories were patterned after the California Survey. Eventually, other state surveys, as well as the U.S. Geological Survey (whose first director would be the unpaid geologic assistant on the Brewer expedition, Clarence King), emulated many practices of the California Survey.

Josiah Whitney, director of the California Geological Survey, was a good scientist and manager but not a diplomat. He refused to modify the program of the Survey so as to more closely support the mineral industry of the state. Money became tighter, and eventually, in 1874, the Survey was abolished.

In 1880 a new earth-science agency, the California State Mining Bureau, was established to conduct studies and collect statistics on the state's mining industry. It was headed by a state mineralogist appointed by the governor. In 1927 it became the Division of Mines and Mining, a part of the Department of Natural Resources, and in 1929 the name was shortened to Division of Mines. Olaf P. Jenkins was appointed chief geologist under W. W. Bradley, state mineralogist. Jenkins, a man of great energy, was the force behind the first comprehensive geologic map of the state of California, published in 1938.

Exalting the Wilderness: John Muir

John Muir (Fig. 2.24), born in Scotland in 1838, emigrated with his family to a pioneer farm in Wisconsin when he was 11 years old. The tasks of clearing and plowing the land, and raising and harvesting wheat and corn, were a constant struggle for the family. During this period young John developed a keen interest in wild things, the plants and animals of the Wisconsin wilderness. But his father, a deeply religious man, was a hard taskmaster, and the whole family worked long hours eking out a living from the farm. The work even precluded school for the children. John, however, read voraciously and found that his father would not object if he read and studied late into the night, so long as he was attentive to work the next day.

Muir showed an early aptitude for mechanical inventions. He hand-carved elaborate wooden clocks, one with a pendulum 14 feet long and a face so large that it could be read from a neighboring farm. In 1860, at the age of 22, John left home. He exhibited some of his inventions at the state fair in Madison and won a prize for his original, well-crafted designs.

2.24. John Muir, perhaps California's most famous historic figure, as he appeared in 1893 at the age of 55. His insightful studies and eloquent writing gave impetus to efforts to preserve Yosemite, Sequoia, and Kings Canyon National Parks. Courtesy of the John Muir National Historic Site.

While in Madison he met some of the faculty of the University of Wisconsin and was admitted to the freshman class after a few weeks of preparatory studies. He studied Latin, Greek, chemistry, mathematics, physics, and geology, but it was botany that most fired his imagination, and he hiked the neighboring forests and meadows collecting plant specimens for later identification. While at the university, he met Mrs. Jeane C. Carr, wife of his professor of Natural Science and Chemistry, Dr. Ezra S. Carr. She was also passionately devoted to the study of plants, and with her he carried on a lifelong correspondence. Periodically, he had to interrupt his studies to work for his support, and the work he could find included public school teaching.

After about three years as a student, Muir became restless, dropped out of the university, and moved on to Canada, possibly to avoid conscription into the Union Army as the Civil War heated up. He worked at a series of jobs, first at a mill and broom factory near Meaford, Ontario, and then at a wagon shop in Indianapolis, but work of this sort did not satisfy his interest in nature or his love for the outdoors. Finally, breaking free from traditional jobs, he set out on a 1,000-mile tramp to the Gulf of Mexico, focusing on botanical studies. He then went on by ship to Cuba and New York, and eventually, via the Isthmus of Panama, he arrived in San Francisco, in late March 1868.

Muir had read of the renowned Yosemite Valley, and he set out immediately on foot to see it for himself. He traveled south, past San Jose and Gilroy, east over Pacheco Pass to the Central Valley, and on into Yosemite via Snelling and Crane Flat. "Never before had I seen so glorious a landscape," wrote Muir on first viewing the valley, "so boundless an affluence of sublime mountain beauty." He herded sheep for a year and eventually obtained employment in Yosemite, building and operating a sawmill for J. M. Hutchings, an innkeeper. But his compelling interest in the natural beauty of the Sierra Nevada led him to travel and explore extensively on foot, and to study the biology and geology of the region. His youth on the farm in Wisconsin had prepared him for this hard, vigorous regimen. Not only his mechanical talents, but his proficiency as an outdoorsman, experience with animals, exceptional stamina, and indifference to hardship, cold, and hunger served him well.

John Muir came to California at a time when little was known about the Sierra Nevada, yet sheepmen, stockmen, and timbermen were actively exploiting it at an alarming rate. With missionary zeal he set about learning what he could of the range so that he could inform others of its wonders and of the pressing need to protect its fragile environment. The knowledge he gained on his frequent explorations and his enthusiasm for all aspects of the natural world soon made him a prime authority on the region. He was sought out as a nature guide, and frequently conducted visiting scientists on field excursions. And beyond all that, Muir proved to be a gifted writer.

Early in his wanderings in the Yosemite region, Muir found evidence that the main Yosemite Valley, as well as the upland canyons, had been occupied by glacial ice. He plotted the areas of the higher parts of the range that previously had been covered by ice and noted the courses of the glaciers and the effects of the movement of this ice. From these observations he concluded that the Yosemite Canyon had been carved by glacial ice to its present U-shaped form.

Muir's theory of the glacial origin of Yosemite Valley was in direct opposition to that proposed by Josiah Whitney, the director of the California Geological Survey. Whitney believed that the trenchlike form of the valley resulted from the collapse of its floor along two or more fractures or faults. In 1869 he wrote in the *Yosemite Guide-Book* a summary of this view:

> We conceive that, during the process of upheaval of the Sierra, or possibly, at some time after that had taken place, there was at the Yosemite a subsidence of a limited area, marked by lines of 'fault' or fissures crossing each other somewhat nearly at right angles. In other and more simple language, the bottom of the Valley sank down to an unknown depth, owing to its support being withdrawn from underneath during some of those convulsive movements which must have attended the upheaval of so extensive and elevated a chain.

Because Whitney held the twin positions of State Geologist of California and Director of the California Geological Survey, his ideas on the origin of Yosemite Valley were seen to be the more authoritative.

In 1871, Muir made a discovery that supported his case for the glacial origin of Yosemite Valley. He found the first glacier known in the Sierra Nevada, high on Black Mountain in the Yosemite region (Muir, 1873). The finding of this tiny remnant of what had been an earlier, larger glacier was soon followed by the discovery of dozens of other small glaciers on the northern slopes of high peaks.

In order to strengthen his case for the glacial origin of Yosemite, Muir set about expanding his observations and exploring the range south of Yosemite. He wanted to demonstrate that other major canyons in the Sierra had also been occupied and shaped by glaciers. He eventually made at least nine trips to the region of Sequoia and Kings Canyon National Parks, from 1873 to 1912 (Fig. 2.25).

On his first trip, in 1873, in company with Galen Clark, a Yosemite guide and innkeeper, and two others, including Albert Kellogg, botanist with the California Academy of Sciences in San Francisco, he traveled on horseback up the South Fork of the San Joaquin River past Mono Creek. Leaving the party in the main canyon for four days, Muir hiked into a stunning glacial valley known today as Evolution Basin, in the extreme northern part of what is now Kings Canyon National Park. He camped near a long thin lake and wrote (Muir, 1938), "Lake Millar, fourteen hundred yards long, fifty to one hundred and fifty yards wide, has waters of a bright green, and lies along the gothic front of Mount Millar on the south

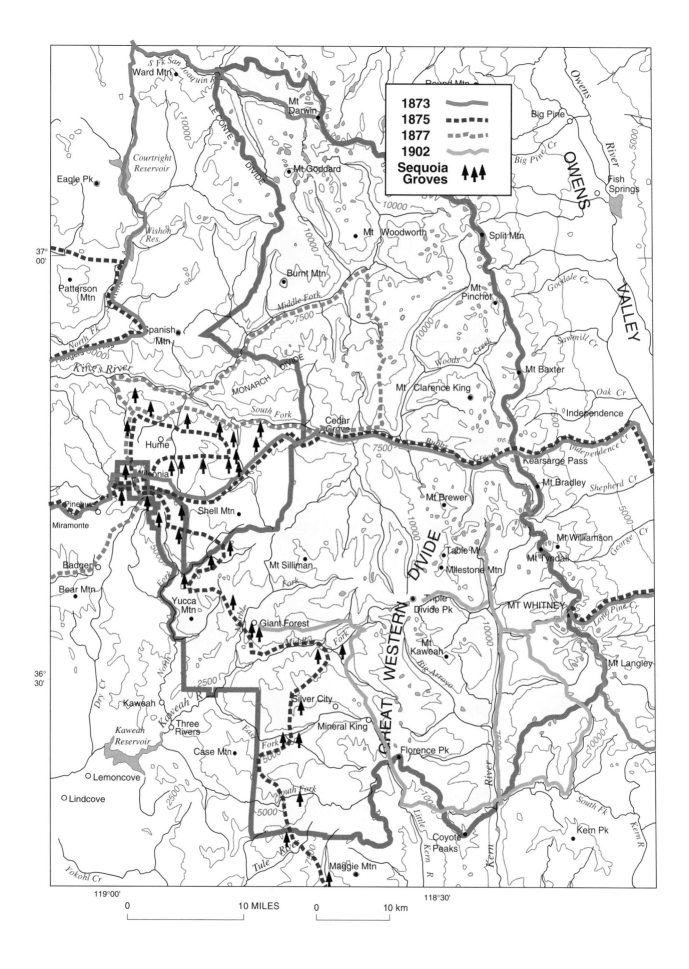

side." The dimensions of the lake define it as what is now known as Evolution Lake. He climbed an unnamed peak (13,385 feet) north of the lake and northwest of Mount Mendel (Fig. 2.26). Muir carried the then just-published Hoffmann map and noted that "The map of the Geological Survey gives no detail of this wild region."

Returning to the main camp, he joined Clark for a climb up a nearby ridge. "The view is aweful—" Muir wrote (Muir, 1938), "a vast wilderness of rocks and canyons. Clark groaned and went home." The party, diminished by one, then climbed the San Joaquin–Kings Divide, followed down the west branch of the North Fork of the Kings River, and " . . . entered a long straight valley remarkable for its many domes on the east side . . . " (Muir, 1938) (probably Dusy Creek and Maxon Dome, just north of Courtright Reservoir) and camped at 8,500 feet. They continued down into "a very interesting yosemite valley of North Fork of Kings River" where a "small mirror-lake occupies the extreme head of the valley where the river issues from a narrow and tortuous gorge in cascades" (this was Granite Gorge). They then climbed several thousand feet by a steep route to the divide between the North Fork and Middle Fork of the Kings River. Apparently from this vantage point Muir caught his first glimpse of Tehipite Valley (Fig. 2.27) on the Middle Fork of the Kings River.

Following a descent of nearly 7,000 feet to the west, down Rodgers Ridge, here marking the crest of the divide, they reached the main river near its confluence with the North Fork, crossed the river, and climbed up the south side of this stupendous canyon, along Mill Creek. They eventually arrived at Thomas Mill, where the Brewer party of the California Geological Survey had begun its explorations nine years earlier. From the mill they passed through Grant Grove, marveled at the giant sequoia trees,

2.25. Four of the ten trips taken by John Muir in the southern Sierra. His first, in 1873, took him from Yosemite to the Sierra Crest near Mount Darwin, then south to the Kings River (leaving and returning to the map at left), through Kings Canyon, across the Sierra Crest at Kearsarge Pass, and on south to climb Mount Whitney by the east face (the first time, from the south, unsuccessfully). His second trip, in 1875, brought him to nearly all of the Sierra big tree groves. His third trip, in 1877, took him to the little-known—and still road-free—Tehipite Valley on the Middle Fork of the Kings River. In 1902 Muir, at the age of 64, made his first venture into the Kern Canyon (and elsewhere) while on the second Sierra Club High Trip.

and eventually made their way into the glacially carved canyon of the South Fork of the Kings River (Fig. 2.28). Muir compared its glacial features with the Yosemite Canyon he knew so well 80 miles to the northwest. After working up the canyon of Bubbs Creek, which was then called the South Fork of the South Fork of the Kings River (Fig. 2.29), toward the main Sierra Crest, Muir left the party and climbed the Kings–Kern Divide, probably at the head of Center Basin. He then retraced his route to Bubbs Creek, crossed the range crest at Kearsarge Pass, and descended into Owens Valley.

While the rest of the party went on to Independence to await him, Muir turned south to make an excursion to Mount Whitney. He reentered the mountains on the Hockett Trail, somewhat to the south of the present site of Lone Pine. But like King before him, guided by the error in the Hoffmann map, he climbed Mount Langley, thinking it was Mount Whitney. Apparently, he had spoken to no one in Lone Pine and had thus failed to learn of the successful ascent of the true Mount Whitney a few weeks earlier. From the summit of Mount Langley he saw a higher peak to the north (Whitney) and set out cross-country in an attempt to gain its summit. After a long walk over rugged terrain and a freezing night without blankets on the exposed Sierra Crest, he turned back only a short distance from the summit and returned on foot to Independence in Owens Valley.

Muir did not give up easily. After one night's rest, he headed south to Lone Pine, but this time went west and climbed up the north fork of Lone Pine Creek. From its upper fork at 10,500 feet, he worked into the broad bowl of its southern branch and climbed to Iceberg Lake and then on up to the main ridgecrest (13,100 feet) in the saddle between Mounts Whitney and Russell (Fig. 2.30). About the subsequent route he wrote, " . . . I pushed direct to the summit up the north flank, but the memories of steep slopes of ice and snow over which I had to pick my way, holding on by small points of stones, frozen more or less surely into the surface, where a single slip would result in death, made me determine that no one would ever be led by me through the same dangers." This was the first ascent of Mount Whitney by the steep eastern route. On the summit he found records showing that Clarence King had finally made the summit a few weeks earlier, after three other parties had climbed the mountain. Later, Muir wrote about this route: "Well-seasoned limbs will enjoy the climb of 9000 feet required for this direct route, but soft, succulent people should go the mule way." (Farquhar, 1965).

Much of the material that Muir gathered during his 1873 exploration

2.26. Looking south across Darwin Glacier to Mount Darwin (13,830 feet) on the left and Mount Mendel (13,691 feet) just right of center. The lakes at the upper left are in the Upper Evolution Valley. The moraines were piled up when the glaciers were somewhat larger than they are today. U.S. Geological Survey aerial photograph by Austin Post.

of the southern Sierra appeared in a series of essays on "Studies in the Sierra" published in the *Overland Monthly* in 1874 and 1875. The article on the "Origin of Yosemite Valleys" closely compared the features of Yosemite, Tuolumne, and Kings canyons (Muir, 1874; Fig. 2.31). The similarity of the overall morphology of Yosemite to that of the other canyons supported Muir's thesis that glaciation—rather than Whitney's faulting-and-collapse hypothesis—was dominant in shaping Yosemite Valley.

In 1875 Muir made his second and third trips to the Kings Canyon region. In July, equipped with riding mules and pack stock, he and several companions entered the mountains from the west through Centerville, and went on to the extensive logging operations among the giant sequoia near Thomas Mill. They then largely retraced the trail taken two years previously, descending into the glaciated canyon of the South Fork of the Kings River (Figs. 2.32, 2.33). Comparing this canyon with the Yosemite Valley to the north, Muir described the similarity of the two with regard to their size, depth, waterfalls, and rocky cliffs. Their pervasive similarity reinforced his concept of the glacial origin of such canyons, and to emphasize this notion he used the word *yosemite* as a general term for all such large, steep-walled, glaciated canyons. The South Fork Canyon, as well as the canyons of the Middle Fork of the Kings (Tehipite Valley), the canyon of the North Fork of the Kings, the Kern Canyon, and even a small glacial canyon on the east side of Mount Whitney, were all termed "yosemites."

Following the South Fork of the Kings River (and Bubbs Creek) upstream, the party crossed the Sierra Crest at Kearsarge Pass and descended to Owens Valley. Continuing south, they climbed Mount Whitney, again by the east-face route that he had used two years earlier. But this time he avoided the hazardous north ridge by descending west from the Whitney–

2.27. Three views looking up the U-shaped glaciated Tehipite Valley on the Middle Fork of the Kings River. *Upper left.* Drawing by Charles Robinson rendered from a sketch by John Muir, as published in *Century Magazine*, following the custom to exaggerate the vertical scale in natural scenes (Muir, 1891). *Upper right.* Modified version of Robinson's drawing in which the horizontal dimension has been optically stretched 200 percent; the proportions of the drawing now closely resemble the natural scene shown below. *Below.* Photograph of the Tehipite Valley from about the same point on the trail where John Muir made his original sketch. U.S. National Park Service photograph by Jim Carson.

2.28. *(Above)* View to the east, up the canyon of the South Fork of the Kings River, from near the foot of the trail that enters the canyon near Cedar Grove. This graphite-and-ink drawing made by John Muir about 1891 overly emphasizes the steepness of the canyon walls. By permission of Holt-Atherton Special Collections, University of the Pacific.

2.29. *(Opposite)* Map of Kings Canyon—the canyon of the South Fork of the Kings River—showing many place names that are now forgotten (Muir, 1891).

Russell Saddle to Arctic Lake (12,200 feet) and then climbing the more gradual northwest flank to the summit.

From September to November of the same year, Muir again visited the Sequoia–Kings Canyon region, this time with the goal of making an assessment of the distribution of the giant sequoia groves on the west slope of the Sierra, following up on his previous observations of how rapidly logging was leading to their destruction. During this long trip he traveled without companions but took a mule, "Brownie," to carry his supplies. He visited a grove of big trees (now called the McKinley Grove) in the drainage of Dinkey Creek, a tributary of the North Fork of the Kings, and then largely retraced his trek of 1873 down the North Fork to the crest of the divide between the North Fork and the Middle Fork (Rodgers Ridge), where he was so stirred by the view of Tehipite Valley to the southeast. Descending 7,000 feet to the main river, he crossed the stream and climbed out of the canyon to Thomas Mill and the sequoia groves.

He then traveled south and explored most of the groves of big trees in

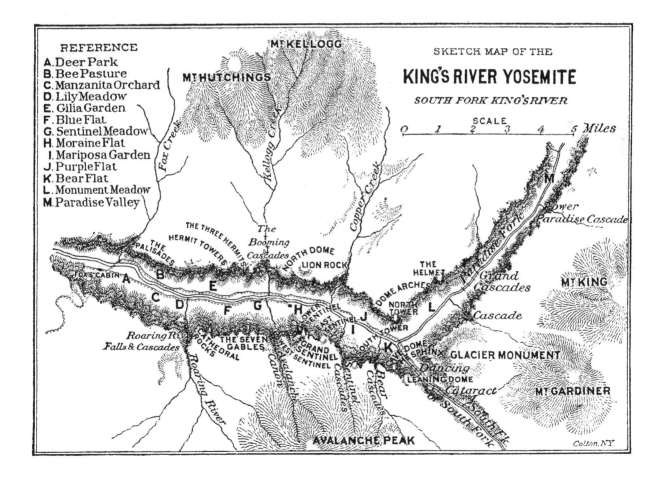

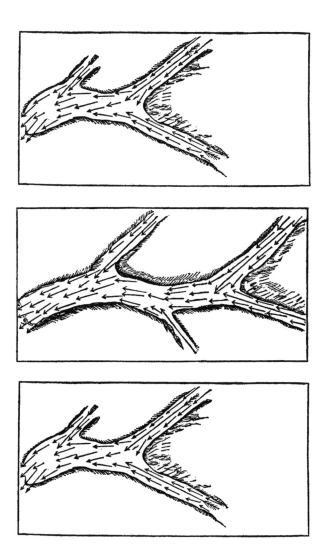

2.30. *(Opposite)* The summit of Mount Whitney (upper arrow) as seen from the east, appearing below the middle of the dark Kaweah Peaks on the skyline. John Muir's ascent in 1873, the first from the east, apparently followed the North Fork of Lone Pine Creek (bottom center) into its southern canyon (left center, with a lake on the left), and then up that canyon to its head, where Iceberg Lake (lower arrow) barely shows above the ridge. From near the lake he ascended up to the saddle and then 1,500 feet up the ridge to the summit. U.S. Geological Survey aerial photograph by Austin Post.

2.31. *(Above)* Maps comparing the outlines of three glacial canyons studied by John Muir: Tuolumne Canyon *(top)*, Kings Canyon (South Fork) *(middle)*, and Yosemite Valley *(bottom)*, the arrows indicating the apparent direction of glacier flow when the canyons were filled with ice. Muir stressed that the overall similarity of these three canyons supports the notion that they had a common origin (glaciation), and that Whitney's theory that Yosemite was special, and had been formed by faulting and collapse, was without merit (Muir, 1874).

the drainages of the Kings and Kaweah Rivers (Fig. 2.25). While in the Giant Forest area, and out of provisions, he met Hale Tharp, who had first visited the forest in 1858 and continued to maintain a summer house in a single large, hollow sequoia tree. After spending several days in the Giant Forest region, Muir continued south through the Tule River Basin to the southernmost limit of the sequoia trees in the drainage of Deer Creek, a tributary of the Kern River.

During the time that John Muir was exploring the region of the Big Trees, his friend and employer from Yosemite, James Hutchings, had un-

2.32. Looking east up the canyon of the South Fork of the Kings River, as seen from the Manzanita Orchard (see map, Fig. 2.29). *Above.* Drawn by Charles Robinson in 1890 while on a trip with John Muir (Muir, 1891). *Below.* Modern photograph showing roughly the same view.

2.33. The south wall of the canyon of the South Fork of the Kings River. Drawing by J. Clement accompanying an article by John Muir (Muir, 1887).

dertaken a trip into the highest Sierra. This party included Dr. Albert Kellogg (botanist), W. E. James, Albert H. Johnson, and several military personnel from Fort Independence. The party traveled to Lone Pine and from there crossed the range crest on the Hockett Trail, and on October 3 climbed Mount Whitney from the west, guided by Johnson, who was a member of the party that had made the first ascent of the mountain two years earlier. The photographer, James, took a series of photographs from the summit, the first taken of and from the mountain.

After returning to Owens Valley, the party again crossed the Sierra Crest at Kearsarge Pass, on October 10, and worked down into Kings

Canyon, where they spent about a week before returning the way they had come. The map that accompanies Muir's 1891 *Century Magazine* article (Fig. 2.29) recognizes the visits of Kellogg and Hutchings to Kings Canyon by the names attached to two peaks north of the canyon. The label on Mount Hutchings was retained by Joseph N. LeConte on his 1896 map (Fig. 3.19), and remains on present maps.

Two years later, in October 1877, Muir made a solitary trip into the parks region on foot. He began by taking the Central Pacific Railroad south to Goshen Junction, near Visalia. The railroad had been completed this far in 1872, just three years after completion of the transcontinental line over the Sierra at Donner Pass, north of Lake Tahoe. Muir walked through the foothills to Hydes Mill in the Kaweah River Drainage on Redwood Mountain, near the present border of Kings Canyon National Park, where he stocked up on provisions. These supplies consisted of flour, tea, sugar, and a piece of dried beef—fare typical of his foot journeys.

Crossing the Kings–Kaweah Divide, he spent several days scouting the Big Trees around the Converse Grove region and discovered several new groves to the east. He then descended to the South Fork of the Kings River, at a point about 2 miles east of Boulder Creek, and walked up the canyon bottom to the spectacular glaciated canyon already visited in 1873 and 1875 (Figs. 2.25, 2.28). From the canyon bottom at 5,000 feet elevation he climbed up the north side and "crossed over the Monarch Divide to the Middle Fork by a pass 12,200 feet high" (Bade, 1924). The exact route that took him to this altitude is not known, but he probably crossed Cirque Crest and descended Windy Creek to the main canyon of the Middle Fork of the Kings. He then followed the Middle Fork down to the isolated Tehipite Valley (Fig. 2.27), which was first visited by sheepman Frank Dusy in 1869. Muir marveled at the massive, craggy granite walls of this gem of a Yosemite-like glaciated valley, flanked by Tehipite Dome on its north side. He continued down the Middle Fork Canyon to near its junction with the North Fork and climbed up the south wall of the canyon, as he had in 1873.

The 1877 trip was Muir's last trip of discovery in the parks region. Thereafter, much of his time was taken up in writing and pushing his agenda for the preservation of the Sierra ecosystems. He returned again in June 1890 to the canyon of the South Fork of the Kings River to gather more facts in his effort to see this region, too, included in a national park. On this trip he once more compared the glacial features of this canyon with those of Yosemite Valley and wrote "A rival of the Yosemite," an article in the November 1891 *Century Magazine*, which compared

2.34. North Tower, on the north wall of the canyon of the South Fork of the Kings River, as seen from the foot of Glacier Monument (see map, Fig. 2.29). *Left.* Drawing by Charles Robinson while on a trip with John Muir in 1890 (Muir, 1891). *Right.* Modern photograph.

the two canyons, stating that they "are wonderfully alike, and they bear the same relationship to the fountains of the ancient glaciers above them." In fact, apparently to further emphasize this comparison of the two canyons, many of the place names he selected for the Kings Canyon—as set out in the map in the *Century* article (Fig. 2.29)—are nearly identical to those of Yosemite Valley. On this map the names North Tower (Fig. 2.34), Grand Sentinel (Fig. 2.35), Cathedral Rocks, Glacier Monument, Dome Arches, and Leaning Dome are all near duplicates of place names on the walls of Yosemite Valley, and they commonly occur in comparable places on the walls of Kings Canyon. Unique names, however, such as Seven Gables, the Sphinx, Lion Rock, and Paradise Peak (Fig. 2.36), were also selected.

The artist Charles D. Robinson traveled with Muir in 1890 and prepared the illustrations for the *Century* article from sketches made in the field. Some of Robinson's artwork was based on sketches by Muir of areas the artist did not visit, such as the Middle Fork Canyon. These illus-

2.35. Grand Sentinel, on the south wall of Kings Canyon, which rises 3,470 feet above the valley floor. *Left.* Drawing by Charles Robinson while on a trip with John Muir in 1890 (Muir, 1891). *Right.* Photograph taken by A. B. Clark in 1895.

trations utilized the style of that period, which was to exaggerate the vertical scale to make the scenery more picturesque (Figs. 2.27, 2.28, 2.33, 2.34, 2.36). The vertical scale in the Tehipite Dome drawing was rendered at twice the true horizontal scale, thereby depicting mountains and valleys as extraordinarily sharp and steep-walled (Fig. 2.27). Likewise, Muir's drawing from nature of Kings Canyon (South Fork; Fig. 2.28) clearly exaggerates the steepness of the canyon walls. This style was no doubt appealing to Muir because the steepness of the valley walls supported the glacial-origin theory.

In the summer of 1901 Muir took his two daughters on a trip to the southern Sierra, and in July of 1902, at the age of 64, he joined the second Sierra Club High Trip and made his first trip into the upper Kern Canyon region. The trip began at Giant Forest, followed up the Middle Fork of the Kaweah River to Bearpaw Meadows, and reached the mining town of Mineral King by way of Timber Gap. From Mineral King they went

2.36. Paradise Peak, looking east from the slopes at the foot of the Helmet (see map, Fig. 2.29). *Left.* Drawing by Charles Robinson while on a trip with John Muir in 1890 (Muir, 1891). *Right,* modern photograph.

south, crossed the Kaweah–Kern Divide at Farewell Gap, and headed thence down the Little Kern River to cross the Great Western Divide at Coyote Pass and drop into the canyon of the Kern River, which they followed north for 25 miles. They camped at Junction Meadows on the Kern for several days and made side trips to the headwaters of the Kern and up the Kern-Kaweah River. They then ascended Whitney Creek to Crabtree Meadow and climbed Mount Whitney. From the camp at Crabtree Meadow they traversed around to Golden Trout Creek, probably via Siberian Pass and Big Whitney Meadows, and then descended the creek back to the Kern River. Muir, drawing on geologic insights gained on an 1885 trip to Yellowstone, deduced that the natural bridge near Golden Trout Creek was produced by hot-spring deposits. He praised the alpine majesty of this region, the highest of the High Sierra.

Muir made two more trips into the parks region. In 1908 he joined the Sierra Club High Trip to the Kern Canyon region and rode on horseback up the Canyon from Kern Lakes to the Big Arroyo region. From there the group traveled over the Kern–Kaweah Divide to Mineral King and on to Giant Forest. He made his last trip to the southern Sierra in 1912; two years before his death at 76, he traveled to Giant Forest by car.

Creating and Enlarging the National Parks

The writings of John Muir in the 1870s had focused public attention on the rapid felling of the giant sequoias as well as the grazing damage wrought by hordes of sheep and cattle in the high mountain meadows. His was an early voice warning of the dangers of unchecked exploitation and of supposing the resources of the mountains to be boundless. In 1878 George Stewart, editor of the *Visalia Delta*, began calling for an end to the rapid destruction of the giant sequoia in the mountains east of Visalia.

As the recreational potential of the southern Sierra began to be realized, others joined in the call for conservation. Residents of the Central Valley found that they could escape the summer heat at the Mineral King mining camp or at camps located near the mountain lumber mills. Mount Whitney, the highest peak in the country, became a popular destination. In 1878 a group of 30 hikers from the Porterville area climbed Mount Whitney with William Crapo as guide. Four women in the group were among the first to reach the summit.

In the summer of 1881 a scientific expedition set up a camp at 12,000 feet on Mount Whitney to measure the amount of solar energy received at high altitude. The party was led by Professor Samuel Langley, director of the Allegheny Observatory in Pennsylvania. The group established their instruments on the summit of the mountain and spent several days and nights making observations. Through the efforts of Captain Otho Michaelis of the U.S. Army Signal Corps, who was attached to the expedition, the Mount Whitney Military Reservation was established in 1883; it included a part of the Sierra Crest embracing Mounts Whitney, Williamson, and Langley (Fig. 2.37). Little use was ever made of the reservation, however, and the land reverted to the forest reserve after the turn of the century.

During the same summer that the Langley expedition was active, a group of three Tulare men, William Wallace, Frederick Wales, and J. W. A. Wright, entered the mountains from the Hockett Trail with the

2.37. Part of the mineralogic and geologic map of California by William Irelan, State Mineralogist, published January 1, 1891, a few months after the creation of Sequoia National Park. Note that the park is named but its boundaries are not shown. Commodity symbols: 1, gold; 2, silver; 11, bismuth; 13, chromic iron. The numbered squares on the map are townships, 6 miles on a side. Longitude on this map uses Washington, D.C., as the prime meridian (Irelan, 1891).

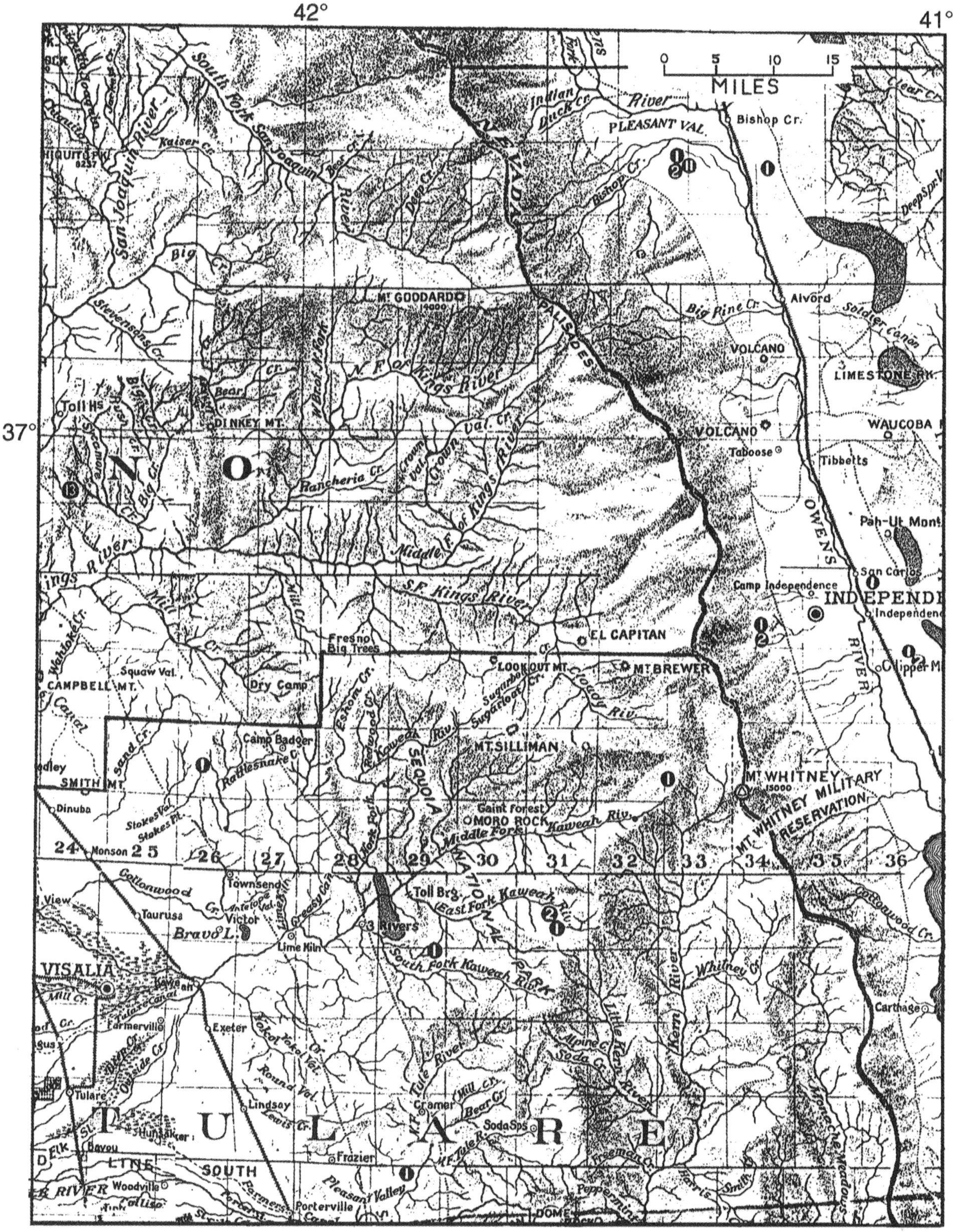
42°
41°
37°
0 5 10 15
MILES
South Fork San Joaquin River
San Joaquin River
Kaiser Cr.
Bear Cr.
Big Cr.
Stevensons Cr.
Indian
Duck Cr.
River
Bishop Cr.
PLEASANT VAL.
Deep Cr.
NEVADA
PALISADES
Clear Cr.
Deep Spr. V.
Big Pine Cr.
Alvord
Soldier Canon
VOLCANO
LIMESTONE RK.
WAUCOBA
Mt. GODDARD
N. F. of Kings River
Toll Hs.
DINKEY MT.
Big Cr.
Rancheria Cr.
Crown Val. Cr.
Middle F. of Kings River
Taboose
Tibbetts
OWENS
Pah-Ut Mont
San Carlos
Camp Independence
INDEPENDE
Independence
RIVER
Clipper M
N
O
Kings River
Mill Cr.
S.F. Kings River
Fresno Big Trees
EL CAPITAN
LOOK OUT MT.
Mt. BREWER
Squaw Val.
CAMPBELL MT.
Dry Camp
Canal
Eshom Cr.
Redwood Cr.
Kaweah Riv.
SEQUOIA
Sugarloaf Cr.
MT. SILLIMAN
Camp Badger
Rattlesnake Cr.
Sand Cr.
SMITH MT.
Dinuba
Stokes Val.
Stokes Mt.
Gaint Forest
MORO ROCK
Kaweah Riv.
Middle Fork
Mt. WHITNEY
MT. WHITNEY MILITARY RESERVATION
24 Monson 25 26 27 28 29 30 31 32 33 34 35 36
Cottonwood Cr.
Townsend
Taurusa
Victor
Bravo L.
Lime Kiln
3 Rivers
Toll Brg.
East Fork Kaweah Riv.
NATIONAL PARK
South Fork Kaweah Riv.
VISALIA
Kaweah
Mill Cr.
Farmerville
Exeter
Yokol Val.
Whitney Cr.
Carthage
Cottonwood Cr.
Tule River
Alpine C.
Soda C.
Little Kern River
Kern River
Round Val.
Tulare
Lindsay
Cramer
Mill Cr.
Bear Cr.
Soda Sps.
T U L A R E
Frazier
Freeman Cr.
LINE
SOUTH
Woodville
Porterville
Pleasant Valley
Peppermint
Harris
DOME

intent of climbing Mount Whitney. They traveled up Volcano Creek (now Golden Trout Creek), climbed the mountain from the west side, and met the Langley expedition. The three then turned north and explored the headwaters of the Kern River and descended into the Kern Canyon. West of the canyon they made the first ascent of Kaweah Peak, on the Great Western Divide. This group of peaks is a landmark that had been noted earlier from the Central Valley, as well as by the Brewer party, from Mount Brewer. In 1883, as a result of this excursion and from access to other available data, Wright published what was then the most authoritative map yet compiled of the southern Sierra region, from the Palisades to Olancha Peak (Fig. 2.38). This map, published in a volume on the history of Fresno County, proved to be important in defining the land for possible inclusion in a national park.

Efforts to protect parts of the forests and alpine country of the southern Sierra did indeed intensify, as timbering, grazing, and destruction of fish and game became more widespread and more wanton. George Stewart, editor of the *Visalia Delta*, was joined by the members of the Wallace group, the Langley expedition, and other interested citizens in supporting the creation of a national park patterned after Yellowstone, which had been established as a national park by act of Congress on March 1, 1872.

In May 1889, the U.S. General Land Office opened an area of one township (36 square miles) of timbered land along the Mineral King Road for public land purchase. Since it became apparent that other townships would follow, Stewart sought support from Congressman William Vandever of California, and Vandever introduced a House bill for the creation of a national park. The bill gained support from many quarters, including S. W. Waterman, the governor of California, and following congressional approval it was signed into law by President Benjamin Harrison, on September 25, 1890 (Dilsaver and Tweed, 1990). This act set apart two

2.38. Map of the general area of the parks by J. W. A. Wright, compiled from data collected during his travels of 1881. Longitude designations on the bottom are given in degrees west of Washington, D.C., rather than the modern convention of degrees west of Greenwich, England (which are shown at the top of the map). This is one of the first maps to place Mount Whitney in its correct position. Sheep Mountain (now called Mount Langley) is the peak 5 miles southeast of Mount Whitney that Clarence King climbed in 1871 believing it to be Mount Whitney (Wright, 1883).

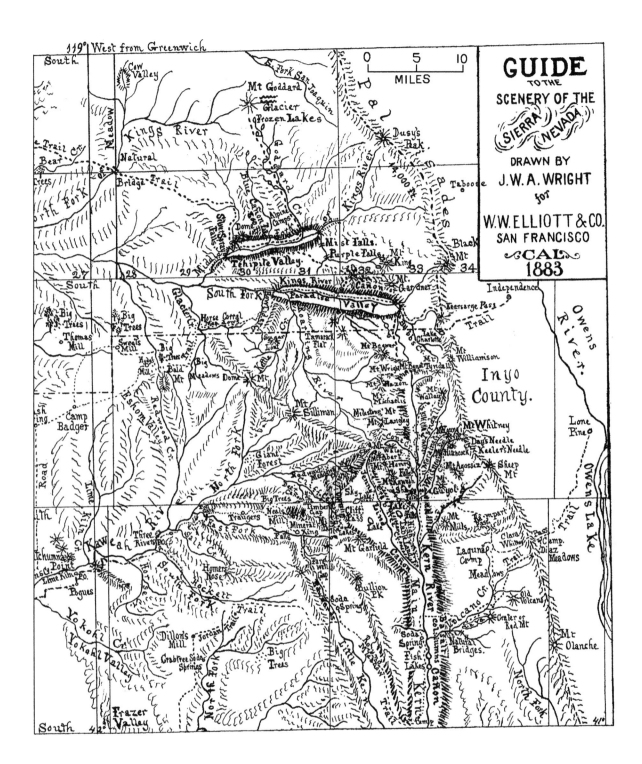

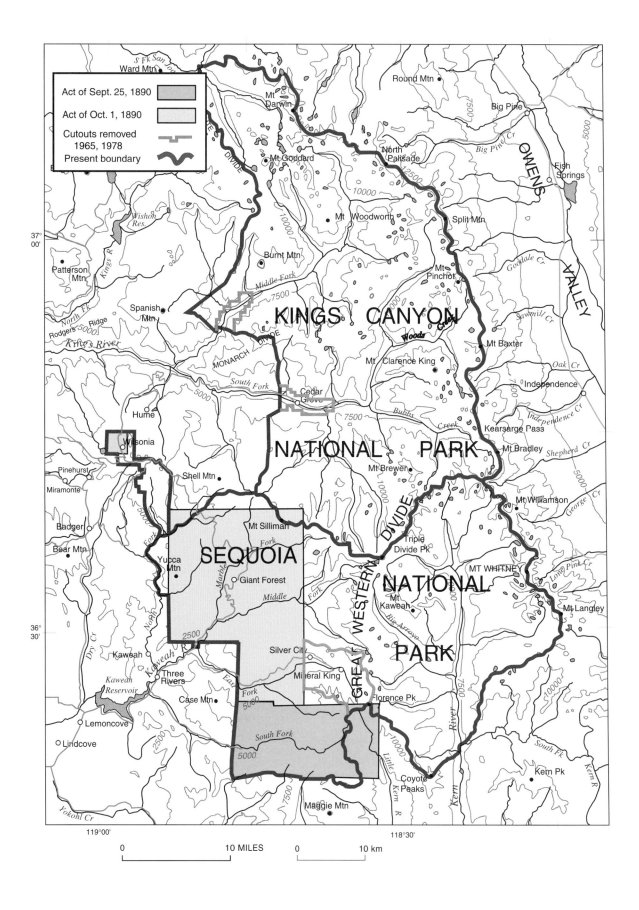

townships of land (72 square miles; Fig. 2.39) in what is now the southwest corner of Sequoia National Park. The tract embraced several sequoia groves, including the Garfield Grove. This was the second national park in the nation, and the first in California.

At that time, somewhat ineffective state reservations protected Yosemite Valley and the Mariposa Grove of big trees, and unknown to Stewart, another bill, for the creation of Yosemite National Park, was moving through the federal legislature. Included in the Yosemite bill before it was put to a vote was a proviso that five more townships be added to Sequoia National Park and that another national park of 4 square miles be created to protect the magnificent General Grant Grove (Fig. 2.39). To the surprise of all, this expanded bill passed the Senate and was signed into law by President Harrison just six days after passage of the initial bill creating Sequoia National Park. This new bill, enacted basically to create Yosemite National Park, more than tripled the area of Sequoia National Park, to seven townships (252 square miles), so that it then occupied approximately the western half of the present park. One of the first maps to identify the new park, without, however, indicating its boundaries, was that of Irelan (1891; Fig. 2.37). The new small park to the north, to be called General Grant National Park, would protect that grove of big trees from logging. Considerable later expansion would extend the park to the north and east, but about 30 miles of the original park boundary still limits the southwest extremity of the park today.

The task of supervising and patrolling the new national park fell to the U.S. Army. In the spring of 1891, a detachment of cavalry from the Presidio of San Francisco, under the command of Captain Joseph H. Dorst, arrived at Red Hill, near the settlement of Three Rivers. The group included 58 men, 60 horses, and 20 mules. Their task was difficult, because

2.39. Map showing the evolution of the park boundaries. The dark color is the area of the first Sequoia National Park, consisting of two townships, which was created by act of Congress September 25, 1890. The lighter color indicates an addition to that park of five more townships and the creation of General Grant National Park of 4 square miles six days later, by the act of October 1, 1890. Subsequent legislation expanded Sequoia National Park to essentially its present size in 1926 and created Kings Canyon National Park in 1940. The heavy colored lines indicate cutouts from the parks reserved for reservoirs (Middle and South Forks of Kings River) and for a national game refuge (Mineral King). These exclusions were removed and the areas added to the parks in 1965 and 1978, respectively.

only two wagon roads gave access to the park area, the Colony Mill timber road up the North Fork of the Kaweah River, and the Mineral King mine road up the East Fork of the Kaweah. Perhaps worse, the park boundaries were not marked on the ground, and only a few local mountaineers were familiar with the region.

In the face of considerable local opposition, the Army explored the park area in order to control the timbering and grazing interests. The military park supervisor arrived with a troop of cavalry each spring and sent detachments to patrol various parts of the park, including the General Grant Grove area. Park boundaries were located and marked, and by 1896 a sketch map of the park by one Lieutenant Davis was completed. In 1898 the troops did not arrive until September, because of the exigencies of the Spanish-American War, and by that time multitudes of sheep had been driven into the park illegally and had laid waste to considerable areas of meadowland.

During the summers of 1898 and 1899, the military had some assistance from a civilian forest ranger, and by 1900 the first full-time forest ranger assumed his post at the park. As road access improved and year-round operations became possible, more visitors made use of the park, and the need for civilian rangers grew. The military realized increasingly that the administration and protection of the park should be the responsibility of a civilian ranger corps supported by the Department of the Interior. Troops nonetheless continued to arrive each summer until the spring of 1914, when the Department of the Interior appointed Walter Fry first superintendent of Sequoia and General Grant Parks. After 23 years, military control of the parks had ceased.

Initially, the national park superintendents all reported directly to the Department of the Interior, but as more parks were added to the system, this plan of supervision became more and more unwieldy. In 1916, therefore, the National Park Service was created, with Stephen Mather as its first director. Mather, supported by preservationists and groups like the Sierra Club, pushed hard for an enlargement of Sequoia National Park to include the high alpine country to the east. Finally, in 1926, the park was more than doubled in area, to 604 square miles. Expanding east to the main Sierra Crest and north to the Kings–Kern Divide, the park now included the Great Western Divide, the upper Kern River Drainage, and the main peaks along the crest, among them Mounts Whitney and Tyndall. The newly established park boundaries are essentially those extant today, except that the Mineral King region, with its mining potential, was then excluded from the park.

Ever since the early success in creating Sequoia National Park, preservationist groups strove to do the same for the Kings Canyon region, with its stupendous canyons, vast forests, and high alpine terrain. John Muir had been defeated in his efforts to bring this about in the early 1890s. Opposing these efforts were powerful timber and grazing groups as well as those who saw opportunities to dam the rivers for both irrigation projects and hydropower. Finally, in 1940, Congress acted to create Kings Canyon National Park, a nearly roadless area of about 709 square miles adjacent to Sequoia National Park on the north. The new park was bounded on the east by the Sierra Crest and incorporated the General Grant National Park, which then no longer existed as a separate national park. Soon after passage of this the Redwood Mountain Sequoia Grove area was incorporated into the park, by proclamation of President Franklin Roosevelt. The new park boundaries, however, excluded the main glaciated canyons of both the Middle and South Forks of the Kings River, to leave open the possibility for later reservoir development (Fig. 2.39).

These boundaries remained unchanged for another 25 years. During this period, the road-accessible canyon of the South Fork of the Kings River, one of the scenic wonders of the Sierra, remained in the Department of Agriculture's National Forest, but was administered by the Department of the Interior's National Park Service. This awkward administrative situation prevailed until August 6, 1965, when the reservoir exclusions were lifted. The park boundaries were adjusted so that both the Cedar Grove region of the South Fork Canyon and the Tehipite Valley region of the Middle Fork Canyon became parts of Kings Canyon National Park. In 1978, Sequoia National Park was again enlarged, this time to include the 25-square-mile Mineral King Game Refuge, a region encompassing the old mining district that had previously been administered by the U.S. Forest Service (Fig. 2.39). These additions raised the combined area of the two parks to 1,350 square miles and established their current boundaries.

Journey over all the universe in a map, without the expense and fatigue of traveling, without suffering the inconveniences of heat, cold, hunger, and thirst.

—*Don Quixote*, MIGUEL DE CERVANTES (1547–1616)

MAPPING

In response to the gold fever of 1849 a growing tide of immigrants had flooded into California. They came overland by wagon, on horseback, or on foot, by sea around Cape Horn, and by sea, land, and sea, two voyages linked by a trek across the Isthmus of Panama. Across the land, pressure was building for the construction of a transcontinental railroad, and by 1869 one would be completed. When mining of the precious metals slowed, prospectors spilled out from the gold country in the northwestern Sierra, and settlers established themselves throughout the Central Valley of California. Eventually, fortune seekers of all sorts moved into the Sierra foothills and mountains.

The rapid pace of development had its impact on the land. Vast herds and flocks soon laid waste to mountain meadows. Booming lumber mills exploited the forests for timber and fuel. And from the travels of the stockmen and sheepherders evolved a network of trails. But along with these exploiters of the land came a few who saw beyond quick profit. They admired the serenity and unique qualities of the great mountain range and sought to learn more about it. They recognized that unbridled exploitation was quickly leading to the destruction of this remarkable wilderness, and that much would be lost forever if not quickly protected.

The primary task of the early explorers was to learn what was there, to chart the overall features of the land, and to identify the more likely routes for trails and roads. Some, like John Muir, did very little mapping, but raised the public's concern for the wild lands, and its awareness of the issues involved. Increasingly, however, the need for detailed mapping became apparent. Most of the earliest reconnaissance mapping in the region had been done by the early government surveys, principally those of the Army Corps of Engineers under the command of Lieutenant John C. Fremont, Lieutenant Robert Williamson (railroad surveys), and Captain George Wheeler (Surveys West of the 100th Meridian), as well as those of the California Geological Survey under the leadership of Josiah Whitney. The need for good maps became more pressing when Sequoia National Park was created in 1890, and members of the Sierra Club, established in 1892, began a program of extensive reconnaissance mapping. Such maps were needed in planning and carrying out excursions into the back country, and members of the club urged the government to develop detailed maps of the region. Finally, about 1900, the U.S. Geological Survey began a systematic topographic-mapping campaign that produced a series of remarkably good maps (initially at a scale of 1:125,000) that showed all the features of the land in previously unattainable detail.

The Territorial Surveys: King and Wheeler

A week after returning to San Francisco from the 1864 expedition in the southern Sierra, Clarence King, James Gardiner, and Richard Cotter left for Yosemite. Yosemite Valley had just been granted to the State of California as a park in perpetuity by act of congress, signed by President Lincoln, and the surveyors were directed to collect data for a map, fix the boundary lines, and study the geology of the area. Their work was barely completed before they were driven out of the mountains by snowstorms.

In 1865, because of financial constraints on the California Survey, King and Gardiner were assigned on loan to a U.S. Army survey of the region from the Mojave Desert to the Colorado River, under the command of General McDowell. The following summer they returned to the Survey and worked in the central Sierra, in the upper San Joaquin–Mount Ritter region.

King, who had more ambitious plans, resigned from the California Geological Survey in 1866. Carrying strong letters of recommendation from Colonel Williamson and William Brewer, he traveled to Washington and proposed to Edwin Stanton, Secretary of War, the need for a survey of the entire west, one that would develop information on the routes to be

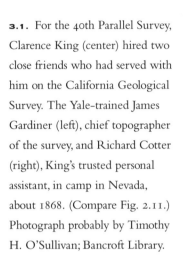

3.1. For the 40th Parallel Survey, Clarence King (center) hired two close friends who had served with him on the California Geological Survey. The Yale-trained James Gardiner (left), chief topographer of the survey, and Richard Cotter (right), King's trusted personal assistant, in camp in Nevada, about 1868. (Compare Fig. 2.11.) Photograph probably by Timothy H. O'Sullivan; Bancroft Library.

taken by the three railroads and several wagon roads then under construction. He emphasized to Stanton that the lack of surveying during the Civil War and the increasing numbers of settlers crossing this region had placed a heightened priority on this work. The proposal was supported by the Chief of Engineers, General A. A. Humphreys, and with the help of friends in Congress, the plan was approved and attached to an appropriations bill.

In March 1867, King, then only 25, was named geologist in charge of the U.S. Geological Exploration of the 40th Parallel, reporting to the Secretary of War. He hired his friends from the California Geological Survey, James Gardiner as chief topographer and Richard Cotter (Fig. 3.1) as his personal assistant. He was also fortunate in obtaining the services of Timothy H. O'Sullivan, a photographer who had helped Matthew Brady document the Civil War.

In addition to his mapping with the 40th Parallel Survey, which he pursued with vigor, King maintained his abiding interest in the geologic features of the West. In 1870 he described the first glacier to be discovered in the United States, on Mount Shasta in northern California. By 1872, most of the field work and some of the reports for the 40th Parallel Survey were completed, but other tasks dragged on until 1878. Gardiner stayed with the Survey until 1873 and then left to join the Hayden Survey. O'Sullivan moved to the Wheeler Survey in 1871.

In June 1871, while his 40th Parallel Survey group was at work in Wyoming, King made a trip to San Francisco to obtain supplies and hire packers, and took the opportunity to make another attempt on Mount Whitney. No doubt his interest was stimulated by the new information from Charles Hoffmann regarding Hoffmann's efforts to refine the position of the mountain by making observations from Owens Valley the previous summer. Hoffmann would utilize these measurements in preparing his topographic map of the central part of the state, which was published in 1873 (Fig. 2.15). But unbeknownst to him (and to King), the readings were taken on the wrong mountain, leaving Mount Whitney placed on the map some 4 to 5 miles south of its true location.

After a jolting three-day stage ride from Carson City to Lone Pine, King would later write (1935), "I left a Green barometer to be observed at Lone Pine, and carried my short high-mountain instrument, by the same excellent maker." In Lone Pine, King gained a companion, Paul Pinson, and the two proceeded west through a range of low hills, where King noted that "the granite was riven with innumerable cracks, showing here and there a strong tendency to concentric forms, and I judged the immense spheroidal boulders which lay on all sides, piled one upon another, to be the kernels or nuclei of larger masses." This is clearly a description of the spheroidal weathering in the Alabama Hills, a small fault-block range that protrudes above the great debris slope rising westward to the Sierra escarpment.

Taking a route he expected would bring them to Mount Whitney, King was guided by the position supplied by Hoffmann, a position based on the two 1864 magnetic bearings to the mountain (by Hoffmann from Mount Brewer and by King himself from Mount Tyndall), and on the 1870 bearings taken from Owens Valley, which were erroneously sighted not on Mount Whitney, but on Mount Langley. King and his companion began climbing a "southern canyon," probably Tuttle Creek, but soon encountered stormy, wet weather and made camp the first night at 10,000 feet in an alpine grove under an overhanging rock. After a meal of beef, toast, and tea, they bundled up in their overcoats for the night before a blazing fire.

The next morning having dawned clear, they strode up a glaciated valley toward timberline. As they pushed upward, mist began to develop, and toward the top of the mountain they found the region largely in clouds that prevented a clear view of the greater environs. At the summit was a small cairn in which was fixed an Indian arrow pointing west. King was overjoyed to have finally reached the summit of the peak he had failed to attain on two previous occasions. He hung the barometer from the summit cairn and noted (King, 1935) that they had attained "the summit of the United States, fifteen thousand feet above two oceans." Assuming that only Indians had climbed the peak before, he left at the summit a half dollar on which he had inscribed his name as a record of his first ascent of Mount Whitney. On his return to Lone Pine, he proclaimed his success to the local press.

In early 1871, King published several articles in the *Atlantic Monthly* recounting his exploits in the mountains of California. These were combined, others added, and the whole published as a book, *Mountaineering in the Sierra Nevada*, in early 1872. This popularized account, widely acclaimed at the time, has been reprinted and remains a favorite (King, 1935). The altitude King's book gave for Mount Whitney, 15,000 feet, is perpetuated on later maps, including the Irelan map of 1891 (Fig. 2.37).

In 1872 King was headquartered in San Francisco, but was seldom still for long. Early in the year he spent several weeks in Hawaii on a leave of absence, and visited the ongoing volcanic activity at Kilauea volcano on the big island. He was most interested in examining small streams of red-hot molten lava and studying the lava after it had cooled and solidified.

Following his return to California, he set off in April to investigate glaciation in the Sierra Nevada, between the Fresno and Merced Rivers, but was unable to accomplish much because of the heavy snow that had fallen that year. His two assistants remained in the range, intent on moving into the higher country when the snow melted. King, meanwhile, made a tour of the 40th Parallel Survey groups working in Nevada and Wyoming. Returning then to San Francisco, he found the mining world agog over reports of a discovery of diamonds at an undisclosed spot in the West, probably in Arizona.

In August, King left again for 40th Parallel Survey business, and spent three weeks in mining camps in Nevada. In September he crossed the Sierra near Yosemite to Owens Valley, and there met, as planned, a friend and accomplished landscape painter, Albert Bierstadt. They traveled up Bishop Creek and met King's glacier-study group in the High Sierra, joining them for several weeks in the vicinity of Mount Humphreys and Evolution Valley. Returning at length to Owens Valley, they observed the effects of the great earthquakes that had occurred there earlier in the year, and then reentered the range at Kearsarge Pass, west of Fort Independence. While the group retraced the trail King had taken in 1864 with the

Brewer Party, Bierstadt filled his sketchbook with drawings of the mountains. They all then traveled down Bubbs Creek to Kings River Canyon and eventually emerged from the mountains at Visalia, where King and Bierstadt took the train to San Francisco.

There they found the diamond fever running high, and learned that several companies had been formed to promote the mining of the gems. After ferreting out the discovery location, in western Colorado, King, with trusted aides, visited the region and located the site on a sandstone mesa, thus in a geologic environment unfavorable for diamond formation. They nonetheless found not only diamonds, but rubies, sapphires, garnets, amethysts, spinals, and emeralds, as well. But the presence of gems only in recently disturbed ground and in holes apparently poked by a stick led King to deduce that the discovery site had been salted. Hastening back to San Francisco, and to the offices of the company holding options on the property, he reported that the strike was obviously a fraud, and was clearly the hero of the hour. In 1876 King was elected to the National Academy of Sciences.

That same year King published the atlas for his 40th Parallel Survey, a document that included colored geologic maps and cross sections as well as topographic maps. The topographic maps, prepared by James Gardiner, were among the first in the American West to use contour lines rather than hachures to portray the topography. The *grade curves* (contours), each maintaining a constant elevation, were spaced at vertical intervals of 300 feet. The use of such lines of equal elevation to depict

3.2. Index map from Clarence King's "Geological and topographical atlas accompanying the report of the Geological Exploration of the Fortieth Parallel" (1876). Here King uses the name "Mt. Clarence King" for the mountain named for himself, in place of the name "Mt. King" that Hoffmann, in his earlier 1873 map (Fig. 2.15), had used. King added his first name, apparently, because he wanted to remove any uncertainty about whom the mountain was named for. Later maps, including the Wright map of 1883 (Fig. 2.38), the Muir map of 1891 (Fig. 2.29), and the LeConte map of 1896 (Fig. 3.19), continued using the name "Mt. King." When the U.S. Geological Survey's 30-minute Mt. Whitney Quadrangle was published in 1907, the name "Mt. King" was used, and it continued to be used on the editions of 1919, 1921, and 1927. Finally, the edition of 1933, and those after, honored King's wish and the mountain was labeled "Mt. Clarence King." Inexplicably, the name "Mt. Clarence King" is located on King's own index map 20 miles too far south. The mountain is actually slightly north of Independence.

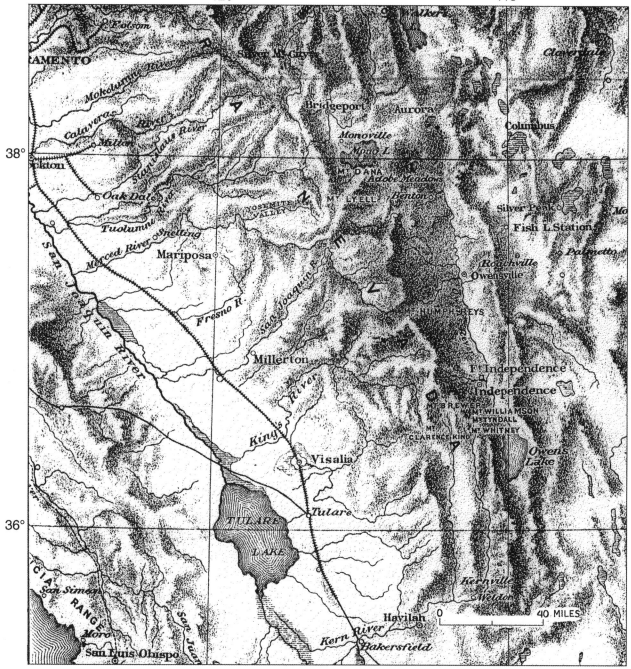

topography was considered especially important for the planning of future irrigation projects.

King's index map of the 1876 atlas covered much of the western states, including the Sierra Nevada, and provides a good look at the state of knowledge at that time (Fig. 3.2). The course of the Kern River was adapted from the Hoffmann map of 1873 (Fig. 2.15), and the shape and size of Tulare Lake roughly match those on the Goddard Map of 1857 (Fig. 2.8). The railroad had at this time reached nearly to Bakersfield.

A curious feature of the map is that Mount Clarence King (which King himself discovered and named) is wildly misplaced. This drafting error put the mountain south (rather than north) of Mount Brewer and the town of Independence. This is also the first map to use King's full name (both first and last) for this mountain. Perhaps he anticipated some confusion with the biblical Three Kings for whom the Kings River was named, and he wanted others to know for whom the mountain was named.

All these maps ultimately required knowledge of precise latitude and longitude, because this was the means by which one mapped area would be connected with another, and the method by which any spot could be uniquely defined. The lesson of the eastern California boundary was clear. In 1850, the intent of the California legislature, in defining part of the boundary as following the 120° meridian, was to include all of the northern Sierra Nevada, and its eastern watershed, in the state of California. They used the best available map (Fremont's 1850 map, Fig. 2.3) in defining that boundary—not knowing that the map was in error and that the 120° meridian passed through Lake Tahoe (named Bonpland Lake on the map) rather than 25 miles east of it as shown on the map. Thus, much of the land intended to be included in California became a part of Nevada.

Consequently, every effort was made to locate a few principal points with the greatest precision obtainable. Because precise readings of time of day were crucial to precision, and because these readings were available only via telegraph lines, only points adjacent to a telegraph line were selected, and similar methods were employed by the various territorial surveys, including King's 40th Parallel Survey. The field astronomers employed a zenith telescope and an astronomical transit generally set on brick, stone, or concrete piers for stability (Figs. 3.3 and 3.4).

Latitude determinations were made by measuring the altitude of stars with the zenith telescope. Latitude can be approximated by measuring the angular height of the North Star (Polaris), which is about one degree removed from the celestial North Pole. At the Equator, Polaris will always be about on the horizon, and at the North Pole it will always be almost directly overhead. In actual field work, the astronomer fixed the positions of many different stars and compared their altitudes with known star positions previously recorded and published in a star catalog (an *ephemeris*).

The most precise method then in use for the determination of latitude was to measure the altitude of each of a pair of stars, both of which were

3.3. Astronomical station of King's 40th Parallel Survey, near Ruby Mountains, Nevada, about 1870. Two instruments for astronomical observations are mounted on piers. On the left is an astronomical transit, used for the measurement of transit times of stars across a north-south line (the meridian) for longitude determination. On the right is a zenith telescope, for measuring the altitude of stars at known times for latitude determination. Accurate time was obtained by telegraph; telegraph equipment and a lantern are visible on the box/table behind the observer. The astronomical engineer (probably James Gardiner) is holding a chronometer. Photograph by Timothy H. O'Sullivan; U.S. Geological Survey Photograph Library.

3.4. Astronomical transit used for the accurate determination of longitude. Note the small bull's-eye oil lamps (the upright cylinders on the right and the left), each of which shines a beam of light down the hollow horizontal axis of the instrument to illuminate the field, so that the cross hairs are visible at night. The field setup of the instrument is shown in Fig. 3.3 (U.S. Coast and Geodetic Survey, 1882).

selected for their position in the heavens, the two both near the zenith but on opposite sides of it. Stars near the zenith—straight overhead—were preferred because the corrections necessary for refraction (the bending of light rays by the atmosphere) are minimal there, since starlight from near the zenith passes through the least atmospheric thickness. Stars from opposite sides of the zenith were selected because that choice allowed correction for inaccuracies in determining the horizontal. The measurement procedure was commonly repeated on five to ten pairs of stars, and for

TABLE 3.1
The turning of the globe: Conversion of time, arc, and longitudinal distance

Time	Arc[1]	Distance at Equator[2]
1 day	360°	21,600 nautical miles or 24,870 statute miles
1 hour	15°	900 nautical miles or 1,036 statute miles
1 minute	15'	15 nautical miles or 17.3 statute miles
4 seconds	1'	1 nautical mile or 1.1515 statute miles
1 second	15"	¼ nautical mile or 1,520 feet
0.066 second	1"	101 feet
0.01 second	0.15"	15.2 feet

[1] A degree (°) of arc is divided into 60 minutes (60') of arc, and a minute (') of arc is divided into 60 seconds (60") of arc.

[2] All distances between meridians of longitude are less north or south of the Equator because of the curvature of the globe and the consequent convergence of the meridians toward the poles. At 37° north latitude, that of the highest Sierra Nevada, the distances between meridians are about 80% of those at the Equator.

several nights of observation. With careful work the latitude of a given location could be determined to within 0.1" of arc, equivalent to less than 10 feet on the ground (Table 3.1).

Measurement of longitude is much more difficult than that of latitude, because of the need for precise time (see Sobel, 1995). If one has the means to determine the difference in time required to bring a star to the same relative position in the heavens at two different sites, one of them a site of known longitude, the other at a point not yet fixed, then the difference in longitude between the two sites is known, as the following may make clear.

The Earth is divided into 360 degrees of longitude, and it turns on its axis once every 24 hours. Hence, in one hour it will turn 15° of longitude, which is equivalent to 900 nautical miles on the ground at the Equator (see Table 3.1). For navigation at sea, the timing of observations to an accuracy of one minute (equivalent to 15 miles of longitude) or of one second (equivalent to one-fourth mile of longitude) is often sufficient. But for land-based mapping, which will fix state boundaries and ultimately establish land-property grids, a far greater precision is required. Hence the astronomers strove for a precision of $1/100$ of a second in timing star transits, which is equivalent to 15.2 feet on the ground at the Equator.

For longitude measurements the transit instrument would be set, by the use of astronomical observations, so that its telescope pointed precisely in the plane of the meridian (toward due north or due south). The task was then to determine the time of transit of a given star, that is, the moment that it crossed the meridian, as defined by the instant that it crossed the vertical crosshair in the telescope. Commonly, the telescopes were equipped with five evenly spaced vertical hairs, so that by averaging five readings more precise times were obtained.

For the determination of longitude at a new station, to be used thenceforth as a primary triangulation point, precise time was obtained via telegraph from an observatory whose position had already been determined. Hence such new stations, and their astronomical instruments, were required to be on, or very near, an existing telegraph line. Surveys undertaken in Nevada and California relied on base stations at observatories in Ogden, Salt Lake City, or Colorado Springs. The longitudes of these base stations had in turn been determined by time signals transmitted from the Lakes Station in Detroit or from the U.S. Naval Observatory in Washington, D.C.

Time was recorded at both the base station and the new station by means of a chronograph (Fig. 3.5), a mechanical device that records "time by the yard." The time signal was recorded as a penned line on a rotating paper-covered drum. The drum was driven by a mechanical clock motor that revolved it once per minute, and the individual seconds were recorded as spikes about 1 inch apart by a mechanically driven pen. The beginning

of each minute was recorded as a missing second mark. These time marks were compared and standardized with the recorded telegraphed time signals received from the observatory base station, which were superimposed on the paper record at both stations.

At both the base station and the field station, an astronomer would register the transit of a given star as a spike on the revolving paper. Each astronomer would tap a telegraph key at the precise instant that the star crossed the vertical hair in his telescopic field of view. This signal was recorded simultaneously on chronographs at both the observatory and the field station by telegraph. By measuring the distance between the two star spikes (at either station) on the chronograph paper, the time difference between the transits of the same star at the two stations could be measured with a precision much less than one second. Generally, since the base station was east of the field station, the time of transit of a given star would be earlier at the base station, and the delay in time of the star's transit at the field station would be a measure of the increase in west longitude of the field station relative to the base station.

During a single night's work, when skies were clear, the transit times of from six to a dozen stars could be measured. The astronomical work was repeated nightly, sometimes for weeks, at both the base station and the field station, and transits of dozens of stars would be measured, so as to accumulate as much data as possible. By averaging a large number of measurements, the difference in star transit times between those at the base station and those at the field station could be held to a precision within a few hundredths of a second.

Of particular concern was the so-called "personal equation," the fact that one of the two astronomers may have a reaction time different from that of the other, and for example tap the electrical key with a consistently greater delay than that of his partner at the other end of the tele-

3.5. Chronograph that records "time by the yard." The time signal is recorded on a paper-covered drum that revolves once per minute, so that tick marks are scribed every second, about 1 inch apart. The astronomer triggers an electric signal at the instant that a given star passes the vertical crosshair of the telescope. Measurement of the position of the signal on the chronograph paper permits time determination as precise as $1/100$ second for the passage of the star. Such precision can fix the location of the station to within 10 to 20 feet of longitude (see Table 3.1). The chronograph was used only at primary astronomical stations adjacent to a telegraph line (U.S. Coast and Geodetic Survey, 1882).

graph line. If both in fact had the same delay in response time, then the time difference of measured star transits between the two stations would be accurate. But to correct for a difference in response times and thus to reduce errors, the reaction times of the astronomers was carefully measured by a special device so that corrections could be made for different reaction times between the two. Another system in use was to switch the field station astronomer with the base station astronomer halfway through the observations. When the slower operator was at the more westerly (later) station, the time difference would be too long, and when the slower operator was at the more easterly (earlier) station, the time difference would be too short. By averaging together all the readings of time difference for the astronomical transits, errors arising from personal differences of this sort could be minimized. Transferring the astronomers across this great distance was facilitated by the use of the transcontinental railroad, which was operational from 1869 on.

The surveyors also developed a method for eliminating the inaccuracies caused by the time required for the electric signal to pass through the telegraph wire and activate the needle on the chronograph, even though this factor is small because of the great velocity of electromagnetic waves (186,000 miles per second under ideal conditions). When the western observer measured the time difference of the two signals (one his, the other his partner's from the distant base station) for the same star on the chronograph's paper record, his own signal would show little delay because the distance of travel through the few inches of wire to his chronograph would be minimal, but the signal from the earlier, eastern station would be delayed about $1/100$ second if that station were 1,860 miles east. Hence the time difference would be $1/100$ second too short. In contrast, when the eastern observer measured the time difference of the two signals on his chronograph, his own signal would show negligible delay, but the signal from the later western station would arrive $1/100$ second too late, making the time difference $1/100$ second too long. Hence by averaging the time differences determined by the chronographs at the two stations, one at each end of the telegraph line, the bias caused by the transmission time of electric signals through the telegraph wire could be eliminated. George Wheeler (1889) noted that the mean probable error of longitude at his primary stations was 0.27 arc seconds (22 feet) and of latitude 0.08 arc seconds (8 feet) (compare with Table 3.1).

At some of the major astronomical field stations that lay on flat ground, the distance from the astronomical bench mark to a new bench mark about 4 miles distant was measured with the greatest precision possible.

This baseline was measured by means of a 20-foot plank (with engraved brass rulers affixed to the ends), which was used to measure the distance between flat-headed nails set in the ground. After tediously carrying this plank over the ground, and leapfrogging from nail to nail until the distant bench mark was reached, the crew repeated the process in the opposite direction and remeasured the line until a sufficient accuracy was attained. From each endpoint station, the transit was set up and angles were then measured to neighboring mountain peaks to fix their location by triangulation. These peaks, in turn, would then become secondary stations for the location, by further triangulation, of still more stations.

At stations away from the telegraph line, timepieces (chronometers) were used to measure the time of star transits so as to establish longitude. The chronometers were checked for accuracy whenever the party visited a telegraph station. Another method of making longitude measurements relied on the establishment of time by astronomical means alone. Such methods were far less accurate than those obtained by the use of telegraphed time, and they required lengthy observations, comparison with star catalogs, and the use of complex corrections. One of the more common such methods utilized fixing the time of culmination (highest point) of the moon. This method of determining time was generally accurate to within about 5 seconds of time, equivalent to more than a mile on the ground at the latitude of the Sierra.

When King's 40th Parallel Survey was well under way, Congress authorized several other groups, funded by the Army's Corps of Engineers, to conduct geographic and geologic surveys of the arid West. These surveys explored and mapped large areas across the Rocky Mountains, the Colorado Plateau, and the Great Basin of Utah and Nevada. During the 1870s, Major John Wesley Powell continued his work on the Colorado River Basin after leading the historic initial expedition down the river in 1869. Dr. Ferdinand Hayden conducted surveys in Colorado, Wyoming, and Nebraska under the auspices of the General Land Office and then the Department of the Interior. Lieutenant (later Captain) George Wheeler (Fig. 3.6) commanded the U.S. Army Geographical Surveys West of the 100th Meridian in Nevada, Colorado, Arizona, and California (Dawdy, 1993). The rationale for assigning surveying to the U.S. Army Corps of Engineers was the long tradition of Army surveys of the West, including the Lewis and Clark Expedition in 1805 and the Pacific Railroad Surveys of 1853–54. Good maps are essential for military success in the field.

Of the surveys undertaken by Powell, Hayden, and Wheeler, only those led by Wheeler worked near the region of the highest Sierra. Wheeler,

3.6. Lieutenant George Wheeler, who commanded the U.S. Army Geographical Surveys West of the 100th Meridian. U.S. Geological Survey Library.

sixth in his West Point class of 1866, began his Army career as engineer in a survey of Point Lobos, near Monterey, California. In 1869 he was assigned to head a reconnaissance from Camp Ruby (northeast of Elko), in east-central Nevada, with the mission of investigating the White Pine Mining District, and then if possible to investigate navigation in the lower Colorado River. This survey resulted in the publication of a map and short text in 1869.

In 1871, the U.S. Army Engineer Department noted that several government-funded surveys managed by civilians, including Clarence King's, were conducting geographic and geologic surveys of the West in a realm where mapping traditionally had been done by the Army. The Army Corps of Engineers, therefore, formally established the United States Geographical Surveys West of the 100th Meridian and placed Lieutenant Wheeler in charge. This survey was directed to prepare maps of the area south of the Central Pacific Railroad in Nevada and Arizona, to note the numbers and disposition of the Indians, to survey potential rail and road routes, to describe the mineral deposits, climate, and vegetation, and to note the availability of wood and water along the way. In order to formulate these maps, the expeditions were charged with determining as accurately as possible the latitude and longitude of principal points within the assigned area.

Wheeler was a strong-minded leader. He showed little mercy in dealing with unfriendly Indians, and he readily invoked the articles of war to keep his men in line. He helped organize the Lyons and Wheeler Mining Company early in 1872 so as to acquire profitable mining properties in southeastern California while he was engaged in government work. He was antagonistic toward the civilian-managed federal surveys, criticizing their emphasis on geologic and other scientific studies rather than the strict geographic-topographic mapping that he and the Army favored.

Despite his anti-science stance, Wheeler's 1871 100th Meridian surveying party did enlist many well-qualified surveyors and scientists, including Grove Karl Gilbert (Fig. 3.7) as chief geologist. The expedition began on the railroad line at Halleck Station in central Nevada, northeast of Elko, and traversed to eastern California via Carlin, Eureka, and Bel-

3.7. Grove Karl Gilbert, whose modest entry into government science as geologist with the Wheeler Survey belied his multitude of later accomplishments. He became chief geologist of the U.S. Geological Survey and was widely regarded as the country's foremost geologic thinker. U.S. Geological Survey Library.

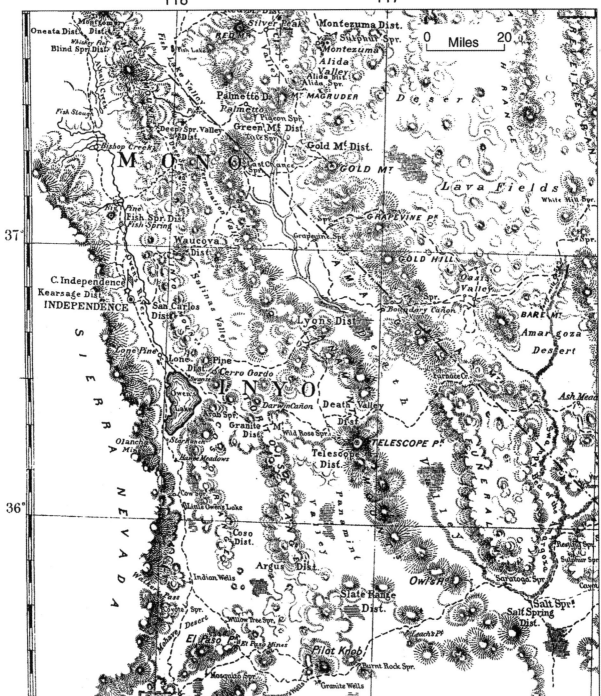

mont. The corps visited sites along the east front of the Sierra Nevada while working for several days out of Camp Independence. Astronomical measurements were made at the camp, and longitude was determined by measurement of lunar culminations. This work led to the preparation of a map published in 1872 (Fig. 3.8).

This map (Wheeler, 1872), very much a sketch map, with a scale of 24 miles to the inch, was limited on its west side by the crest of the Sierra Nevada and extended from the southern limit of the range to a point north of Bishop Creek. The map was prepared quickly from rather skeletal data, to be included in a proposal for further funding. It carries the disclaimer "This Map is a hasty and partial compilation from the topographical data already received, many still remaining en route. Upon it is projected in Skeleton the ground-work of the area examined. All points except the prominent astronomical positions are subject to a slight modification upon the final Map." No peaks in the Sierra are labeled, but the map is quite detailed in its depiction of the mining districts active at that time.

The map was later refined, utilizing all available sources, and published in 1874 in a folio as Atlas Map Sheet No. 65, which covered the Sierra from 35° 40' to 37° 20' north latitude (Fig. 3.9; Wheeler 1874). The Sierra part of this map made use of and credited the Hoffmann map, which was based on the data collected by the California Geological Survey from 1864 to 1870 but was not published until 1873. The 1874 Wheeler map (like the Hoffmann map) also positions Mount Whitney incorrectly, at the latitude of Kaweah Peak—at the present position of Mount Langley. The true Mount Whitney was first climbed in 1873 and the error recognized, but too late to be entered on the Hoffmann or Wheeler maps.

In 1875 Wheeler fielded eight distinct parties, one of which again visited the region of Owens Valley and headquartered at Camp Independence. Astronomical observations and triangulation determined the position of several important peaks in the Sierra. Peaks located by the Wheeler surveys in the southern Sierra were North Palisade, Split Mountain (South Palisade), Black Mountain, Mount Williamson, Mount Whitney, Mount Langley (also called Sheep Mountain, Mount Corcoran, and old [or false] Mount Whitney), Mount Denels, and Olancha Peak. These astronomically based points were critical in the later preparation of accurate maps of the region. The group also worked up Independence Creek and investigated the Kearsarge Mining District. At 10,500 feet on the Kearsarge Pass trail, Timothy O'Sullivan photographed Flower Lake as an example of a typical Sierra glacial lake (Fig. 3.10).

3.8. Part of Wheeler's preliminary sketch map of the 1871 survey (Wheeler, 1872). The focus of the map is the mining districts east of the Sierra Nevada. No new work was done in the Sierra, and no peaks were located. This map was rushed into publication so as to provide justification for continued military funding of the mapping program.

3.9. *(Opposite)* Part of atlas sheet number 65 (1874), Wheeler Survey of 1871, original scale 8 miles to the inch. This map is an elaboration of the 1872 map shown in Fig. 3.8, to which material from other sources has been added, but for which no additional field work had been done. The Hoffmann map published the year before (1873; Fig. 2.15) provides all data for the Sierra Nevada, and is incorporated unchanged, even including the trail of the Brewer party. Both maps label Mount Whitney at the present site of Mount Langley, south of Whitney's actual position. The true Mount Whitney lies north of the north margin of Owens Lake. Clarence King mistakenly climbed (and labeled) the wrong peak in 1871, believing it to be Mount Whitney, and this error was incorporated on both the Hoffmann map and this map.

3.10. *(Above)* View across Flower Lake (10,500 feet) near the Kearsarge Pass Trail. The main Sierra Crest in the background is about 1 mile south of Kearsarge Pass. From a lithograph rendered from a photograph by Timothy O'Sullivan when with the Wheeler Survey in 1875 (Wheeler, 1889).

The Scaling of Mount Whitney: Errors Compounded

Although visible from the Lone Pine area in Owens Valley, Mount Whitney is not conspicuous in outline from that vantage point, because it is flanked by closer peaks that appear higher. For this reason the exceptional height of the peak long went unnoticed from Owens Valley. As we have seen, it was observed in 1864 from Mount Brewer, to the northwest, by the party of the California Geological Survey. They sighted a peak to the southeast on the main crest and stated, "The other high point, eight miles south of Mount Tyndall, and, so far as known, the culminating peak of the Sierra, was named by the party Mount Whitney" (Whitney, 1865).

As recounted previously, Clarence King first attempted to climb this peak from the northwest, in 1864, but found it to be too far off. He climbed Mount Tyndall instead and from it took bearings to Mount Whitney, which is six miles to the south-southeast. A few days later King again attempted to attain the summit of Mount Whitney, this time from the southwest via the Hockett Trail. He traveled for three days north of the trail to the base of the mountain and struggled up the west side but was turned back just 300 to 400 feet from the summit when he encountered a sheer precipice on the southwest side. King measured an altitude of 14,740 feet at his highest point and estimated that the peak was about 15,000 feet high.

In 1871 King set out to climb the peak again, this time from Owens Valley. He directed his efforts toward the position of the peak as it had been determined in 1870 by Hoffmann. This incorrect position, combined with poor visibility in cloudy weather, caused King to believe that he had scaled Mount Whitney, whereas actually he had climbed Mount Langley. He hung his barometer on the summit and reported its height at about 15,000 feet.

As a result of King's reports and book, this peak came to be identified as Mount Whitney by settlers in Owens Valley, but when others attempted to climb it, a curious issue unfolded. In 1873 (two years after King's reported success) Watson Goodyear (Fig. 2.12), who had accompanied Hoffmann on the California Geological Survey of 1870 to Owens Valley, presented a talk at the California Academy of Sciences, in San Francisco. In his talk, he described how he and a companion rode mules to the summit of King's mountain in July of that year. They found King's coin there, yet they saw a peak 5 or 6 miles to the north that was considerably higher. This higher peak was the one seen in 1864 from Mount Brewer by the Brewer party and the one the party named Mount Whitney at that time. King had mistakenly climbed the wrong mountain, had not seen the higher mountain to the north because of cloudy weather, and thereby had completely confused the issue. The mountain that King climbed and took to be Mount Whitney was locally called Sheep Mountain, was later called Mount Corcoran, and is now named Mount Langley.

In this same year, 1873, several other parties, spurred on by Goodyear's report, did achieve the summit of the true Mount Whitney by climbing the western slope. On August 18 a group of Lone Pine residents, Charles D. Begole, Albert H. Johnson, and John Lucas, made the first known ascent of Mount Whitney, which they approached from the west. They called the mountain Fisherman's Peak. Later in August a second party, consisting of William Crapo and Abe Leyda climbed the mountain, and on September 6 the third ascent was made by Crapo, William L. Hunter, Tom McDonough, and Carl Rabe (Farquhar, 1965). Rabe, who had been employed as a cook by the Geological Survey (Fig. 2.12), learned about the handling of surveying instruments as an assistant to the topographers. He was entrusted with a barometer (Fig. 2.14), which he carried to the summit during this ascent. An identical mercury barometer was set up in the town of Lone Pine, and readings were made every half hour during the day, so that corrections could be made later for natural, weather-related variations in air pressure. When the readings were compared, and corrections applied, the summit elevation was made to be 14,898 feet (Fig. 3.12).

Clarence King was excited by Goodyear's report and again raced to the southern Sierra with the goal of climbing Mount Whitney. He and Frank Knowles, a settler in the Tule River region, finally climbed the mountain from the west on September 19, 1873, one month after the first ascent. In the second edition of his book, *Mountaineering* (King, 1935), King added a 16-page section in which he described his belated climb of the mountain and admitted his mistaken identification of the mountain in 1871, owing both to cloudy weather during the ascent and to magnetic aberrations that had affected his compass reading from Mount Tyndall in 1864. He eloquently (but somewhat sarcastically) stated in referring to his climb of 1871,

> My little granite island was incessantly beaten by breakers of vague impenetrable clouds, and never once did the true Mount Whitney unveil its crest to my eager eyes. Only one glimpse and I should have bent my steps northward, restless till the peak was climbed. But then that would have left nothing for Goodyear, whose paper shows such evident relish in my mistake, that I accept my '71 ill-luck as providential. One has in this dark world so few chances of conferring innocent, pure delight.

King absolved Hoffmann of all blame in the matter and noted that the incorrect mapped position of Mount Whitney had resulted partly from an erroneous magnetic bearing that he measured from the summit of Mount Tyndall to Mount Whitney in 1864. This incorrect reading, which apparently resulted from a local magnetic irregularity on the summit of Mount Tyndall, in turn misled Hoffmann in his identification (and measurement) of the supposed Mount Whitney as seen from Owens Valley in 1870. The mislocation was recorded on two maps, those of Hoffmann of 1873 (Fig. 2.15) and Wheeler of 1874 (Fig. 3.9), which were published before the error came to light. Hence, King's incorrect compass reading from Mount Tyndall in 1864 had come back to haunt him. It led to Hoffmann's incorrect angle measurement from Owens Valley in 1870, which was incorporated in the information Hoffmann supplied to King, who went on to climb the wrong mountain in 1871.

King described his feelings at the summit in his inimitably colorful prose:

> This is the true Mount Whitney, the one we named in 1864, and upon which the name of our chief is forever to rest. It stands, not like white Shasta, in a grandeur of solitude, but about it gather companies of crag and spire, piercing the blue or wrapped in monkish raiment of snow-storm and mist. Far below, laid out in ashen death, slumbers the desert.
>
> Silence reigns on these icy heights, save when scream of Sierra eagle or loud crescendo of avalanche interrupts the frozen stillness, or when in symphonic fullness a storm rolls through vacant cañons with its stern minor. It is hard not to invest these great dominating peaks with consciousness, difficult to realize that, sitting thus for ages in presence of all nature can work, of light-magic and color-beauty, no inner spirit has kindled, nor throb of granite heart once responded, no Buddhistic nirvana-life even has brooded in eternal calm within these sphinx-like breasts of stone.

In October 1873 John Muir, carrying the Hoffmann map—and apparently unaware of the revised location of Mount Whitney—climbed Mount Langley and noted the higher peak to the north. He attempted to traverse to it, but was thwarted by a high intervening ridge, and spent the night blanketless dancing about to keep from freezing. The next morning he turned back and returned to Lone Pine. He then climbed Whitney solo from the east, making the summit a few weeks after King's ascent.

In 1875, parties from the Wheeler Survey climbed the mountain, and attempted to straighten out the confusion caused by King's 1871 ascent and his assigning of the name Whitney to the wrong mountain. William A. Cowles ascended the peak first climbed by King (the present Mount

3.11. The true Mount Whitney. 4.8 miles distant, as viewed to the northwest from the summit of the false Mount Whitney (Mount Langley). Sketched September 22, 1875, by William A. Cowles when with the Wheeler survey (Wheeler, 1889).

Langley) and from there sketched the true Mount Whitney standing off to the northwest (Fig. 3.11). The true Mount Whitney was climbed by four Wheeler Survey men on September 24 and by three others on October 13. From the summit they made extensive barometric measurements and measured angles to many peaks and other points, including vertical angles of depression to a point in Lone Pine and to the flagstaff at Camp Independence. To improve their accuracy, they later reversed these vertical angles, that is, they measured vertical angles of elevation from the same valley sites back to the peak of Mount Whitney, which by now could be correctly identified from the valley.

The altitude of the peak as determined by these paired vertical angles was 14,470 feet, and that determined from the barometers the party had carried to the summit was 14,471 feet. The accuracy of the vertical-angle measurements was of course dependent on the altitude determinations of Lone Pine and Camp Independence, which themselves had been determined by barometer. But the barometric measurements at these settle-

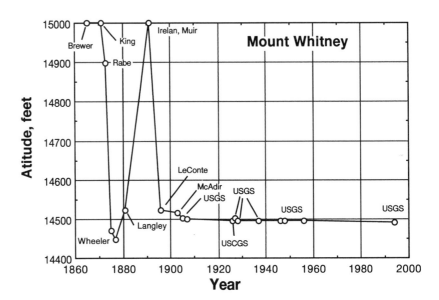

3.12. Reported altitudes of Mount Whitney since its discovery in 1864. Rabe was the first to measure the altitude with a barometer carried to the summit, and later compared his readings with a stationary barometer maintained at Lone Pine in Owens Valley. Wheeler made the first measurements by triangulation, and the U.S. Geological Survey later ran the first level line to the summit.

ments were considered to be of high reliability because many repetitive measurements had been made. Considering these problems, the elevation determined by the Wheeler group was remarkably good, just 20 to 21 feet short of the 1994 value of 14,491 feet (Fig. 3.12).

Wheeler's horizontal-angle measurements also led to the first reliable determination of the geographical coordinates of Mount Whitney, the highest point in the land. Its position was determined at 36° 34' 32" north latitude and 118° 17' 30" west longitude. This position was a basis for subsequent maps for many years, including the LeConte map of 1899 (Fig. 3.20).

Measuring the Ground: The General Land Office

Through numerous expeditions, the explorers and mappers slowly improved their depiction of the shape of the land and the precise location of natural features. But as the settlers moved in, another kind of surveying became necessary, that of marking and dividing the land on a grid, so that it could be allocated to the various purposes of homesteaders and railroads and the like. The General Land Office, under the U.S. Department of the Interior, carried out this function.

The primary system of land division adopted in the western states, including California, was based on measurements east or west from a designated meridian of longitude, the north-south *principal meridian*, and north or south from a designated parallel of latitude, the east-west *baseline*. The country was divided into a grid made up of quadrangles 6 miles on a side called *townships*. Each township was then divided into 36 squares of 1 square mile each called *sections*. The sections each contained 640 acres, and the quarter-sections (a common land unit taken as a homestead) each contained 160 acres.

The surveying was generally contracted out to local surveyors, who concentrated on surveying the land divisions and marking them on the ground, although, for them, depicting these parcels accurately on a map was a secondary concern. Later, when the township and section corners were reoccupied on the ground and resurveyed onto a more precise map, these land divisions were found commonly to be somewhat irregular in shape, rather than regular squares. By that time, however, the land was taken up, and the boundaries were generally legally fixed in their original surveyed and marked positions.

The numbering of the townships began at the intersection of the principal meridian and the baseline. The east-west divisions are called *ranges* and the north-south divisions, *townships*. The ranges are designated E or W and the townships N or S, depending on which side of the meridian and the baseline they fall. Hence a given township is defined as, for example, "township 11 S, range 14 E" indicating that it lies, uniquely, 11 townships south of the baseline, and 14 ranges east of the principal meridian.

The main surveying tools used for this work were compass, transit, and chain. The chain used for measuring distance was made up of 100 links of steel wire. Each link was 7.92 inches long, and the total length of the chain, and the unit of measure thus designated a *chain*, was 66 feet, or 4 rods, in length. Every tenth link in the chain had a brass tally mark, notched to indicate its number. One mile is 80 chains and an acre is made up of 10 square chains. Hence a quarter section (one quarter square mile) contained 40 x 40 = 1600 square chains = 160 acres. The link chain as an instrument of land measure was superseded by the steel tape before the turn of the century. The tape, which was generally ¼ or ⁵⁄₁₆ inches wide and 100 feet long, was graduated to tenths of a foot. It proved to be more accurate than the link chain because the chain tended to lengthen as the links stretched, owing to the tension normally applied while measuring. Moreover, the steel tape could be dragged easily over irregular ground, whereas the chain tended to catch on rocks and brush.

California, which had never been organized as a territory, was admitted to the Union September 9, 1850, and as part of the terms of admission, two sections of land in every township (out of the total of 36 sections in a township) were awarded to the state. A program of land surveying was sorely needed to stimulate settlement and commerce, and in March 1851, the first Surveyor General of California, Samuel King, was appointed; he traveled to Washington, D.C., where he was briefed on mapping procedures and picked up three solar compasses, a transit, and other equipment.

Upon King's return, he set up an office in San Francisco and contracted for surveyor Leander Ransom to establish a baseline and a prime meridian, both of them to pass through the summit of Mount Diablo (3,849 feet), a prominent, well-defined peak in the coast ranges 25 miles east of San Francisco. The peak proved to be so rugged that Ransom could not run survey lines down its flank, but had to place his starting point some 12 miles south of the peak. From here, through a series of offset lines to the east and north, he established the corner of townships 1 N and 1 S and ranges 2 E and 3 E (that is, 12 miles east of the summit). The baseline was then continued back to the peak. This initial Mount Diablo survey is the basis for all subsequent land division in central California and in the entire state of Nevada.

The Act of Admission that added California to the Union in 1850 also ratified the state's boundaries. The principal obtuse angle midway along the state's eastern boundary was defined as that point where the 120th degree of west longitude intersects the 39th degree of north latitude. When this boundary was decreed, it was not well known where this boundary angle would occur. The Fremont map indicated that it would lie considerably east of the Sierra Nevada, and some 25 miles east of Lake Bonpland, the name Fremont used for Lake Tahoe (Fig. 2.3). As a result of his work in 1855, Goddard was one of the first to determine that the principal bend lay within the range and in the middle of Lake Bigler (now Lake Tahoe) (Uzes, 1977), and consequently that Fremont's longitudes were in error. Thus, the intent of those who defined the boundary—to include all of the northern Sierra Nevada and its eastern watershed in the state of California—was frustrated and considerable land became part of Nevada. In addition, this mismapping complicated later surveying and the location of the boundary, since the primary boundary angle occurred not only within the Sierra, but in the middle of Lake Tahoe, and could neither be occupied by surveyors nor marked by a boundary monument.

In 1855 A. W. Von Schmidt was commissioned to survey the public lands east of the Sierra Nevada. Von Schmidt carried his survey over all of Owens Valley, working from the Mount Diablo Baseline, which passed a few miles south of Mono Lake to the region south of Owens Lake. Included in his survey were townships 1 S to 12 S and ranges 31 E to 35 E, as numbered from Mount Diablo. Von Schmidt's crew consisted of the party chief, a compassman, three chainmen to measure distance, and two axmen to clear the lines.

In 1873 the now Colonel Von Schmidt was commissioned to make an accurate survey of the California–Nevada boundary line. He found that the Colorado River, which had been on the east side of the river valley during a survey in 1861, had migrated almost three miles west. This required that he modify the location of the line. The position of the line was further corrected in 1899 when the U.S. Coast and Geodetic Survey resurveyed and monumented the line. In 1901, the California Legislature declared this line to be the correct and legal boundary between Nevada and California.

By the 1880s, land surveys had reached into the Sierra Nevada, and townships and ranges—all relative to the Mount Diablo Baseline and Meridian—were set out on maps (Figs. 2.37, 2.38, 3.13). The original Sequoia National Park, created in 1890, occupied two of these townships, and a week later five more townships were added to the park by federal statute (Fig. 2.39).

One of the first maps to show the new park name (but not the park's boundary) was the Irelan map, published January 1, 1891, a few months after the creation of Sequoia National Park. This map, a colored mineralogic and geologic map of California (Fig. 2.37; Irelan, 1891), indicates mines by commodity symbols, and shows eight mapped geologic units. It is one of the few maps that show the Mount Whitney Military Reservation. The Irelan map depicts the North Fork of the Kings River extending much too far east, so that it, in effect, heads south of Mount Goddard and drains the main Sierra Crest to the east. This aberration was not shown on the 1883 Wright map (Fig. 2.38), and, as we shall see, was corrected on the 1893 LeConte map. The Irelan map also referred to the stream west of Mount Brewer as Cloudy River, though the current name, Roaring River, had been used by Wright (1883) and would be used by LeConte (1893).

In August and September, 1890, John Muir published articles in the nationally read *Century Magazine* supporting the proposed Yosemite National Park. After the park was created he wrote another article for *Century* spelling out the case for enlarging Sequoia National Park to include the Kings River canyons. His article in *Century* in November 1891 included two maps, one of them a detailed large-scale map of the South Fork canyon of the Kings River (Fig. 2.29), the other a small-scale map showing a proposed enlargement of Sequoia National Park to include

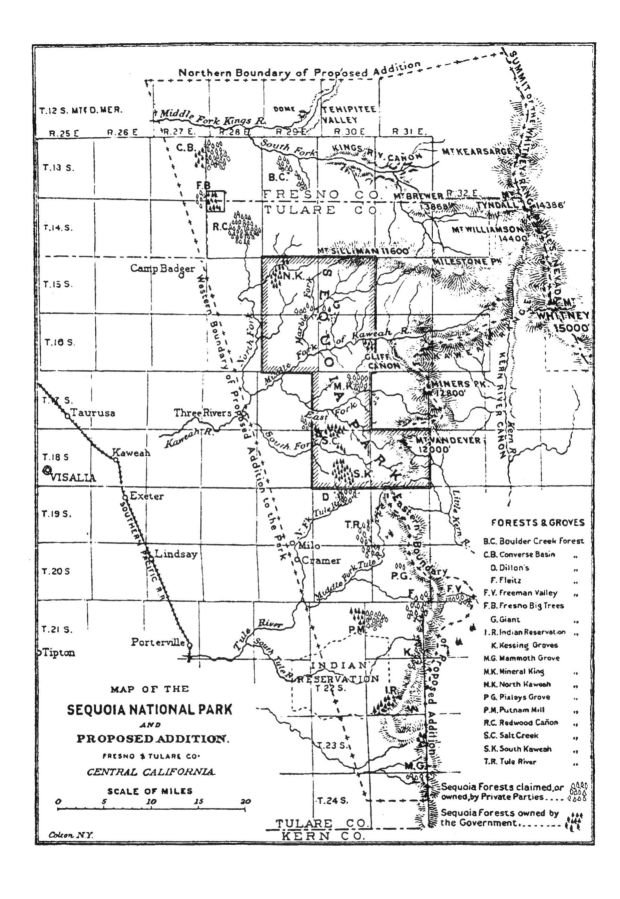

both the Middle Fork and South Fork canyons of the Kings River (Fig. 3.13). This is one of the first published maps showing the boundaries of the park as then constituted.

The Wright map (Fig. 2.38; 1883), the Irelan map (Fig. 2.37; 1891), and the Muir map (Fig. 3.13; 1891) all show no upper south-trending segment of the South Fork of the Kings River, a segment that drains what is now called Paradise Valley. Wright refers to the main east-trending segment of the canyon as Paradise Valley and also as Kings River Canyon, and the Muir map labels it Kings River Canyon. Apparently, the first map to show the south-trending segment (shown actually as a southwest-trending canyon) labeled as Paradise Fork is the map published with the Muir article in November 1891. Both LeConte maps (Fig. 3.17; Dyer, 1893, and Fig. 3.19; LeConte, 1896) designate Paradise Valley as on modern maps, along the south-trending canyon.

The Irelan map (Fig. 2.37; 1891) shows in large type the place name El Capitan on the south side of the upper South Fork of the Kings River Canyon in a place equivalent to what is now called the Grand Sentinel. Grand Sentinel seems to appear first as a place name in the Muir map (Fig. 2.29; 1891) and is on both LeConte maps, and the name El Capitan was apparently dropped after the Irelan map appeared. Both the Irelan map (Fig. 2.37; 1891) and the Muir map (Fig. 3.13; 1891) report an altitude of 15,000 feet for Mount Whitney, the altitude reported by Whitney from the estimates of the Brewer expedition and the altitude reported by King for his 1871 barometric measurements of Mount Langley, which at the time he thought was Mount Whitney. LeConte's 1896 map gives an elevation of 14,522 feet for Mount Whitney, the elevation reported by Langley.

Both the Wright map (Fig. 2.38; 1883) and the Irelan map (Fig. 2.37; 1891) use a coordinate system showing longitudes in the parks region between 41° and 42°, in contrast to both older and more recent maps, which show longitude between 118° and 119°. In the late 1800s, the General Land Office required surveyors to employ a system in which the prime meridian passed through Washington, D.C., rather than Greenwich, England. The Greenwich longitude at Washington (as measured at the U.S. Naval Observatory) is 77° 03' 02.3" W, and if we subtract 77° 03' 02.3" from 118° we get about 41°. In 1884, the International Meridian Conference, meeting in Washington, D.C., agreed to adopt the "meridian passing through the center of the transit instrument at the Observatory of Greenwich as the initial meridian for longitude." The United States standard for marine navigation was fixed at Greenwich thereafter, but for some years land surveyors continued to use the Washington standard.

3.13. One of the first maps to show the boundaries of Sequoia National Park. The map appeared in the November 1891 issue of *Century Magazine*. John Muir was intent on including the Yosemite-like canyons of the Kings River, the Sierra Crest, including Mount Whitney, and additional sequoia groves within the park. Here he shows the proposed boundaries of an enlarged park intended to include these features (Muir, 1891).

Mapping the High Trails: LeConte, Solomons, and the Sierra Club

The Sierra Club was founded in 1892 by a group of 27 citizens of the San Francisco Bay region (Fig. 3.14). John Muir (Fig. 2.24) was one of the founders of the club and served as its first president, until his death in 1914. The club was founded in response to growing public interest in the recreational aspects of the Sierra; its mission was to explore and enjoy the mountains of the Pacific region and to further the preservation of their natural features. Two hundred and fifty members and friends attended the first general meeting in September 1892 at the California Academy of Sciences in San Francisco, and between five and six hundred attended the second general meeting in October.

One of the founding members of the club was Joseph LeConte, the

3.14. John Muir was president of the Sierra Club from its founding in 1892 until his death in 1914. The club's first seal featured a pine tree and the motto *Altiora Peto* (I seek higher things). It is surprising that Muir endorsed this motto, since it nearly duplicated *Altiora Petimus* (we seek higher things), the motto of the California Geological Survey headed by Josiah Whitney, with whom he had considerable scientific disagreement. In any event, Willis Polk designed a new seal with no motto in 1894. This seal, with the giant sequoia tree replacing the pine tree, Half Dome and alpine peaks in the background, and a rustic border, was used up to 1940. The seal with the broad border and without the cones and branches was used during the war years (1940–46), after which a modified seal with a more simplified scroll was adopted, the seal still used today. Though the seal has changed through the years, the club's objectives—to explore and enjoy the mountains of the Pacific region and to promote the preservation of their natural features—have not changed.

first professor of geology at the University of California, which had been established in 1870. He had studied at Harvard under Professor Louis Agassiz, an eminent geologist well known for his work on glaciation. At the end of the first session of the university, LeConte joined a group of his students for an excursion into the High Sierra. During their visit to Yosemite Valley, they met John Muir, then working at a sawmill near Yosemite Falls. Muir accompanied the group over the Sierra Crest to Mono Lake. It may have been during this trip that LeConte and Muir discussed ideas concerning the importance of glaciation in sculpting the High Sierra and Yosemite Valley.

Professor LeConte's son, Joseph N. LeConte (Fig. 3.15), also a founding member of the Sierra Club, would come to have a more lasting im-

3.15. Joseph N. LeConte and party celebrating with a smoke at Millwood, near General Grant, at the end of their trip of 1894. Left to right: John Pike, Ed Hutchinson, LeConte, and Jim Hutchinson. LeConte, whose belt is at its last hole, is carrying an aneroid barometer in a leather case over his shoulder. Bancroft Library.

pact on the exploration of the High Sierra than had his distinguished father. The young LeConte was trained early as a mountaineer. At the age of four he climbed—and was partly carried—up Mount Tallac, near Lake Tahoe. At eight he made his first camping trip to Yosemite.

In 1890, at the age of 20, "Little Joe" (a nickname that stuck because of his short stature) made an extended trip into the Kings Canyon region with three fellow students from the University of California. At that time these mountains, which had been traversed by the Brewer party 26 years earlier, were still little known, and largely unmapped. On this trip they took horses into the main South Fork of the Kings River Canyon, camped for several days near Cedar Grove, climbed the walls of the canyon, and explored the neighboring mountains. They then went over Kearsarge Pass, south to Lone Pine, and up Mount Whitney from the Hockett Trail and the western approach. Turning north, they traveled up Owens Valley, over Mono Pass to Tuolumne Meadows, and on to Yosemite Valley and Hetch Hetchy Valley, and returned to Berkeley by way of Angels Camp and Altaville. On this trip LeConte carried a surveyor's transit and made measurements defining the location of various peaks and watercourses, with a view toward improving the existing maps.

The young LeConte began teaching mechanical engineering at Berkeley in 1892 but was able to spend his summers in the mountains. He was a good photographer and took many of the first photographs of remote areas in the High Sierra, some of which were published in the *Sierra Club Bulletin*. These photographs were taken primarily on dry glass plates, which had superseded wet glass plates about 1879. The earlier wet plates were prepared wet, exposed wet, and processed while still damp, and their development required a portable darkroom. LeConte and other photographers of this period carried the heavy, fragile glass plates and developed them later. George Eastman first produced cellulose nitrate (film) negatives in 1889, but film was not generally utilized until about 1895. "Never intentionally 'arty,' most of his compositions reveal a sensitive reaction to the finest moments of the mountain scene," said the great wilderness photographer Ansel Adams (1944), in commenting on LeConte's photographs. "It is this quality that differentiates between a mere *record* and a *creative, sympathetic statement.*"

LeConte's chief interest was maps, however. As a result of his 1890 trip to Kings Canyon, and as travel in the mountains increased, he realized that there was a compelling need for reliable maps, especially of the high country. He assembled all available mapping of the range and set about systematically improving the few sketchy extant maps, most of which

3.16. Surveyor's compass used for measuring bearings to sighted targets. The compass circle is graduated to one-half degree.

were based on the Hoffmann map made during the Brewer party survey and published in 1873.

Joseph N. LeConte determined the position of unmapped peaks by taking compass bearings (Fig. 3.16) to points whose geographic coordinates had been surveyed previously, such as those reported in Wheeler's *Geographical Surveys West of the 100th Meridian*. From the summit of each new peak he occupied, he took bearings to the known points so that the peaks "were located as accurately as a small surveyor's compass and the local variation of the needle would allow." From each peak "the intervening country [was] sketched, the sketch being afterwards corrected by numerous compass bearings."

The very first issue of the *Sierra Club Bulletin* contains one of Joseph N. LeConte's maps (Fig. 3.17). Volume 1, number 1 of the bulletin, published in January 1893, includes an article by Hubert Dyer (one of the students accompanying LeConte on the 1890 trip) on the Mount Whitney Trail. Attached to this article is a tip-in map of the Kings Canyon–Mount Whitney region signed only "J. N. L." and given no title or further accreditation. Neither this map nor its origin is mentioned in the article, but the map is clearly the work of the younger LeConte. Apparently, the publication of this map inspired Little Joe with a missionary zeal to map the highest Sierra.

In the next issue of the *Sierra Club Bulletin* (volume 1, number 2, June 1893), an announcement entitled "Maps of the Sierras" mentions, "The Sierra Club has recently distributed to its members two maps of portions of the Sierras. These maps were compiled for the Club by Mr. J. N. LeConte, of the University of California." One map, which covers the

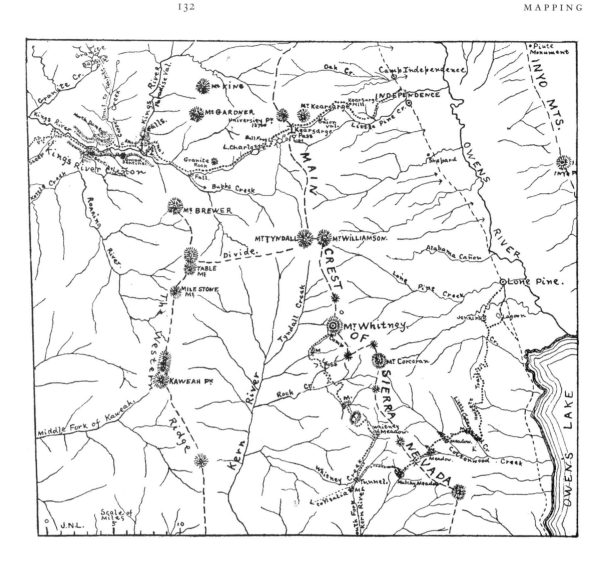

3.17. J. N. LeConte's first published map, that of the Kings Canyon–Mount Whitney region, appeared in 1893 in the first issue of the *Sierra Club Bulletin* attached to an article by Hubert Dyer on the Mount Whitney Trail. The map, identified only by the initials J. N. L. and lacking title or further accreditation, marks the beginning of LeConte's dedication to the study and mapping of the entire High Sierra.

Yosemite region, makes use of atlas sheets of the Wheeler survey, the Eighth Annual Report of the U.S. Geological Survey, the official Fresno County map, and the notes of T. S. Solomons, who visited the region in 1892.

The other map, which covers the Kings River–Mount Whitney region, is an extension and enlargement of the map published earlier in the year with the Dyer article. This map is based on the official county maps of Fresno, Tulare, and Inyo Counties and on the notes of LeConte's trip of 1890. These new maps were among the first to attempt to show in some detail the position of all known peaks and drainages in the highest stretch of the Sierra. The maps, drawn at a scale of 4 miles to the inch, were widely used by Sierra Club members and all others interested in travel in the High Sierra.

Theodore S. Solomons (Fig. 3.18) was another young mountaineer who developed an early interest in the remoter parts of the Sierra (Sargent, 1989). As a teenager Solomons had visited his uncle's farm near Fresno and had quickly become enamored of the unknown and unmapped snowy peaks visible to the east. In 1888, at the age of 18, he made a trip into the better-known northern Sierra and toured from Lake Tahoe through Alpine and Calaveras Counties.

His first major excursion into the High Sierra was in the summer of 1892. The first leg of the trip led from Donner Lake to Yosemite Valley. There he joined J. N. LeConte, and the two explored the Ritter Range

3.18. Theodore Solomons, who pioneered a near-crest route from the San Joaquin River to the Kings River drainage, and who conceived a crestal trail through the length of the High Sierra. His early efforts were extended by several others, and the route they pioneered eventually became the high-altitude trail that today reaches from Yosemite to Mount Whitney. It was named the John Muir Trail in honor of the great naturalist. Bancroft Library.

and the east escarpment near Mono Lake. Then, leaving LeConte, who had to prepare for his teaching duties at the University of California, Solomons and a companion headed south with pack mules. They ascended Mount Ritter a second time, climbed Mount Lyell, explored the upper reaches of the Middle Fork of the San Joaquin River, and visited Fish Valley, near the southern extremity of the newly created Yosemite National Park.

Solomons, like LeConte, had made an extensive study of all existing maps that included the region. "In desperation, I raided the Surveyor-General's Office," he noted many years later (1940), "and almost swooned with delight when they handed out plat after plat of official land surveys of nearly every township in Fresno and Tulare counties. Apparently the southern High Sierra was not only explored but meticulously surveyed. . . . " When he traveled over the mapped terrain he found "that these gorgeous specimens of the draughtsman's art were pure fabrications, the products of an imagination unsullied by the slightest acquaintance with the Sierra Nevada. We discovered it in the field—bitterly."

In the summer of 1894, after visiting Tuolumne Canyon and Yosemite Valley, he and a companion pushed south on September 1 with two burros and a horse. They went up the Middle Fork of the San Joaquin River to Fish Valley and found themselves surrounded by the huge flocks of sheep that overran the mountain meadows. They searched out a high route to the south, crossed a rugged pass (the present Silver Pass) into the drainage of the South Fork of the San Joaquin River, and descended to (and named) Vermilion Valley, now occupied by the Lake Edison Reservoir. They traveled up Bear Creek and climbed a 13,075-foot peak that they called Seven Gables, from its fancied architectural shape (their peak is not to be confused with the cliff of that name on the Muir map; Fig. 2.29). "Roughly speaking, one might say that the sight was sublime and awful . . . ," he wrote, "I believe I was then looking upon the finest portion of the crest of the Sierra Nevada. . . . "

Their trek was cut short when they encountered near-impassable terrain and were overwhelmed by an early snowstorm. Describing a night without shelter in the storm, Solomons writes: "My companion was taken by a chill, and I stood up hour after hour alternately warming the two rags of blankets, and making hot coffee." With 4 feet of snow on the ground, they abandoned their stock and most equipment and walked out of the mountains, arriving at Fresno several days later.

Solomons had set himself the compelling task of pioneering a 200-mile high-elevation trail paralleling the main crest, but remaining just west of it, from the Yosemite region to Mount Whitney. He had already, in 1892, reconnoitered the route from Yosemite Valley on the Merced River through the Lyell–Ritter country into the upper drainage of the Middle Fork of the San Joaquin River. In 1894 he explored the Silver Pass route between the upper reaches of the Middle and South Forks of the San Joaquin Rivers. And in 1896, he began near the endpoint of the 1894 trek in the canyon of the South Fork of the San Joaquin River, with a view toward attacking the next obstacle, a route over the San Joaquin–Kings Divide. "Such a high mountain route practicable for animals between the two greatest gorges in the Sierra," he had written in 1895, "has never been found—nor, indeed, sought—so far as my researches have revealed."

At Jackass Meadow (below the present Florence Lake Reservoir), Solomons and one companion left their stock and set off with heavy backpacks up the South Fork of the San Joaquin River. Thus equipped—and able to travel over far rougher country than was possible with pack animals—they began one of the epic backpacking adventures of the Sierra. At the confluence of the South Fork with a larger stream coming in from the east (which Solomons called the Middle Fork of the San Joaquin and was later named Evolution Creek), they turned south and worked up the creek, with the hope of finding a route over the San Joaquin–Kings Divide. Solomons named a mountain west of the creek Emerald Peak because of its distinctive color. This mountain acquires its coloration from multihued metamorphic rocks. As they moved higher up this glaciated valley, they found themselves surrounded by a group of high granite peaks. He named the peaks Darwin, Fiske, Haeckel, Huxley, Spencer, and Wallace, and the lake at their base, Evolution Lake. Solomons wrote, regarding his selection of names for this group, "I could think of none more fitting to confer upon it than the great evolutionists, so at-one in their devotion to the sublime in Nature" (Solomons, 1940). They spent a few days climbing the peaks in the region, but like LeConte during the same summer, failed to recognize the utility of the pass at the head of the basin (the present Muir Pass), which opened to the upper Middle Fork of the Kings River.

They then went west on the Goddard Divide and climbed Mount Goddard, from which they had a view of the possible route for pack stock over the San Joaquin–Kings divide at the head of Evolution Basin, but a snowstorm changed their plans. Instead, they headed for shelter, south off the mountain and down into the dauntingly rugged Disappearing Creek region. This creek was so named because its water disappears beneath the

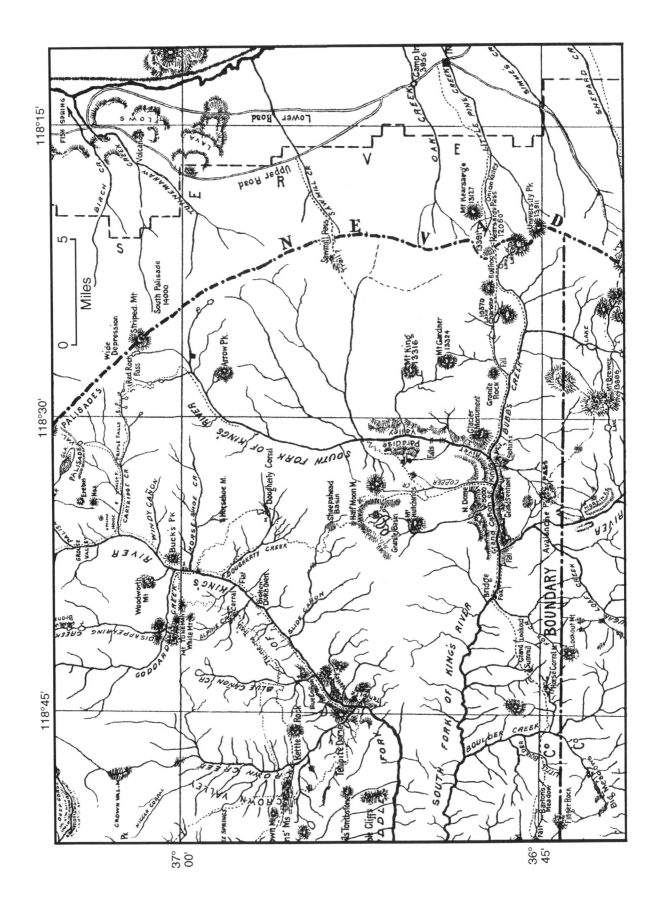

massive piles of loose blocks and boulders that fill the canyon bottom. It is unlikely that a foot trail will ever be built in this forbidding country.

After two days in this gorge, the travelers emerged in the canyon of the Middle Fork of the Kings River, near Simpson Meadow, thus pioneering a route over the High Sierra segment of the South Fork San Joaquin–Middle Fork Kings Divide. They descended the Middle Fork of the Kings River to marvel at the Yosemite-like canyon of Tehipite Valley and spent a hard day climbing Tehipite Dome, after which they slept near the summit.

Returning to Simpson Meadow by means of a sheep trail high on the north canyon wall, they spent several days climbing south over the high divide between the Middle and South Forks of the Kings River. On August 28, exhausted but triumphant, the pair finally reached the canyon of the South Fork of the Kings River and civilization. Although they had followed a particularly rough course from the head of Evolution Creek to the Middle Fork of the Kings River, Solomons did define a high route from Yosemite to Kings Canyon, and he was the first to sketch in the general topography of much of this region.

In that same year, 1896, LeConte revised and combined his earlier maps and added new information to produce a single large map of the high part of the range, from the Yosemite region south to Mount Whitney. He used previously published information as well as new data of the unsurveyed regions of the High Sierra that he and several other explorers, including Solomons, L. A. Winchell, and Nathaniel F. McClure, had gathered. This reconnaissance map was not alleged to be accurate or complete, but it was the best that could be done in depicting the topography of this wild and remote region. It was published at a scale of 4 miles to the inch (1:250,000) and was designated as publication number 12 of the Sierra Club (Fig. 3.19).

In the announcement of this map (in the May 1896 issue of the *Sierra Club Bulletin*), J. N. LeConte appealed to club members for new information, particularly on the region of the divide between the headwaters of the Kings and Kern Rivers, about which little was known. In order to improve the map he requested information on the major streams and "the compass bearings of at least two known peaks from each of the principal forks, turnings, and lakes, together with sketches of its general course between these points, and of its tributaries up to their sources." One goal of this effort was the search, begun by Solomons, to find a High Sierra route, passable to pack stock, from the Yosemite region to Mount Whitney, a trail that eventually would become the John Muir Trail. But even as LeConte continued his efforts toward more topographic information, the

3.19. Part of J. N. LeConte's map of 1896 (Sierra Club publication no. 12). This map remained the most authoritative map of the High Sierra from Yosemite to Mount Whitney for about a decade, until the 30-minute quadrangle maps of the U.S. Geological Survey were published.

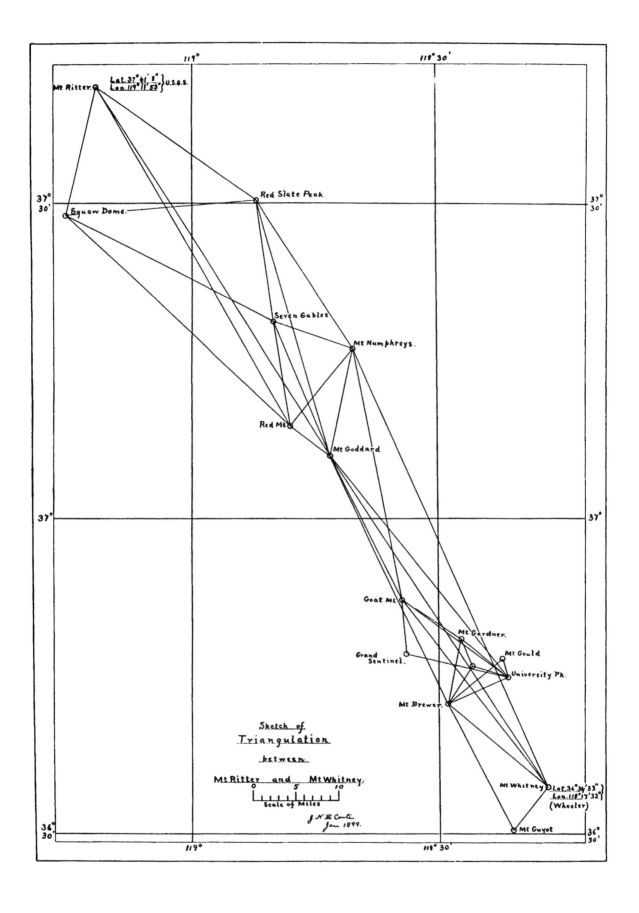

U.S. Geological Survey was preparing to embark on a program of systematic mapping of the Sierra.

LeConte, meanwhile, continued his interest in upgrading the maps. In 1899, he released the results of a triangulation survey of the Sierra from Mount Ritter to Mount Whitney, in the course of which he had occupied some 16 major peaks (Fig. 3.20). And in 1904, in the company of Grove Karl Gilbert (now with the U.S. Geological Survey; Fig. 3.7), the Hutchinson brothers, James and Ed, and others, he made a journey into the region of the South Fork of the San Joaquin River. They went by stage to Fish Camp, and then on horseback to Blaney Meadows and up Evolution Creek, in what is now the northern end of Kings Canyon National Park. While Gilbert examined evidence of glacial action on the rocks around Evolution Lake, and the others went off to climb Mount Humphreys, LeConte traveled alone up to the head of Evolution Creek and on to the crest of the Goddard Divide. This saddle at the drainage divide between the upper reaches of the South Fork of the San Joaquin River and the Middle Fork of the Kings River is the present site of Muir Pass on the John Muir Trail. "At exactly 12 o'clock the top was reached," LeConte (1905) wrote.

> The other side, as I had feared, broke down in the savage black gorges of the Middle Fork region, which were choked with snow and frozen lakes far down below. It would certainly be an impossibility to get an animal down anywhere along this part of the divide when the snow was deep, and even late in the season the success of such an undertaking would be very doubtful. Another thing which strengthens my opinion that the place could not be used as a pass is that no signs of sheep are to be found there.

The torturous route and heavy snow on the south side of the pass apparently obscured to LeConte the practicability of this saddle as the main

3.20. In order to determine precise locations, J. N. LeConte (1899) performed triangulations spanning 90 miles from Mount Ritter to Mount Whitney, along the central Sierra Nevada Crest. For this work LeConte used a telescopic instrument mounted on a mapping plane table. He climbed the 16 major peaks shown by circles in order to fix their position relative to one another (as well as to locate an additional 200 other points) by the method of intersecting lines. The validity of the entire map (and its scale) hangs on the locations of Mount Ritter and Mount Whitney, which were previously determined by the U.S. Geological Survey and the Wheeler Survey, respectively.

high-altitude route between the San Joaquin and Kings Rivers. George R. Davis of the U.S. Geological Survey first crossed this pass three years later with a pack train and named it Muir Pass, and LeConte himself took pack mules over the pass in 1908.

Thwarted in their attempt to climb out of the Evolution Basin country, the party went back down Evolution Valley and headed west to climb into the drainage of the North Fork of the Kings River, some 7 miles west of Muir Pass. From there they went through Crown Valley, dropped down the steep trail into Tehipite Valley on the Middle Fork of the Kings River, and worked up the river to the site of Simpson Meadow. From there they climbed over the Monarch Divide at Granite Pass and descended to the Canyon of the South Fork of the Kings River.

Finally, in 1908, LeConte met the challenge of tracing in a single trip a high-mountain route with pack animals close to the range crest from Yosemite to the South Fork of the Kings River. This route had been pioneered in sections by LeConte and others over more than a decade, but this, the first attempt to make the whole trip, was a major achievement. His trip chiefly followed the route of the present John Muir Trail. Like Solomons before, the party went up Evolution Creek, on over Muir Pass, to the upper Middle Fork of the Kings River. From there the route led over the Monarch Divide at Granite Pass and down to the canyon of the South Fork of the Kings River.

During his active teaching career at Berkeley, the younger LeConte found time to maintain his interest in the mountains and the Sierra Club. For 42 years he served as a director of the club and was president from 1914 to 1916. He was appointed honorary president in 1931 and died in 1950.

Another charter member of the Sierra Club was William E. Colby, who in 1894, at the age of 19, climbed Mount Dana and traveled down the Tuolumne Canyon. In the summer of 1901, with the encouragement of John Muir, Colby led the first Sierra Club Outing. These trips, spanning two to three weeks in the mountains, generally included several dozen participants. The emphasis was on an appreciation of the natural wonders of the wilderness, and the outings generally included several scientists who could instruct participants in botany, zoology, and geology. Many new routes were pioneered, and ascents of unclimbed peaks were made by mountaineers connected with these expeditions. Colby continued to be an inspirational leader of these trips for several decades.

The greatest mountaineer of the Sierra Nevada, however, was Norman Clyde (Fig. 3.21). He was born in Philadelphia in 1885, and lived in Ohio and Canada until he was 17 (Clyde, 1962). After graduating from Geneva College in Pennsylvania, he taught at several small rural schools and eventually went to the University of California at Berkeley for graduate study in education. He left the university before obtaining his master's degree and spent the next several years teaching in small schools in northern California. His marriage having been brought to an end after just three years, when his wife died of tuberculosis, he became a loner and spent increasing amounts of time in the mountains.

Clyde joined the Sierra Club in 1914 after joining a club outing near Yosemite. That same summer he traveled with a pack train the length of the Sierra Crest and made his first ascent (of fifty!) of Mount Whitney.

3.21. Norman Clyde, "the pack who walks like a man," a Sierra mountaineer who made more first ascents of major peaks than any other climber in any mountain range. Bancroft Library.

He accompanied the Sierra Club High Trip to the Evolution area in 1920 and made several first ascents on that trip. In 1924 he became principal of the high school in the town of Independence, in the shadow of the east escarpment of the Sierra Nevada. Here he could get away on short notice and spend time in the mountains that he loved with a passion that persisted the rest of his life. In 1925 alone, he climbed 48 major peaks, half of which no one had previously climbed.

In time, the Independence townspeople grew increasingly dissatisfied with Norman Clyde, the principal, because managing the school was clearly not his highest priority. On Halloween in 1927, Clyde drove off—with several shots from his revolver—a group of rowdy students intent on vandalizing the school. Later, bowing to public pressure, he resigned from his post with the school and thereafter spent most of his time in the mountains. In the summers he commonly joined Sierra Club High Trips, and in the winters he acted as caretaker of mountain lodges, where spare time was devoted to reading classics in the original Greek and writing about his travels.

Some called Clyde "the pack that walks like a man" because his packs dwarfed him. Clyde, a small man weighing about 140 pounds, commonly carried a 75-pound pack, a pack that contained every conceivable item he might need on his arduous trips into the backcountry. Wisely, he was not one to refuse excess food or equipment offered by others leaving the mountains. Such happenstances enabled him to extend his stays and ramble deeper among the high mountains.

Among his first ascents of major peaks within the parks are Dragon Peak, Triple Divide Peak, Diamond Peak, Mount Lippincott, Gray Kaweah, North Guard, Mount Genevra, Mount Russell, Mount Goethe, Deerhorn Mountain, Mount McAdie, North Palisade, Thunderbolt Peak, Mount Stewart, Goodale Mountain, Kid Mountain, and Cardinal Mountain. Norman Clyde made more than 100 first ascents in the Sierra, and in the process pioneered innumerable new routes. No one else—in any mountain range—has remotely approached this achievement.

Detailed Topographic Mapping: The U.S. Geological Survey

Criticism of the overlapping work and the unnecessary expense of operating the several separate federally funded surveys had led to a movement to consolidate them into one U.S. Government survey. In 1878 a Special Committee on Scientific Surveys of the Territories of the United States was appointed by the National Academy of Sciences, with the mission of

3.22. Clarence King signed on as an unpaid volunteer geologic field assistant with the California Geological Survey in 1864 (Fig. 2.11). At the age of 25, in 1867, he became Geologist-in-Charge of the Geological Exploration of the 40th Parallel (Fig. 3.1), and in 1879 was appointed the first director of the U.S. Geological Survey. U.S. Geological Survey Library.

recommending the future course for the surveys. After due deliberation the committee recommended the establishment of a single national geological bureau within the Department of the Interior. On March 3, 1879, the U.S. Geologic Survey was created by act of Congress and signed into law by President Rutherford B. Hayes. Clarence King (Fig. 3.22) was confirmed as its first director a month later.

The Geological Survey was charged with classifying the public lands and examining the geologic structure and mineral resources of the national domain. King directed the Survey during its formative period and established the basic principles that guided its work for many years. He divided it into two general units: a Mining Geology Division, headed by Raphael Pumpelly, and a General Geology Division, headed by Major John Wesley Powell. Allen D. Wilson was the first Chief Topographical Engineer. King remained as director of the Survey for only two years, before embarking on a career as a mining geologist. He was succeeded by

Powell. King had a lifelong attraction to dark-skinned women but remained unmarried until the age of 45, when he met a young African-American woman, a nursemaid in the home of a friend. Shortly thereafter, in 1888, he and Ada were secretly married, and King lived a double life for the rest of his days. Not only did he keep his marriage a secret from family and friends, but he married Ada under the name James Todd, without revealing his true name to her. To cover his exhaustive traveling, he told her he was a railroad porter. Ada bore him five children.

Thereafter, King sought his fortune in dozens of promising mines over the western part of the country, and in Canada, Alaska, and Mexico, but financial success eluded him. His last years were spent in almost constant travel, examining mining properties and testifying as an expert witness in cases dealing with mining disputes. In 1900 he testified in a lawsuit involving the Copper Kings at Butte, Montana, then went on to the goldfields of Alaska, and early in 1901 he made an appraisal of lead deposits at Flat River, Missouri. Learning that he had tuberculosis, he traveled to Chicago, then on to Florida and the West Indies for a vacation. After a brief trip to Washington, D.C., he visited his mother in Newport, Rhode Island. Turning west again, he traveled to Prescott, Arizona, then on to a special clinic in Pasadena, California, but his sickness was too far advanced to respond to treatment. Returning then to Phoenix, Arizona, he wrote at length to Ada, revealing his true identity.

King maintained his sense of humor to the end (Wilkins, 1958); when, during a break in his delirium, he heard the doctor mention that the heroin must have gone to his head. "Many a heroine," he is reported to have joked, "has gone to better heads than mine is now." King died December 24, 1901.

Two decades after its founding, topographers of the U.S. Geological Survey undertook the prodigious task of systematically mapping the highest Sierra. Many of the techniques used were those that had been perfected by Hoffmann in his Sierra work with the California Geological Survey. As reported in the *Sierra Club Bulletin* in May 1897, 41 topographic atlas sheets of areas in eastern California, reaching from Oregon south to about the region of the parks, were ready for distribution. The article noted that "The topographic maps are prepared from actual surveys, to serve as a basis for the geologic maps, and show surface elevations, roads, and all necessary physical and cultural details. . . . They will be sold at the rate of five cents per sheet. Orders for one hundred sheets, or over, whether for the same or different sheets, will be sold at the rate of two cents per sheet."

In the early 1900s this surveying program was extended into the high southern Sierra, including the area of the two parks. The work produced a set of 30-minute topographic maps that showed in unparalleled detail the lay of the land. The new maps eliminated large tracts of terra incognita and truly filled in the blank spaces in the California map. These quadrangle maps, bounded by parallels of latitude and meridians of longitude 30 minutes (one-half degree) apart, are 34.5 miles (exactly 30 nautical miles) north-south and about 27 miles east-west at this latitude. The quadrangles were published in three colors at a scale of 1:125,000, or about 2 miles to the inch, and each includes about 930 square miles. The six principal quadrangles, Mt. Goddard, Bishop, Tehipite, Mt. Whitney, Kaweah, and Olancha, were surveyed during the seasons of 1902 to 1909.

Mapping the terrain required four steps, each calling on somewhat different skills and equipment: astronomical observations, triangulation, leveling, and topographic mapping. The astronomical observations were made to determine the exact location of a few critical points, each of which was an easily accessible bench mark on flat ground near a telegraph line. Determinations of accurate latitude and longitude (as described in Chapter 2) fixed the position of each map in its correct place on the planet.

The next step in mapping was triangulation, or the determination of the horizontal as well as the vertical position of an array of secondary points that cover the area of interest. Triangulation was accomplished by first measuring the angles from the fundamental, astronomically determined stations to easily identifiable points (usually high peaks). From those of these peaks that could be climbed, surveyors would then measure the bearing and vertical angles to still other distinctive points. The instrument used for this work was a transit (Fig. 3.23), an instrument capable of measurement to within 1 to 10 seconds of arc (0.0003 to 0.003 degree). When these same distinctive points were sighted from one or more different predetermined points, and the angles to them measured, then the position and the altitude of these new points could be calculated. By this method an array of precise interconnected triangles was constructed, with a peak or other distinctive point occupying one apex of each triangle. The triangulation method requires that at least one of the group of many interconnected triangles have one side of precisely known length. This was ensured by connecting the triangles back to the astronomically determined starting points (Fig. 3.20). Alternatively, the distance could be measured on the ground between two points. For such measurement a line was chosen on flat ground, as in Owens Valley, and the line length

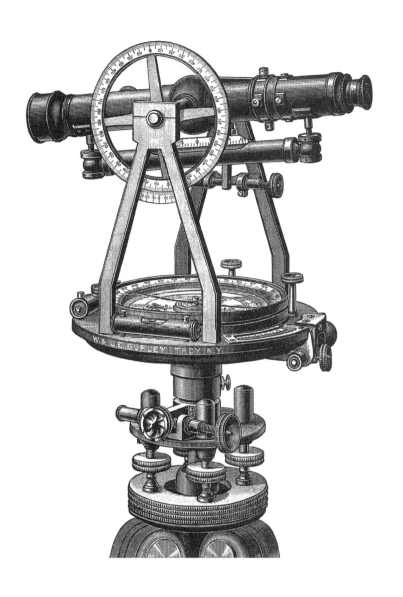

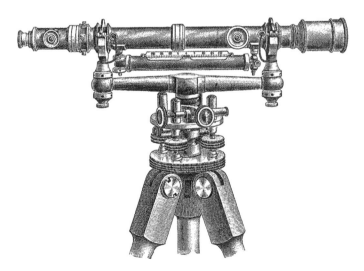

was determined by *chaining*, that is, by physically measuring the distance between two instrument stations with a steel tape.

Many corrections were necessary to improve the accuracy of the triangulation (and leveling) measurements. The unavoidably rough handling of the instruments required that they be tested daily, so that they could be adjusted for alignment. One of the tests involved first sighting at a distant object, specifically, placing the crosshairs of the telescope on the tip of a mountain peak and taking careful readings of horizontal and vertical angles to the peak. Then the instrument was rotated precisely 180° about its vertical axis and then around its horizontal axis so that the tip of the peak touched the crosshairs again. If the instrument was in adjustment the vertical hair would pass through the peak and the vertical reading (above the horizontal) would be the same as the initial reading. Corrections were also made for the offset of targets resulting from light refraction by atmospheric effects, and for the curvature of the Earth. First-order leveling surveys, in which the highest precision was sought, required that the instrument be shaded by an umbrella at all times, so as to minimize the un-

3.23. *(Opposite, top)* A turn-of-the-century surveyor's transit used for measuring azimuth, or horizontal angles (indicated on the horizontal circle outside of the compass), and elevation, or vertical angles (indicated on the vertical circle by the telescope). The instrument is leveled by adjusting four thumbscrews at the base (as indicated by the two spirit levels on the horizontal circle outside the compass). Vertical and horizontal angles can be measured to one-half minute by means of engraved vernier scales. The instrument can also be used as a level by using the large spirit level attached to (and beneath) the telescope (Wentworth, 1903).

3.24. *(Opposite, bottom)* Surveyor's Y level used to measure the relative elevation of two bench marks along a survey line in combination with a graduated staff or rod held vertically on each bench mark. The telescope is leveled by means of the thumbscrews, as verified by the large spirit level affixed to the bottom of the telescope. The level of the instrument can be checked by first leveling it (centering the bubble in the spirit level) and then sighting a distant object that aligns with the telescope's horizontal crosshair. Next, the telescope (with its attached bubble level) is released and lifted out of its two Y-shaped supports or clamps near each end of the telescope, then reversed and placed back in the supports. Then, when the complete instrument is swung 180° around its vertical axis, and the same distant object is sighted again, the horizontal crosshair should fall on the same position on the object, if the instrument has been properly adjusted (Wentworth, 1903).

even expansion of metallic parts caused by temperature changes under direct sunlight.

The leveling surveys, the third stage in mapping, were employed to establish precise elevations of critical points, since the barometric altitudes previously used in reconnaissance mapping were not sufficiently accurate for this new generation of mapping. A leveling survey provides a precise difference in elevation between two points, each then generally marked with a permanently emplaced brass tablet, or bench mark. The elevation difference was measured by sighting a horizontal line through an instrument and taking a reading off a graduated rod or staff placed on a bench mark.

The leveling instrument (Fig. 3.24) was constructed so that it would "shoot" only along a level line. As such, the instrument could not be elevated or depressed, as a transit can. It would be set up midway between a point of known elevation and one to be measured. It was then adjusted with thumb screws to accord with a spirit level (a device similar to the bubble in a carpenter's level), so that it could be sighted only in a level or horizontal direction. A vertical staff or rod, some 12 feet high, calibrated with accurate divisions to hundredths of a foot (or the metric equivalent) was placed sequentially on each of the two ground points to be measured, the instrument placed roughly midway between the two points. The elevation was carried forward by first backsighting to the point of known elevation, reading the rod scale at the central horizontal crosshair in the instrument, then turning the instrument forward and foresighting to read the rod scale on the new point of unknown elevation. The difference between the backsight reading and the foresight reading determined the elevation of the new forward bench mark, whether it be above or below the first bench mark. The instrument was then moved forward, the previous foresight point now became the backsight point, and a new forward point was selected for the foresight. The survey leapfrogged ahead in this fashion. When leveling up or down a steep slope the foresights and backsights had to be placed much closer together, or the level line would sight above the top or below the base of the rod, and no reading could be obtained. Consequently, moving up or down a steep slope was slower and more laborious, because the instrument had to be set up more often, at closer spacings.

The level in use at that time, the Y level (Fig. 3.24), was constructed so that, after sighting at a distant object, the entire telescope portion, with attached level bubble assembly, could be lifted out of its Y-shaped clamps and reversed, and then, after the instrument was rotated 180°, sighted to see if the same distant object was still aligned with the cross hair. If it were not, adjustments were made to bring the instrument into alignment. This procedure can be compared to a builder's first adjusting a timber so that it lies horizontal, by use of a carpenter's level, then reversing the level to see if a level reading is still obtained on the timber. Generally, line segments were run twice, forward and backward, and if the difference between the two was outside an established range of tolerance (about ½ inch per mile), the lines were resurveyed until agreement within that tolerance was reached.

Level lines were always surveyed in from previously leveled regions, and ultimately from the Pacific Coast. At the coast the elevation of sea level was established by multiyear observations of tide gauges at San Francisco, San Pedro, and San Diego. The tidal bench marks were the fundamental source from which all elevations were derived, even in midcontinent.

The first level lines to the eastern foot of the range in the parks area were made in 1905, by topographer R. A. Farmer and levelman M. D. Shannon of the U.S. Geological Survey (Gannet and Baldwin, 1908). They carried a leveling survey from a previously measured datum at Mojave along the stage road to Keeler on the east side of Owens Lake in Owens Valley, and then north along the Carson and Colorado Railway to Laws, east of Bishop. In the same year Farmer also ran a line from near Lone Pine to the crest of the range at Whitney Pass and on to the summit of Mount Whitney. From this survey they measured the elevation of the peak at 14,501 feet above sea level, and this is the elevation that appears on the Mt. Whitney Quadrangle sheet published in 1907, as well as on the editions printed in 1919, 1921, and 1927. In addition, C. H. Semper, levelman, ran a line from Mount Whitney west along the trail to Crabtree Meadow, northwest to the Kern River, and south down the Kern to Rock Creek.

Twenty years later, in 1925–28, the leveling was repeated from Lone Pine to the summit of Mount Whitney by a crew from the U.S. Coast and Geodetic Survey. In this first-order survey, they used more sophisticated techniques than had been used by the U.S. Geological Survey. The crew spent the summer of 1925 on the upper part of the Lone Pine–Mount Whitney Trail, carrying the level line from near the summit of the mountain to Whitney Pass at 13,300 feet. Their work was hampered because the horse trail had been closed due to heavy rockslides, and all equipment had to be backpacked up the mountain. The work was finally terminated because of high winds, freezing temperatures, and snow and sleet storms. Three years later, in the summer of 1928, the line was completed by leveling from Whitney Pass to Lone Pine. The elevation of Mount Whitney determined by this measurement was found to be 14,496 feet, as shown on the 1936 topographic quadrangle. This value is just 5 feet lower than the first leveling measurement that the U.S. Geological Survey had made. Later adjustments have been made, not only of the final leveling leg from Lone Pine to the summit of Mount Whitney, but also of the various elevations, reaching all the way back to the tide gauges on the Pacific Coast. The adjusted elevation of Mount Whitney as shown on the 1937 30-minute quadrangle map is 14,495 feet, and that on the 1956 15-minute Mt. Whitney Quadrangle is 14,494 feet. A final adjustment, in 1988, makes the elevation 4,416.9 meters (14,491 feet), as shown on the 7.5-minute quadrangle map.

The instruments used during much of the topographic-mapping stage were a *plane table* and an *alidade* (Fig. 3.25). The table, a solid wooden board about 2 feet square with extendable, adjustable legs, was set up directly over the bench mark on one of the previously triangulated peaks, and on it was fixed a map of the triangles previously established. After the plane table was leveled by means of a ball-and-socket joint, it was rotated until the map on the table was accurately oriented, that is, so that the north arrow on the map pointed directly north. The alidade, a small telescope attached to a straightedge base mounted parallel with the tele-

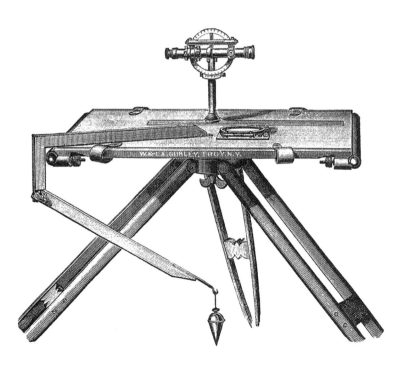

scope's line-of-sight, was then placed on the map with one edge of the straightedge on the point that represents the bench mark above which the plane table stands. With the alidade in this position, when another peak of predetermined position is sighted through the telescope, the straightedge will lie exactly on the map point representing that peak. With everything thus in alignment, further sightings can be made.

The topographer then sights all visible points such as peaks, lake outlets, and stream junctions through the alidade, which is resting on the plane table, and draws lines (rays) along the straightedge extending from the sighting point on the map (the bench mark) to each of the target features. The features are sketched in even though their exact position along the ray (their distance from the observer's position) is not yet known. But when the plane table is later set up on a *different* known point, at *its* bench mark, and rays are drawn to the same targets as before, the sketch map can be upgraded, because the precise position of each target has now been defined by the intersection of the two rays that have been drawn to that target, one from each of the two bench marks. When each target is first sighted, the vertical angle to it (either up or down, as measured on the alidade circle) is recorded, and from this angle and the distance to the target, the elevation of the target relative to the bench mark can be determined.

The map is then sketched by reference to the newly determined points. This evolving sketch map, showing the position, course, shape, and elevation of each feature, is repeatedly refined as "shots" of the same features are made from new plane-table positions situated on still other triangulated

3.25. Plane table with alidade and compass. A map showing the position of known triangulation points is affixed to the table. The table is leveled and oriented by first placing the straightedge of the alidade (which is mounted parallel to the telescope's line of sight) so that it connects two points on the map, one designating the occupied station, the other designating a distant known point. The leveled board (and map) is then swiveled around its vertical axis to bring the known distant point into alignment with the vertical crosshair of the telescope, and the now-oriented board is locked in place. Other selected targets are then viewed through the telescope, and rays drawn to them (as well as vertical angles measured to them), thus defining their precise direction from the occupied station. When the plane table is then carried to another known vantage point, and the process repeated, with rays drawn to the same selected targets (and vertical angles measured to them), the intersection of the two rays to the same target defines its location and elevation (Wentworth, 1903).

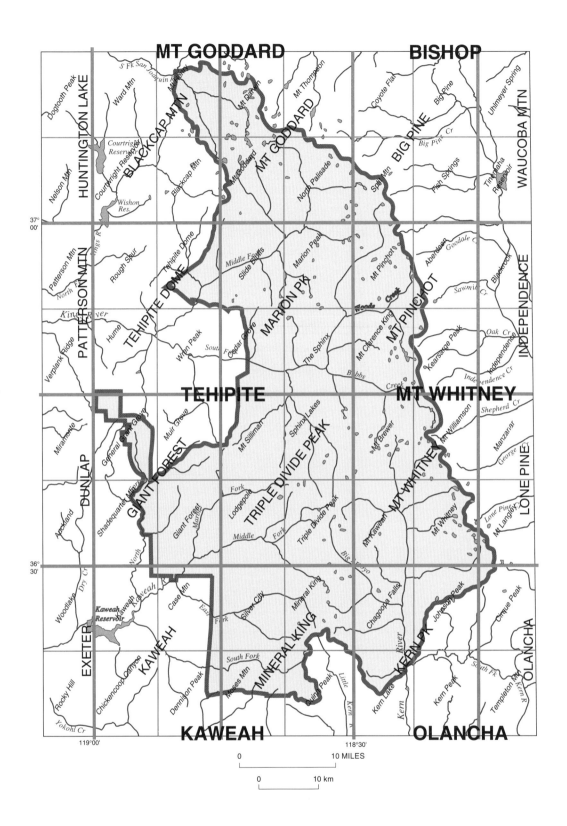

TABLE 3.2
Mapping of the 30-minute quadrangles
in the Sequoia–Kings Canyon region

Quadrangle	Surveyed	Published	In Charge	Topographers
Kaweah	1902	1904	R. U. Goode	E. C. Barnard, A. I. Oliver, R. B. Oliver, W. C. Guerin
Tehipite	1903	1905	E. M. Douglas, R. B. Marshall	R. B. Marshall, G. R. Davis
Mt. Whitney	1905	1907	E. M. Douglas, R. B. Marshall	G. R. Davis, C. L. Nelson, S. N. Stoner
Olancha	1905	1907	E. M. Douglas, R. B. Marshall	S. N. Stoner, J. P. Harrison,
Mt. Goddard	1907–09	1912	R. B. Marshall, T. G. Gerdine	G. R. Davis
Bishop	1910–11	1913	R. B. Marshall, T. G. Gerdine	G. R. Davis, B. A. Jenkins

3.26. U.S. Geological Survey topographic quadrangle maps of the parks area. Large, horizontal boldface type marks the center of 30-minute quadrangles, diagonal uppercase type marks the 15-minute quadrangles, and diagonal lowercase type marks 7.5-minute quadrangles.

peaks. The topographers in fact had to occupy (set up shop on) a large number of peaks in the region in order to make the multiple shots needed to draw the elevation contours and shape of every mapped feature of the landscape. Mountaineering skills were essential to the men who traversed this alpine region and occupied the many high peaks that would need to be included in the surveying plan. Technical and graphic skills, intuition, and survival skills were all important in this work, and some topographers became masters of the enterprise.

To cover the region of the highest Sierra (Fig. 3.26 and Table 3.2), these men produced six 30-minute quadrangle maps. These were the first contour maps made of the area, because accurate elevation control had not previously been available. The maps show the details of the topography by means of contour lines (drawn generally 100 feet apart) each of which is drawn at constant elevation and thus consistently perpendicular to the down-slope direction. All previous maps showed topography by hachures, fine lines drawn downslope, which, while suggesting the lay of the land artistically, convey a good deal less topographic and elevation information.

Most of this surveying was accomplished by a small team of U.S. Geological Survey topographers (Table 3.2). Two of the more active were

3.27. Robert Marshall, U.S. Geological Survey topographer. Of the six 30-minute Quadrangles that cover the region of the highest Sierra, Marshall was in charge of the mapping of five. U.S. Geological Survey Library.

Robert B. Marshall (Fig. 3.27) and George R. Davis (Fig. 3.28). Of the six quadrangles that cover the region of the highest Sierra, Marshall was involved in the mapping of five, and Davis, of four. Davis was the first to take pack stock over Muir Pass, and his name can still be found in the summit registers of many of the isolated high peaks, including the first ascents of Black Mountain, Milestone Mountain, and Mount Baxter.

With these 30-minute, 2-miles-to-the-inch, topographic contour maps the U.S. Geological Survey had taken a giant step forward in Sierra mapping. A comparison of these maps with the best mapping previously available, the 1896 J. N. LeConte map (Fig. 3.29), shows what a tremendous advance these new maps represent. In a 10-minute quadrangle of the main Sierra Crest extending south from 37° north latitude, an area that encompasses more than 100 square miles, we see that the LeConte (1896)

3.28. *Above.* George Davis with a plane table and alidade set up next to the summit cairn of Mount Whitney. Mount Langley ("False Mount Whitney," 14,027 feet) is 5 miles southeast, in the background at right. Note that the alidade telescope is depressed; all other peaks in the range are lower. U.S. Geological Survey Library. *Below.* Davis, U.S. Geological Survey topographer on four of the six quadrangles that cover the area of the highest Sierra, including the Mt. Whitney and Mt. Goddard Quadrangles. He was the first to take pack stock over Muir Pass, and his name can still be found in the summit registers of many of the isolated high peaks; he made the first ascents of Black Mountain, Milestone Mountain, and Mount Baxter.

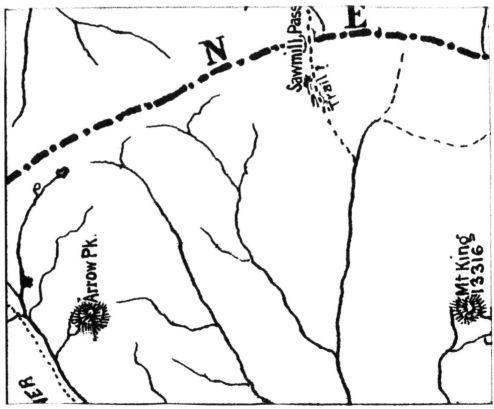

map shows only two named peaks, Mount King and Arrow Peak, and only two mapped lakes, whereas the U.S. Geological Survey 30-minute Quadrangle (surveyed in 1905) shows 13 named peaks and dozens of mapped lakes. In addition, the new map depicts countless details of topography, drainage, and elevations, by the use of contours.

With the new U.S. Geological Survey maps now available, the era of terra incognita ended. Trips were no longer harrowing journeys of exploration into the unknown, and travel entered a new phase. Recreational use of the parks increased. Construction of the John Muir Trail, a major undertaking, began in 1915, and the trail was finished in 1938. The High Sierra Trail, from Giant Forest over the Great Western Divide at Kaweah Gap to the Kern River, was built between 1928 and 1932. With the expansion of the parks and the building of trails, access for hikers and mountaineers was eased considerably. New routes were pioneered over still more remote passes, and climbers scaled increasingly challenging peaks and more difficult routes.

In the 1950s the Survey undertook a new series of topographic quadrangle maps drawn at twice the scale and detail of the first-generation 30-minute maps (scale 1:125,000). The 15-minute quadrangle map series (Fig. 3.30; scale, 1:62,500, one mile to the inch) was the first to take advantage of new aerial-photography technology. This process employed overlapping vertical (straight down) aerial photographs of the terrain. They were taken on cloud-free days in strips along parallel flight lines several thousand feet above the ground. To prepare the composite photogrammetric map, projectors supported on racks above a table are set up in such a way as to mimic the position of the aircraft above the land surface. The projectors then project overlapping photo images down onto the table, and the images are adjusted for relative position by means of a small number of identifiable points visible on each of the two photos, points that had previously been located in the field by triangulation. Projection of two overlapping photographic images, one with red filters and the other with blue filters, produces a stereoscopic image when viewed through colored glasses, with a red lens for one eye and a blue lens for the other.

The resulting three-dimensional model projected onto the table is then employed to trace each of the many topographic contours on a map fixed to the table. This is done by use of a miniature, movable table a few inches high which can be slid around on the large table through the projected stereoscopic image. It is moved so that a reference spot at its middle is just on the land surface of the image. Then the small table, with an affixed pencil, or stylus, below the spot, is moved around the projected landscape,

3.29. (*Broadside*) *Left*. Part of the LeConte map of 1896 (Fig. 3.19), showing the state of knowledge of a segment of the Sierra Nevada Crest (dash-dot line) at that time. The region is between 36° 50' and 37° 00' north latitude in the upper drainage of the South Fork of the Kings River. *Right*. The same area, at the same scale, as published 11 years later on the 1907 U.S. Geological Survey 30-minute Mt. Whitney Quadrangle map.

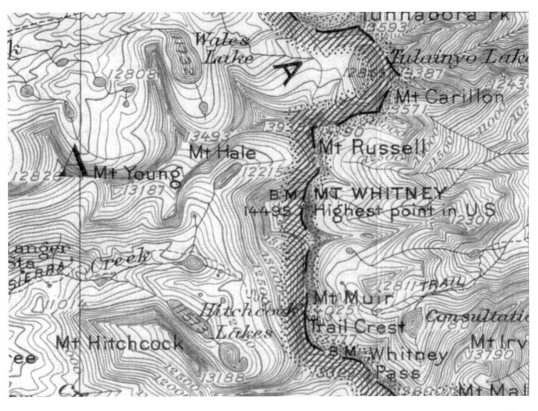

always keeping the spot on the land surface of the image (avoiding flying into air outside the image or digging into the hill inside the image), and drawing the first contour on the map. When a contour of a given elevation has been drawn over the whole map, then the small table is raised a predetermined amount and the next uphill contour is followed on the stereoscopic image and drawn on the map. In addition to contours, all other features visible on the photograph—including lakes, streams, main roads, and peaks—can be transferred from the image to the map. Later field checking of the photogrammetric compilation allows the addition of roads, trails, boundaries, and features that are not visible on the photographs.

These modern maps are thus remarkably accurate in the depiction of peaks, lakes, watercourses, and elevations. Rarely is any important feature omitted, and the contours faithfully represent the details of the land surface, such as the irregularities of slopes, the position of craggy terrain, and the merging of contours on vertical cliffs. But a detailed comparison of the photogrammetrically prepared, 15-minute maps with the original 30-minute quadrangles does emphasize the high quality of the 30-minute maps. The on-the-ground surveyors of the original 30-minute quadrangles had performed a remarkable job under extraordinarily difficult conditions. Their work had truly removed the veil of uncertainty from our knowledge of the highest Sierra (Fig. 3.30).

What is currently the final chapter in the topographic mapping of the Sierra Nevada began in the 1970s with the production of a new series of 7.5-minute quadrangle maps designed to replace the 15-minute series. Each of the 7.5-minute maps (scale 1:24,000) covers one-fourth the area covered by a 15-minute map (scale 1:62,500) on a larger piece of paper (22 x 27 inches as compared to 17 x 21 inches). These new maps were made with still more advanced technology. Commonly, new sets of aerial

3.30. *Above.* Part of the 30-minute U.S. Geological Survey Mt. Whitney topographic map edition of 1907, made by plane table and alidade. I have doubled the scale here, from a published value of 1:125,000 to 1:62,500, to match the scale of the map below. The contour interval is 100 feet. *Below.* The same area, shown on the U.S. Geological Survey 15-minute Mt. Whitney topographic map edition of 1956 (scale 1:62,500; 1 mile to the inch). The map utilized aerial photos taken in 1955. The contour interval is 80 feet. A few ponds and minor drainages are missing from the older map, and the more recent map shows more detail in topography, but in general the older map depicts topography, lakes, and drainage extremely well.

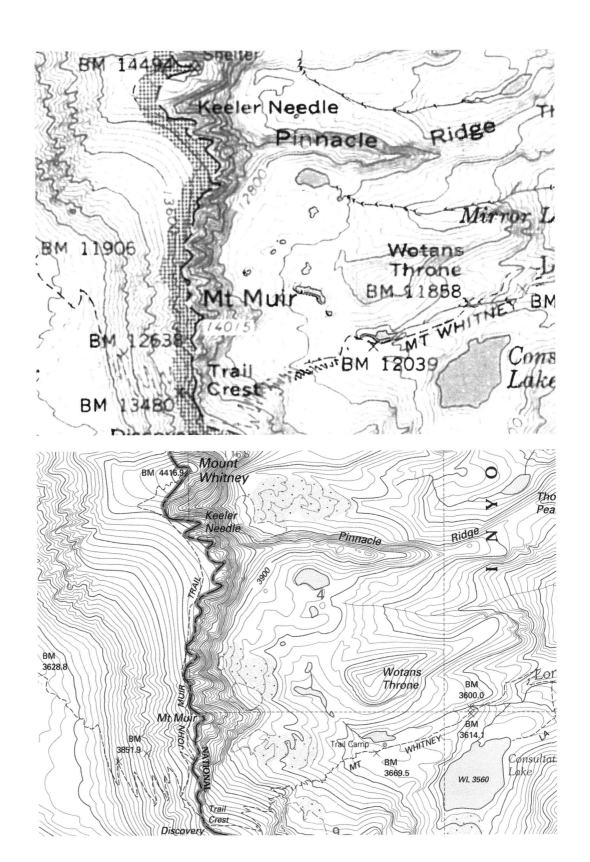

photographs were flown, but some maps utilized the same sets of aerial photographs employed for the 15-minute series. The new technology, however, has not replaced the human cartographer, who must still, in the projection room, actually trace every contour by moving a stylus along the apparent ground surface depicted in the red-and-blue stereoscopic model. Because of their larger scale, the 7.5-minute topographic maps (Fig. 3.31) can show more of the subtle details of topography, and they have more room for names and for cultural features such as buildings in urban areas. The disadvantage of these larger-scale maps is that four times as many map sheets are necessary to cover the same area. And, of course, an area of interest to a user always seems to fall at the boundary between two of the quadrangles, or where four come together. Some of the new maps make use of the metric system of measurement, with contours drawn at 10- or 20-meter intervals of elevation, and about 10 percent of the California quadrangles are metric.

Peaks, Valleys, and Rivers: The Basis of Place Names

The history of exploration of an area is commonly mirrored by the place names given to geographic features such as rivers, streams, lakes, and mountains (Browning, 1986). The names were, of course, assigned primarily by those who first visited the land and had the means to publish or publicize those names. Hence, it was explorers and surveyors who did most of the early naming, particularly of the region's more salient natural features. Names were important in the early years of exploration and mapping as an aid in describing routes; the pioneers established names in part to help those who followed. Later, as land was claimed, timber cut, and animals put out to graze, a second layer of names was added by those who used or lived on the land. And in more recent years, with ever more detailed maps made at ever larger scale, there is more room on the maps

3.31. *Above.* Part of the 15-minute U.S. Geological Survey Mount Whitney topographic map edition of 1956 (see Fig. 3.30). The scale has been increased (for reproduction here) from a published value of 1:62,500 to 1:24,000, to match the scale of the map below. The contour interval is 80 feet. *Below.* The same area as above on the U.S. Geological Survey 7.5-minute Mount Whitney topographic map edition of 1994 (scale 1:24,000). The map utilized aerial photographs taken in 1976 and 1978, and was revised from photographs taken in 1993. The contouring system is metric, with intervals of 20 meters (65.6 feet). The latest accepted altitude of Mount Whitney is 4,416.9 meters (14,491 feet).

for names, and more names are needed. This last generation of names is generally applied to features of less importance, since all of the primary features have already been named.

All through the naming process, the effect of personal and political bias is evident, especially where names of people are used. Just as evident has been the problem of multiple names, two or more different names for the same feature, or of use of the same name for more than one feature, which is particularly confusing when those features are closely spaced geographically.

The U.S. Board of Geographic Names was established in 1890 by executive order to instill some order in the process of naming geographic features. Since then, all new names added to government maps and publications have had to be approved by the board. In 1947, the board was recreated by act of Congress, and its policies were somewhat expanded. The board has always discouraged the practice of honoring living persons in the naming of natural features, and current policy requires that such names not be considered before the honoree has been deceased five years. It is clear that if the Board had existed in the 1860s, we would not have seen the names Mount Whitney, Mount Brewer, Mount Hoffmann, Mount Clarence King, and Mount Gardiner given to first-order features.

In order to explore the origin of place names, I have listed all the names that appear on two of the early 30-minute quadrangles. Within the Mt. Whitney Quadrangle, 104 of the names that appear on the map are within the Sierra Nevada (the map also includes Owens Valley and a part of the Inyo Mountains). Within the Tehipite Quadrangle, which is entirely within the Sierra Nevada, 215 names appear. Although several features, such as a mountain, creek, meadow, or camp, may have been labeled by the same name, I selected from such a redundancy only a single generic name. I then sorted each of the names into one of six categories: descriptive names, plants and animals, local people, famous people, artifacts, and Indian names.

As in most places in the world, the chief group of place names consists of those that are descriptive, and in the southern Sierra about 40 percent of the names describe the object named (Fig. 3.32). They refer to size (Big Meadow, Little Five Lakes, Sevenmile Hill), color (Black Mountain, Red Spur, Emerald Lake), shape (Arrow Peak, Table Mountain, Long Meadow), terrain (Granite Creek, Boulder Creek, Stony Creek), number (Twin Lakes, Triple Divide Peak, Sixty Lake Basin), setting (Summit Meadow, Lookout Mountain, Cliff Creek), fancied architectural similarity (Colosseum Mountain, Window Peak, Carillon Peak), shape (Pear

3.32. Place names of the Sierra Nevada within the Mt. Whitney and Tehipite 30-Minute Quadrangles, sorted into six general categories. As in most areas, descriptive names are the most widely used. The more common use of names of famous people in the Mt. Whitney Quadrangle probably results from the conferring of place names in this high alpine country by the early college-trained scientists and mappers. In the lower country of the Tehipite quadrangle, settlers and herdsmen tended to emphasize local people in their place names.

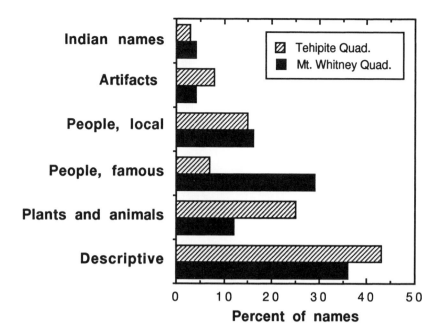

Lake, Lake South America, Slim Lake). A further set of names, hard to classify, also seems to fit best in this group; they are names that inspire an emotion or abstract thought (Inspiration Point, Consultation Lake, Scenic Meadow).

A second important category of place names comprises those that invoke particular plants or animals. Plants so honored include trees (Tamarack Lake, Yucca Creek, Cedar Grove), and bushes (Brush Canyon, Bunchgrass Flat, Heather Lake). Animals may be large (Buck Creek, Little Bearpaw Meadow, Deer Ridge) or small (Rattlesnake Creek, Bullfrog Lake, Gnat Meadow). Domestic animals (Goat Mountain, Kid Creek, Horse Creek) and animals not native (Moose Lake, Lion Rock, Whaleback) are represented. Many names are derived from birds (Ouzel Creek, Big Bird Lake, Wren Peak).

Many place names honor people, and I have here divided them into two categories: those named for famous, generally nationally or internationally known figures, and those named for local people who used the land. The early explorers were commonly academically trained specialists, and they named features for their professors and friends, or for scientists whom they admired. These names include scientists and academicians (Mount Tyndall, Mount Hale, Mount Muir), surveyors and mappers (Mount Goddard, Mount Clarence King, Mount Brewer), mountaineers (Mount Mallory, Mount Irvine), and administrators and politicians (Mount Whitney, Wilsonia, Corcoran Mountain). Included also are the names of the surveyors themselves. A few names placed in this category honor institutions, primarily universities; these include University Peak (for the University of California), Mount Stanford, Golden Bear Lake, and Mount Caltech.

Local people for whom places were named include homesteaders and settlers (Symmes Creek, Thibaut Creek, Tharps Rock), cattlemen (Collins Meadow, Halstead Meadow, Weston Meadow), sheepmen (Woods Creek, Lewis Creek, Statum Meadow), and members of hiking parties (Mount Gould, Wright Lakes, Longley Pass). Both people categories also include names honoring family members, such as wives, sons, and daughters.

A fifth category embraces those places that are named for an artifact, whether a small object made by people (Cartridge Creek, Bacon Meadow, Rowel Meadow) or a structure (Cabin Creek, Sawmill Canyon, Log Corral Meadow).

The final group of names are those taken from Indian names of particular features (Taboose Pass, Tehipite Dome, Tokopah Valley) or of persons (Sequoia Park, George Creek, Chagoopa Plateau).

An aggregate tabulation of the place names (Fig. 3.32) finds that descriptive names account for about 40 percent of all the names, and names of people for 30 percent. An average of 16 percent of places were named after plants and animals. The remaining 14 percent of names represent artifacts and Indian names.

Famous-people names are about four times more common than local-people names in the Mt. Whitney Quadrangle, whereas they are about equal to local-people names in numbers in the Tehipite Dome Quadrangle. The reason for this imbalance is clear: most of the terrain of the Mt. Whitney Quadrangle is at high altitude, isolated in the winter, and not habitable. Moreover, the high-altitude Mt. Whitney Quadrangle was first explored by scientists who were less likely to know the names of locals, more likely to know of famous people and to be inclined to name features after them. The lower-altitude Tehipite Dome Quadrangle was more accessible to livestock, hunters, and loggers and more suitable for settlement. The greater numbers of settlers there (than in the Whitney Quadrangle) meant that they were more often those who did the naming, and they often chose to apply the names of local people. Similarly, the heavy forests below timberline provide a habitat for a larger population and diversity of living things, accounting for the fact that plant and animal names are about twice as common in the lower western country of the Tehipite Dome Quadrangle.

Indian names are poorly represented in spite of the prior domination of the land by the Indian tribes, who had lived in and adjacent to the mountains for centuries and had used them for trade routes and food gathering. The early American explorers and settlers had little interest in Indian culture, often even antipathy for the ways of Indians, and chose to apply their own names.

The Structures of the Land: Geologic Mapping

The primary purpose of the early maps was of course to show the position of mountains, rivers, and routes, but from the beginning of charting of the High Sierra, different rock types, mineral deposits, and geologic structures were also included on the maps. The first reconnaissance geologic map of a sizable part of California (Fig. 2.7) was that by William Blake, geologist with the railroad survey of 1853 headed by Robert Williamson. This remarkable map defined nine geologic units by color. More than 1,000 copies were printed by the Government, and after printing watercolors were precisely applied to them individually, by hand! Other maps followed: the Wheeler map of 1872 (Fig. 3.8) shows lava fields, salt springs, and dozens of mining districts; and the Wright map of 1883 (Fig. 2.38) shows volcanic features along Volcano Creek (now called Golden Trout Creek), including volcanoes, lava flows, basalt columns, and a natural bridge.

Perhaps the first complete geologic map of the state of California (Irelan, 1891; Fig. 2.37) shows in color eight mapped rock units: granitic rocks; limestone; auriferous slates, mainly Jurassic; volcanic rocks; auriferous gravels, Pliocene; metamorphic rock, Cretaceous; unaltered sandstones, shales, and conglomerates, chiefly Tertiary; Quaternary and Recent; and unexplored and thus unidentified portions. This map also depicts all the mineral deposits—separated by commodity—known at that time.

Aside from these early regional maps, systematic geologic mapping had to await the advent of more detailed base maps. One of the earliest such studies was that of Andrew Lawson, who studied the landforms and glaciation of the upper Kern Basin (Lawson, 1903). He joined a Sierra Club Outing in the summer of 1902 in the company of Grove Gilbert and made use of the latest blueprint topographic-map compilation of J. N. LeConte. In this study he mapped out the limits of the main trunk glacier down the Kern Canyon, but inferred that the canyon owed its origin to a down-dropped block bounded by faults, an idea apparently borrowed from Josiah Whitney's collapse hypothesis for the origin of Yosemite Valley. At that time Lawson did not recognize that the canyon followed a master fault that had offset granitic masses several miles.

In the early 1900s, when the 30-minute quadrangles of the U.S. Geological Survey became available, geologists used this new topographic detail to plot data with increased precision. Adolph Knopf (1905) mapped the metamorphic rocks at the Mineral King mining district, Cordell Durrell (1940) mapped the structure of the metamorphic rocks of the Kaweah foothill belt, and Konrad Krauskopf (1953) mapped the tungsten deposits and general geology of a broad region in the Sierra within Madera, Fresno, and Tulare Counties. Francois Matthes (1960) mapped in reconnaissance the extent of the Pleistocene glaciers in Sequoia National Park. A study that laid the groundwork for the mapping of the individual granitic plutons in Sequoia National Park was that of Donald Ross (1958). Requiring greater detail than was possible with the 30-minute quadrangles available, he mapped the geology on maps he prepared from recently flown aerial photographs.

More detailed geologic mapping was made possible by the advent of the new larger-scale, more detailed 15-minute series of topographic maps in the 1950s. The U.S. Geological Survey began a systematic geologic-mapping program utilizing these maps, and in the early 1950s Paul Bateman and his associates began mapping a strip across the central Sierra one degree (60 miles) wide, and in 1956, I, with Thomas Sisson and our associates, began mapping an adjacent east-trending strip on the south of the Bateman strip, between 36° 15' and 37° north latitude. Most of the quadrangles in the Sierra in this region were mapped and published in the Geological Survey's Geologic Quadrangle Map Series. This mapping was critical for modern studies of the geology of the highest Sierra and has been used to prepare the page-sized geologic maps that follow in later chapters of this book (Fig. 3.33 identifies the quadrangles in question; Table 3.3 lists the sources consulted in each case).

A geologic map is a compendium of information on the nature and history of the rock units in a given area. To begin with, the exposed rocky crust of the area is subdivided into a system of major mappable units that can be consistently inspected and identified, followed in the field, and accurately plotted on a map. In addition to these rock units, the local geologic structures, such as the bearing and inclination of sedimentary beds, metamorphic layering, faults, and fractures, are also mapped. This physical determination in the field of the actual shape and location of rock units and structures is critical to posing questions concerning the relative ages, origin, and history of the area, and to seeking data that might answer these questions.

Concurrent with the mapping is the sampling of the rock units, fossils, and mineral deposits, and the accumulation of numerical data on rocks and structures (for example, the percentages of mineral types in a given rock unit, or the general trends of structural elements in a mass of granitic rock). As geologists, we study this material and these data in a laboratory setting with equipment not available in the field. We analyze samples for their mineral and chemical compositions, experts inspect fossils, specialists select key rock specimens for radiometric dating. The overall synthesis of such data, the geologic map, spells out the sequence of events that unfolded through eons of time to produce what the map tells us is under our feet.

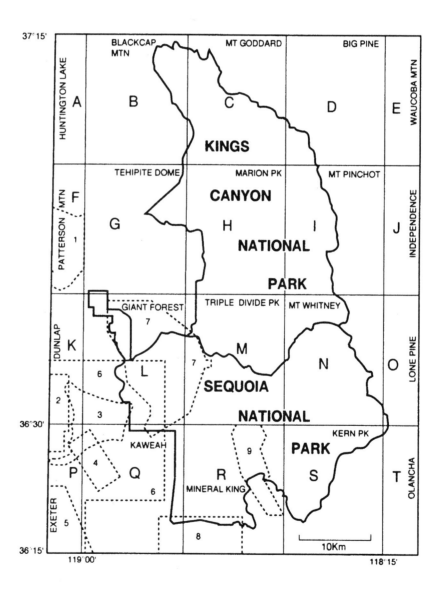

3.33. Sources used by the author in compiling geologic maps of Sequoia and Kings Canyon National Parks and vicinity. (See Table 3.3 for an explanation.)

TABLE 3.3
References used in compiling geologic maps of
Kings Canyon and Sequoia National Parks and vicinity

(See Fig. 3.33 for location of geologic map areas; letters refer to
15-minute quadrangle maps, numbers to special map areas)

Fifteen-Minute Quadrangles

A. P. C. Bateman and D. R. Wones, 1972, Geologic map of the Huntington Lake Quadrangle, central Sierra Nevada, California: U.S. Geological Survey Geologic Quadrangle Map GQ-987.

B. P. C. Bateman, 1965, Geologic map of the Blackcap Mountain Quadrangle, Fresno County, California: U.S. Geological Survey Geologic Quadrangle Map GQ-428.

C. P. C. Bateman and J. G. Moore, 1965, Geologic map of the Mount Goddard Quadrangle, Fresno and Inyo Counties, California: U.S. Geological Survey Geologic Quadrangle Map GQ-429.

D. P. C. Bateman et al., 1965, Big Pine Quadrangle, Geology and Tungsten Mineralization of the Bishop District, California: U.S. Geological Survey Professional Paper 470, 208 pp.

E. C. A. Nelson, 1966, Geologic map of the Waucoba Mountain Quadrangle, Inyo County, California: U.S. Geological Survey Geologic Quadrangle Map GQ-528.

F. J. G. Moore and T. W. Sisson, 1984, Patterson Mountain Quadrangle, unpublished reconnaissance mapping.

G. J. G. Moore and W. J. Nokleberg, 1991, Geologic map of the Tehipite Dome Quadrangle, Fresno County, California: U.S. Geological Survey Geologic Quadrangle Map GQ-1676.

H. J. G. Moore, 1978, Geologic map of the Marion Peak Quadrangle, Fresno County, California: U.S. Geological Survey Geologic Quadrangle Map GQ-1399.

I. J. G. Moore, 1963, Geology of the Mount Pinchot Quadrangle, southern Sierra Nevada, California: U.S. Geological Survey Bulletin 1130, 152 pp.

J. D. C. Ross, 1965, Geology of the Independence Quadrangle, Inyo County, California: U.S. Geological Survey Bulletin 1181-O, 64 pp.

K. J. G. Moore and T. W. Sisson, 1984, Dunlap Quadrangle, unpublished reconnaissance mapping.

L. T. W. Sisson and J. G. Moore, 1994, Geology of the Giant Forest Quadrangle, Tulare County, California: U.S. Geological Survey Geologic Quadrangle Map GQ-1751.

M. J. G. Moore and T. W. Sisson, 1987, Geologic map of the Triple Divide Peak Quadrangle, Tulare County, California: U.S. Geological Survey Geologic Quadrangle Map GQ-1636.

N. J. G. Moore, 1981, Geologic map of the Mount Whitney Quadrangle, Inyo and Tulare Counties, California: U.S. Geological Survey Geologic Quadrangle Map GQ-1545.

O. P. Stone, G. C. Dunne, J. G. Moore, and G. I. Smith, in press 1999, Geologic map of the Lone Pine Quadrangle: U.S. Geological Survey Geologic Investigation Series Map I-2617.

P. J. G. Moore and T. W. Sisson, 1982, Exeter Quadrangle: unpublished reconnaissance mapping.

Q. J. G. Moore and T. W. Sisson, 1982, Kaweah Quadrangle: unpublished reconnaissance mapping.

R. J. G. Moore and T. W. Sisson, 1982, Mineral King Quadrangle: unpublished reconnaissance mapping.

S. J. G. Moore and T. W. Sisson, 1985, Geologic map of the Kern Peak Quadrangle, Tulare County, California: U.S. Geological Survey Geologic Quadrangle Map GQ-1584.

T. E. A. duBray and J. G. Moore, 1985, Geologic map of the Olancha Quadrangle, southern Sierra Nevada, California: U.S. Geological Survey Miscellaneous Field Studies Map MF-1734.

Special Map Areas

1. W. J. Nokleberg, J. G. Moore, M. A. Chaffee, A. Griscom, W. D. Longwill, and J. M. Spear, 1983, Mineral resource potential map and summary report of the Kings River, Rancheria, Agnew, and Oat Mountain Roadless Areas, Fresno County, California: U.S. Geological Survey Miscellaneous Field Studies Map MF-1564-A, one sheet, scale 1:62,500, 13 pp.

2. D. Clemens Knott, 1992, Geologic and isotopic investigations of the Early Cretaceous Sierra Nevada Batholith, Tulare County, California, and the Ivrea Zone, northwest Italian Alps: Examples of interaction between mantle-derived magma and continental crust: California Institute of Technology Ph.D. dissertation, 169 pp.

3. S. E. Goodin, 1978, Metamorphic country rocks of the southern Sierra Nevada, California: University of California, Berkeley, M.S. thesis, 75 pp.

4. D. L. Liggett, 1990, Geochemistry of the garnet-bearing Tharps Peak Granodiorite and its relation to other members of the Lake Kaweah Intrusive Suite, southwestern Sierra Nevada, California, *in* J. L. Anderson, ed., *The Nature and Origin of Cordilleran Magmatism*: Geological Society of America Memoir 174, pp. 225–236.

5. J. Saleeby, Written communication, March 1980.

6. C. Durrell, 1940, Metamorphism in the southern Sierra Nevada northeast of Visalia, California: University of California, Berkeley, Department of Geological Sciences Bulletin, v. 24, no. 1, 118 pp.

7. D. C. Ross, 1958, Igneous and metamorphic rocks of parts of Sequoia and Kings Canyon National Parks, California: California Division of Mines and Geology Special Report 53, 24 pp.

8. M. Sawlan, Written communication, June 1982.

9. C. J. Busby-Spera and J. Saleeby, 1987, Geologic guide to the Mineral King area, Sequoia National Park, California: Society of Economic Paleontologists and Mineralogists, Pacific Section, v. 56, 44 pp.

> The result . . . of our present enquiry is, that we find no vestige of a beginning,— no prospect of an end.
>
> —*Theory of the Earth*, 1788, JAMES HUTTON

> These rocks, these bones, these fossil ferns and shells,
> Shall yet be touched with beauty, and reveal
> The secrets of the book of Earth to man.
>
> —ALFRED NOYES (1880–1958)

TIME, MINERALS, ROCKS, AND PLATES

Geology, the study of the Earth, is a young science that was still rapidly evolving during the period of western exploration. It was in the mid-seventeenth century when a Danish anatomy professor, Nicolaus Steno, studying layered sedimentary rocks, proposed that those layers on the bottom are older than those on top, and that shells and other fossils embedded in the layers were older or contemporaneous with the enclosing rock layers and did not grow in them at a later time.

As scholars in Central Europe and England began to see similarities in rock layers situated great distances apart, it was only a matter of time before the concept of widespread marine deposition was appreciated. Abraham Gottlob Werner, a much-admired professor at the Mining Academy in Freiberg, Germany, popularized the notion that sedimentary layers, or strata, were deposited sequentially from a vast ocean. In 1785 he published a simple classification of these strata, a prelude to later, more elaborate efforts to develop a global geologic column and time scale.

The Geologic Time Scale

Building on these early studies of sequentially layered rock formations and the identification of successive time periods by the use of the fossils

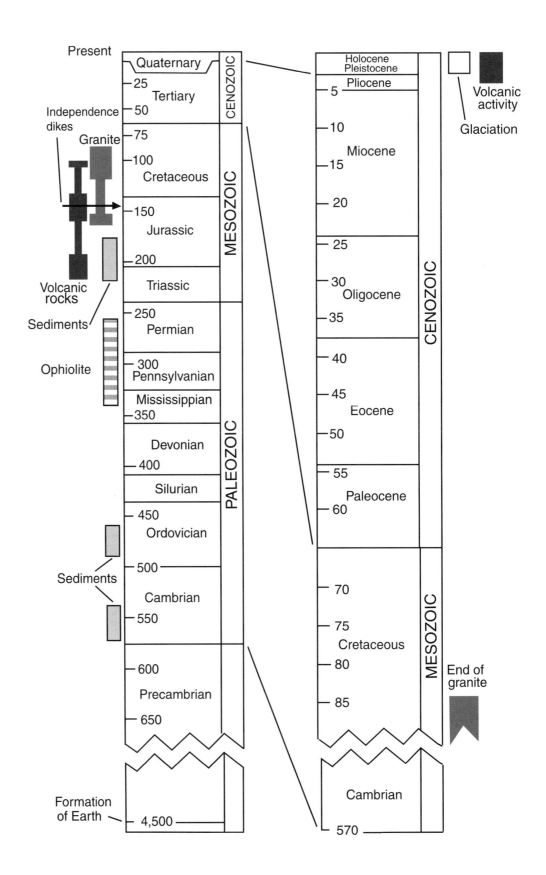

in those rocks, geologists established a geologic "calendar" of eras, periods, and epochs. This scheme rapidly evolved into the *geologic column*—the total worldwide sequence of rocks, from oldest to youngest (Fig. 4.1). In 1823 Adam Sedgwick, professor of geology at Cambridge, coined the term Palaeozoic Series for the widespread lower sequence of bedded rocks. John Phillips, paleontologist of the London Geological Survey and successively professor of geology at King's College, Trinity College, Dublin, and Oxford University, in 1843 named the younger Kainozoic Era (which later became the Cenozoic) and Mesozoic Era. These were felt to be the three major eras of the geologic column, representing rocks that were laid down atop the apparently nonfossiliferous deep strata that were lumped as the Precambrian.

The construction of the column made use of the law of *superposition*, Steno's concept that in a sequence of sedimentary rocks, any given layer or stratum is younger than the layer it overlies. The column also made use of the concept of *index fossils*, the remains of particular organisms (often marine mollusks) that are known to have existed only for a limited time and are thus characteristic of sedimentary rocks of particular, restricted time spans. Although only a small part of the column may be represented at any one place, the careful piecing together of different overlapping sections from different places has established a consistent, workable, and global sequence of age units.

This sequence is divided into eras, periods, and epochs. The primary divisions are the eras, from oldest to youngest, which include the Precambrian (originally thought to have no recognizable traces of life), the Paleozoic ("ancient life"), the Mesozoic ("middle life"), and the Cenozoic ("recent life"). The Precambrian is about seven times longer than the three others combined, reaching from the creation of the Earth, at about 4,500 million years ago, to the beginning of the Paleozoic, at 570 million

4.1. The geologic time divisions, showing ages in millions of years. Important events in the Sierra Nevada are shown on the sides, at their corresponding times. The column on the left shows the four main divisions of geologic time (the oldest at the bottom): the Precambrian, Paleozoic, Mesozoic, and Cenozoic Eras. The column also shows the periods (such as the Cretaceous) into which the eras are divided. Note that the Precambrian Era is about seven times longer than the other three eras combined. The column on the right is an enlargement of the Cenozoic Era and parts of older eras. Note that, for example, the Miocene is an epoch of the Tertiary Period.

years ago (Dalrymple, 1991). The eras are each in turn divided into periods, and the periods into epochs, on the basis of their fossil constituents. For convenience, names have been applied to these smaller age units, as well. The geologic time scale thus arrayed is the primary age classification that geologists the world over employ, and has been for decades (Fig. 4.1).

Working out the chronological sequence of all of the units of the geologic column was a major accomplishment, one that took the better part of 150 years and entailed widespread international collaboration. But the ages of all these units were known only in relative terms—science had only begun to develop means to date them. The next step was to develop means for determining the absolute time period, in years, during which any rock unit was deposited, so that actual dates could be attached to the sequential divisions of geologic time. Many early systems had been devised for this purpose, but invariably they tended to underestimate the actual lengths of geologic time. The ultimately successful tool in dating the divisions of the geologic column is based on the decay rate of radioactive elements in the rocks.

Sedimentary rocks, however, are generally not suitable for radiometric dating because they consist of fragments of other rocks, fragments that may range from clay-sized to boulder-sized. Many of these rock fragments derived originally from igneous rocks (rocks that had previously cooled from molten material), and hence each fragment possesses its own cooling age, and all of the fragments are older than the time of deposition of the sedimentary bed.

Therefore, it is the igneous rocks themselves, which cooled and solidified from magma, that are the rocks most commonly used to date particular levels in the geologic column. In such an igneous rock body, the radioactive clock starts ticking when the rock cools and crystallizes; that is the moment when the products of the radioactive decay begin accumulating within the solid rock body, usually within crystalline mineral material. (Before crystallization, when the rock was still in the magmatic stage, the daughter products of decay, particularly the gaseous ones, would be lost.)

The most commonly dated rock materials are lava flows or volcanic-ash layers that rest on one sedimentary layer and are covered by another. A date from these volcanic rocks provides a minimum age for the layer below and a maximum age for the layer above. An igneous rock that has intruded into another rock from beneath can also be dated. The rock that is cut by the igneous body is necessarily older, and that which is later deposited on top of the eroded and exposed intrusive body is younger. By carefully piecing together the findings of numerous dated rocks from many corners of the world, geologists have assembled a workable time scale divided into the named eras, periods, and epochs (Fig. 4.1). The time intervals signifying the duration of different divisions within the scale are of varying precision, depending on the availability of datable layers, but ongoing geochronological studies are constantly improving it, and evidence of many sorts—including the pace of continental drift—accords well with it.

Dating the Rocks: Radiometric Methods

Radiometric dating methods rely on the fact that some chemical elements occur naturally as radioactive isotopes. An *isotope* is one of two or more forms of the same element that occupy the same position on the periodic table of the elements. These forms have the same atomic number and behave chemically in almost the same fashion, but they differ from one another in the number of neutrons in their atomic nuclei, and consequently differ in atomic weight and in the nature of the radioactive changes they undergo.

Some of the naturally occurring isotopes exhibit radioactive decay and change—from one isotopic form to another—at a fixed rate, called the *half-life*, that is unaffected by "outside" factors such as heat and pressure. The half-life is the time that has ensued when half of the atoms of a given sample of a radioactive isotope have decayed to produce the new *daughter* isotopes. We know what the half-lives of the various radioactive isotopes are. Consequently, when the relative concentrations of a radioactive isotope (the *parent* material) and its daughter isotope in a rock are measured, then the age of the rock can be calculated. The calculated age represents the time that has passed, up to the present, since the point in time at which the rock enclosing the isotope reached such a state (for example, cooled and solidified) that the daughter product could begin accumulating within the mineral structures of the rock. In the case of a lava flow, that initial time is the instant when the lava solidified; in the case of a mollusk shell, or a tree stump that has not yet decomposed, it is the time of the life (or more particularly the death) of the organism.

For radiometric dating, the most commonly used naturally occurring radioactive isotopes (Table 4.1) are radiocarbon (^{14}C), potassium (^{40}K), rubidium (^{87}Rb), and the thorium/uranium series (^{232}Th, ^{238}U, and ^{235}U). Methods that use other isotopes are available, but the isotopes listed in the table are the ones that are most abundant in common Earth materials, and their half-lives are most appropriate for dating rocks.

TABLE 4.1
Common isotopic systems used in radiometric dating

Parent	Daughter	Half-life, years	Dating range, years	Common datable material
^{14}C	^{14}N	5,700	100 to 70,000	Plant and animal remains, water, etc.
^{40}K	^{40}Ar	1.3 billion	50,000 to 4.6 billion	Mica, hornblende, whole rock
^{87}Rb	^{87}Sr	47 billion	10 million to 4.6 billion	Mica, feldspar, whole rock
^{238}U	^{206}Pb	4.5 billion	10 million to 4.6 billion	Zircon, sphene
^{235}U	^{207}Pb	710 billion	10 million to 4.6 billion	Zircon, sphene
^{232}Th	^{208}Pb	14 billion	10 million to 4.6 billion	Zircon, sphene

Elements: C, carbon; N, nitrogen; K, potassium; Ar, argon; Rb, rubidium; Sr, strontium; U, uranium; Pb, lead; Th, thorium

Building Blocks of Rocks: The Common Minerals

A *mineral* is an element or a chemical compound that takes a crystalline form and has formed as a result of geologic processes. Usually, minerals are made up of a few major chemical elements, but other elements, both major and minor, may substitute for the major ones within the crystal structure of the mineral, thereby rendering its composition more complex and altering its character.

Aside from its chemical composition, a mineral generally has a fixed set of characteristics, or properties. For the geologist or mineralogist, the most important properties are color, form, cleavage, hardness, and specific gravity (see Table 4.2). Darker minerals commonly exhibit a characteristic color, but many of the colorless or translucent minerals, such as quartz, can take on various colors, depending on the type and content of their minor impurities. Because virtually all minerals are crystalline, the external form of a given mineral commonly reflects the distinctive shape of its internal crystal lattice. Some crystals are thin and needle-shaped (acicular), others are six-sided (hexagonal) prisms, and still others are short, stubby prisms faced with pyramids. Cleavage, or the manner in which a mineral splits or breaks, is dependent on the internal structure of the mineral crystals. Biotite, for example, one of the mica minerals, has a single perfect cleavage, causing it to split into flat, parallel plates, whereas another mineral may have two or three characteristic cleavages that cause it to break into elongate bodies or blocks. The hardness of a mineral reflects the relative ease with which it is scratched. Hardness is ranked from 1 to 10, talc at 1 being extremely soft, and diamond at 10 being the hardest mineral known. The human fingernail has a hardness of 2.5; a steel knife, about 5.5. The specific gravity of a mineral is its heaviness relative to that of water; a mineral with a specific gravity of 3, for example, weighs three times as much as an equal volume of water.

The granitic rocks that constitute most of the Sierra Nevada are made up primarily of only five rock-forming minerals (Fig. 4.2). All of these minerals are silicates—compounds that include the elements silicon and oxygen. The five are quartz, two types of feldspar (orthoclase and plagioclase), biotite, and hornblende, and the general categories of granitic rocks

TABLE 4.2
Common minerals in igneous rocks

Mineral	Composition	Form and color	Cleavage	Hardness	Specific gravity
Quartz	SiO_2	Six-sided crystals and granular masses; clear, white, gray, or any color	None; conchoidal fracture	7	2.6
Orthoclase	$KAlSi_3O_8$	Prism-shaped crystals, granular masses; pink, salmon, white, gray	Two, at right angles	6	2.6
Plagioclase	$NaAlSi_3O_8$ to $CaAl_2Si_2O_8$	Prism-shaped crystals, granular masses; white to dark gray	Two, at about right angles; striations on cleavage	6–6.5	2.6–2.7
Biotite	$K(Mg,Fe)_3$ $AlSi_3O_{10}(OH)_2$	Stubby six-sided prisms, irregular masses of flakes; black, brown, dark green	One, perfect	2.5–3	2.8–3.2
Hornblende	$(Ca,Na)_2$ $(Mg,Fe,Al)_3$ $Si_8O_{22}(OH)_2$	Long six-sided crystals; black to dark green	Two, intersecting at about 60°	5–6	2.9–3.8
Augite	CA,Fe,Mg $(SiO_3)_2$	Stubby, eight-sided crystals, granular masses; dark green to black	Two, intersecting at about 90°	5–6	3.2–3.9
Olivine	$(Mg,Fe)_2SiO_4$	Stubby crystals; yellowish green to olive green	None; conchoidal fracture	6.5–7	3.2–4.3

4.2. Classification of the common crystalline igneous rocks, on the basis of their mineral content.

are defined by their proportions of these five minerals. Two other minerals that are common in some related rocks are augite and olivine. Other minerals commonly present as small crystals in small amounts in granitic rocks are sphene ($CaTiSiO_5$), apatite [$Ca_5(PO_4)_3(F,OH,Cl)$], zircon ($ZrSiO_4$), magnetite (Fe_3O_4), and ilmenite ($FeTiO_3$).

The rock-forming minerals (and the rocks they form as well) can be classified in two major groups: *felsic* and *mafic*. The term *felsic*—the word is derived from the names feldspar and silica—refers to the light-colored minerals and to rocks that are rich in them. The common felsic minerals are feldspars and quartz. The term *mafic*—a word derived from magnesium and ferric (iron-bearing)—refers to dark-colored minerals and to rocks rich in iron and magnesium. The common mafic minerals are biotite, hornblende, augite, and olivine. Rocks that are about evenly enriched in both felsic and mafic minerals are termed *intermediate*.

Of the major minerals in granite, quartz has the simplest chemical composition. It is silicon dioxide (SiO_2). Quartz comprises about 30 percent of granitic rocks. The mineral itself is a glassy gray (although impurities can give it any of a variety of other shades), and it is the hardest and most chemically resistant of the common rock-forming minerals. Hence, when granitic rocks are chemically and physically broken down by weathering and abrasion in a stream, the quartz is preferentially preserved and becomes a major constituent of the resulting sediment.

The two major feldspars, orthoclase and plagioclase, are aluminum silicates containing various amounts of potassium, sodium, and calcium as integral parts of their composition. This mineral group is the most widespread of all, constituting more than 60 percent of the crust of the Earth. Orthoclase is the potassium feldspar, and plagioclase is a mixture of sodium and calcium feldspars occuring in any proportion. The plagioclase in granitic rocks generally has a ratio of sodium feldspar to calcium feldspar of about 3 to 1; that in the darker gabbro and basalt has a ratio of about 1 to 1.

Biotite and hornblende are the two dominant black minerals in the granites, contrasting with the whitish quartz and feldspar to give these rocks their characteristic speckled, salt-and-pepper look. Biotite is black mica, easily distinguished by its single perfect cleavage. It is a complex water-bearing aluminum silicate that also contains potassium, iron, and magnesium. Hornblende, a member of the amphibole group, is a black (or dark-green), generally elongate (prismatic) mineral that almost always occurs in granitic rocks, although usually less abundantly than biotite. Like biotite, it is a complex water-bearing aluminum silicate, but one that contains sodium, calcium, iron, and magnesium. A small amount of potassium is also present, which permits this mineral to be dated (as can biotite) by the potassium-argon radiometric method.

Augite, a member of the pyroxene group, is an elongate, black to dark-green mineral with two cleavages crossing at about right angles, which distinguishes it from hornblende, which has cleavages meeting at about 60 degrees. Augite, an aluminum silicate containing calcium, sodium, iron, and magnesium, is not common among the predominantly quartz-bearing plutonic rocks, but it becomes increasingly common among the darker plutonic rocks.

Olivine, a glassy green iron magnesium silicate, occurs only in the darkest plutonic rocks, such as gabbro and peridotite. When well formed, olivine, lacking cleavage, occurs in stubby crystals commonly terminated by pyramids.

Solidified Magma: The Igneous Rocks

The word *rock*, as used in the geologic sense, refers to any naturally formed, nonliving, coherent solid material that forms part of the Earth (or any other planet). A rock is usually made up of a mixture of two or more minerals, but in some cases a single mineral may constitute a body of rock. Rock masses are beautifully exposed in the highest Sierra. In fact, geologists who made an early geologic reconnaissance of the High Sierra stated, "We note that the upper parts of the Sierra exhibit, throughout their length, a bare rocky landscape. This is, in fact, one of the broadest exposures of rock known in the world" (Locke and others, 1940).

This statement may be debatable, but it is true that a large part of the Sierra Nevada is above timberline, and that much of it is bare rock with a minimum cover of snow, ice, alluvium, and glacial moraines. Such cover is more common and more extensive on most of the other major mountainous areas on Earth. Anyone hiking the John Muir Trail will be awestruck by the vast expanse of bare rock in the high country, truly a geologist's Valhalla of continuous outcrop.

Much of the granitic rock from which the high peaks are carved is *igneous*, that is, rock that was originally molten and then cooled and crystallized from a high-temperature melt. Such a natural melt, commonly bearing entrained crystals and dissolved gases, is termed *magma* when below the surface of the Earth and *lava* when on the surface. Rock that cooled and crystallized below the surface is *intrusive*; rock that erupted on the surface is *extrusive* or *volcanic* (Fig. 4.2). Most workers now agree that the granitic rock of the Sierra Nevada is dominantly igneous and intrusive, having crystallized from magma that rose from depth and solidified in magma chambers below the surface at a depth of a few miles. An individual body of granite that intruded as a unit and is relatively uniform in composition is termed a *pluton*. A large pluton, or a close assemblage of related plutons larger than 40 square miles in extent, is called a *batholith*. The entire mass of granitic rock in the Sierra is called the Sierra Nevada Batholith. The former magma chambers, now represented by the plutons, also fed volcanoes that once covered much of the Sierra terrain with lava. This volcanic cover, however, has been largely intruded by younger granitic plutons, and in the course of time has eroded away, leaving the plutons exposed. The remaining volcanic rock masses, now deformed and metamorphosed, comprise small patches between the plutons.

Several compositional-classification schemes are in use for the intrusive igneous rocks. The easiest to use are those that consider the proportions of the various minerals included (Fig. 4.2). With a hand lens one can identify most of the minerals and their proportions, especially in the coarse-grained intrusive rocks. The mineral content of volcanic rocks is more difficult to ascertain, because many are very fine-grained, and microscopic examination may be necessary to determine the mineral content before the rock can be classified. Other classifications make use of chemical analyses of the rocks, a system commonly used with volcanic rocks, because of their fine grain size.

The classification schemes equate the mineral proportions of the fine-grained volcanic rocks with those of the coarse-grained intrusive rocks of the same composition (Fig. 4.2). Thus, the volcanics rhyolite, andesite, and basalt are equivalent to the intrusives granite, diorite, and gabbro, respectively. The intrusive rocks granite, diorite, and gabbro trend from light- to dark-colored, as they become richer in the dark minerals biotite, hornblende, augite, and olivine. The rocks become heavier as they become darker-colored, because the individual minerals constituting them become heavier (denser) as the rock samples proceed from light to dark, principally because the darker minerals contain more iron.

The term *granite* is very old. It comes from the Latin *granus* or 'grain,' reflecting the coarse size of its constituent minerals, the units of which are about the size of a grain of wheat. In everyday usage, granite has a broad meaning, the term referring to any fairly coarse-grained crystalline rock. A more restrictive common usage requires that granite contain a significant amount of quartz and feldspar; scientists often use the term *granitic rocks* in this broad sense. But because all such rocks are so abundant, the scientific term "granite" is reserved for a specific kind of light-colored granitic rock that contains 20 to 60 percent quartz and has more than two-thirds of its feldspar in potassium feldspar, relative to sodium-calcium feldspar (plagioclase). Granite, as thus more restrictively defined, is light in color, has a SiO_2 content of about 70 percent by weight (*weight percent*), and is a *felsic* rock.

Gabbro, a dark rock containing primarily sodium-calcium feldspar and augite, with or without olivine, contains about 50 weight percent SiO_2 and is a *mafic* rock. Diorite is intermediate in mineral makeup between granite and gabbro, containing about 60 weight percent SiO_2.

The principal minerals in a granitic rock are generally rather coarse-grained and of similar size, from about ⅛ to ¼ inch. Upon slow cooling under high pressure they have intergrown with one another, forming a dense mass with virtually no open space. In some rocks a given mineral is better formed than the others and is found in decidedly larger crystals (up to an inch or more). Such crystals, markedly larger than their neighbors, are termed *phenocrysts*. In the granitic rocks of the Sierra, the potassium feldspar (orthoclase) commonly forms large phenocrysts.

In the simplified igneous rock classification outlined in Fig. 4.2 the rocks defined as granite include a great volume of rock, the most abun-

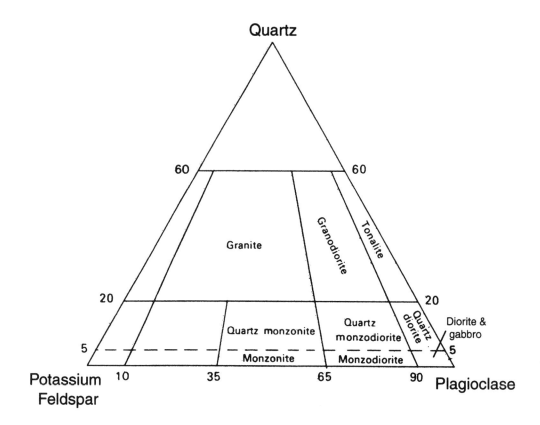
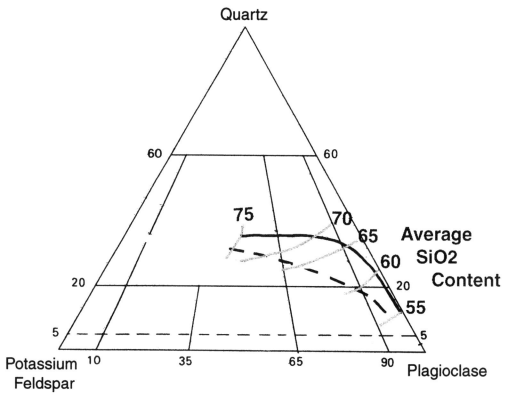

dant "basement" rock on the continents, and surely the most common rock in the Sierra Nevada. For scientific purposes, the general category "granitic rocks" has been subdivided into several subcategories to permit geologists to differentiate compositional changes within and between plutons. The classification recommended by the International Union of Geological Sciences (Streckeisen, 1973) divides the quartz-bearing plutonic rocks according to their proportions of plagioclase, potassium feldspar, and quartz. The proportions are most easily displayed by representing each analysis as a point on a triangular diagram, the three corners representing quartz, potassium feldspar, and plagioclase (Fig. 4.3). Such a diagram thus displays the proportions of only the three most common (light-colored) minerals; it ignores the somewhat less abundant dark-colored minerals, chiefly biotite and hornblende. Using this scheme a rock sample falling at the center of the diagram would contain equal parts of these three constituents. A sample falling at one of the corners would consist of only the one constituent designated by that corner. A sample midway along the quartz-plagioclase side would contain equal amounts of quartz and plagioclase, but no potassium feldspar. And so forth. In this classification the quartz-rich granitic rocks are divided into granite (in the narrow sense), granodiorite, and tonalite. The quartz-poor rocks occupy the fields labeled quartz monzonite, quartz monzodiorite, and quartz diorite. The most quartz-poor rocks include monzonite, monzodiorite, diorite, and gabbro. Natural rocks are rarely of such composition as to fall in the parts of the triangle that are unlabeled, that is, the extremely quartz-rich or potassium feldspar-rich fields.

The mineral makeup of granitic rocks is analyzed by a system of point-counting. In this procedure, the rock is cut with a diamond saw, and a flat face is polished so that a thin, transparent plastic screen with numerous rows of tiny dots can be placed on the rock surface. The rock is

4.3. Diagrams showing volume proportions (numbers in percent) of the minerals plagioclase, potassium feldspar, and quartz in granitic rocks. Equal proportions of the three minerals are represented by a point at the center of the triangle, and 100 percent of each mineral by a point at the corresponding apex. *Above.* Classification of granitic rocks (Streckeisen, 1973). *Below.* Trend lines of analyzed rock samples of the western part of the batholith (solid line) compared with those of the eastern part (dashed line), which are richer in potassium feldspar. Average weight percent SiO_2 (silica) content is shown by the gray trend lines. The Sierra Nevada batholith averages about 68.5 percent SiO_2.

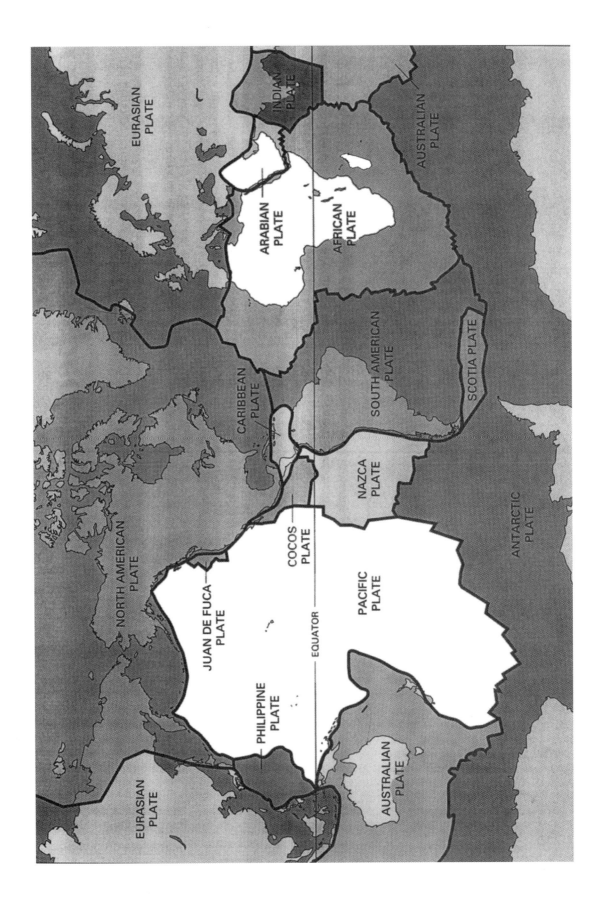

then observed through a microscope, and the mineral that appears beneath each dot is identified and tabulated. When about 1,000 points are identified and counted, the percent in total area of each mineral appearing on the cut rock surface can be determined with some precision. If the rock has no planar structure, which might render the count invalid, and the minerals are distributed at random throughout the rock, then the *area percent* of a mineral can be assumed to be equal to the volume percent.

The analyses of granitic rocks sampled at successive points across rock masses in the Sierra Nevada define trends when plotted on the triangular diagram. These trends can be represented by curved lines that extend from near the plagioclase corner and reach roughly the center of the triangle (Fig. 4.3, lower diagram). These trends yield information on the processes that have produced the variety of related rocks in the Sierra. Most rocks fall in the fields of quartz diorite, tonalite, granodiorite, and granite. Granodiorite is the most abundant rock in the Sierra Nevada Batholith.

The Global View: Plate Tectonics and the Rise of the Sierra

Scientists believe that the Earth's outer shell, which is cool and rigid, is composed of about a dozen major rigid plates (Fig. 4.4), those under the continents approximately 60 miles thick. This is the *lithosphere*, and the lithospheric (tectonic) plates move about over the underlying part of the mantle. The mantle layer directly below the lithosphere, which is called the *asthenosphere*, is somewhat more plastic than the remaining mantle below and the lithosphere above, and accordingly permits the lithosphere to move across it (Figs. 4.5 and 4.6). Below the asthenosphere a somewhat more rigid mantle extends down to the outer core at 1,800 miles (2,900 km) depth. The core, which is fluid in its upper part, extends another 2,160 miles (3,478 km) to the center of the Earth (Fig. 4.5).

The tectonic plates (Figs. 4.4 and 4.6) consist of either oceanic lithosphere, continental lithosphere, or both. Most plates carry both continents and ocean basins. The plates that make up the lithosphere are composed of an upper layer termed the crust and a lower layer termed the lithospheric mantle. In the oceanic part of the plates, the crust is thin (about 4

4.4. About a dozen major tectonic plates constitute the outer surface of the Earth. The heavy solid lines show the boundaries of these plates. Seismic and volcanic activity concentrates largely along these boundaries (Kious and Tilling, 1996).

miles thick), and composed principally of basaltic rocks overlain by about 3 miles of ocean water. In the continental part of the plates, the crust is thick (20 to 30 miles) and composed partly of granitic rocks.

These tectonic plates jostle one another as they move over the plastic asthenosphere in response to extremely slow *convective currents* within the mantle. Theorists envisage hot material rising, due to lower density, then moving laterally as its surface cools, and eventually descending, much as the globules in a lava lamp behave. This slow movement is commonly equated to the rate of growth of one's fingernails (a few inches per year). Through the great duration of geologic time, however, the plates have had ample time to move great distances over the face of the globe and have initiated important geologic events and processes. The present geographic position of the continents—which is not where they *used* to be and not

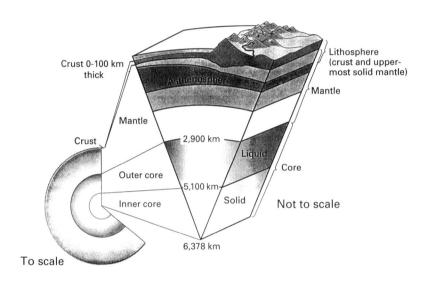

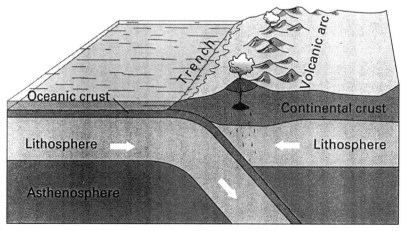

where they *will* be—has resulted from these plate-tectonic movements. Most earthquakes, volcanoes, and many major fractures and faults are located along present or former plate boundaries.

The plate boundaries are basically of three types: divergent, convergent, and transform. The *divergent* boundaries are those where two adjacent plates move apart and new lithosphere is generated, mostly by upward plastic mantle flow in the growing space between them. *Convergent* boundaries are those where two adjacent plates move against one another, causing one to be thrust beneath the other and to descend into the lower mantle. *Transform* boundaries are those, like California's San Andreas Fault, where neighboring plates slide horizontally past one another, and the plates neither grow nor shrink in size. Some plate boundaries cannot be so sharply defined, but are rather broad belts with indistinct boundaries where the effects of plate interaction are distributed.

The Mid-Atlantic Ridge (see Fig. 4.4) is a spreading ridge straddling a divergent plate boundary. Here, the North American and Eurasian Plates are moving apart in the north, and the South American and African plates are moving apart in the south. New oceanic lithosphere is being created (forced up) between them. Most of the plate boundary is beneath the sea, and mantle material moves continuously upward into the growing void between the plates. Magma moves into shallow chambers that feed submarine volcanoes, and these volcanoes repeatedly erupt lava to carpet the slowly spreading seafloor. In places, most notably Iceland, the volcanoes reach the surface, with dramatic effect.

The west coast of South America marks a convergent plate boundary. There, oceanic lithosphere of the Nazca Plate is thrust beneath the conti-

4.5. *(Top)* Cutaway views showing the layers within the Earth. Note that the lithosphere, which is the outermost rigid layer that forms the tectonic plates, includes an upper layer called the crust (Kious and Tilling, 1996).

4.6. *(Bottom)* A plate with thin oceanic crust converging against a plate with thick continental crust, showing the subduction (sinking) of the oceanic lithosphere beneath the continental lithosphere. Above the point where the subducted slab is about 60 miles (100 km) deep, magma is generated, perhaps by the action of the water that is baked out of the downgoing, heating oceanic lithosphere. Water and other volatile constituents lower the melting point of rock that they contact and thus induce melting. The resulting magma collects in chambers that may vent to feed volcanoes, but some will eventually cool and solidify underground and form granitic batholiths, such as those of the Sierra Nevada (Kious and Tilling, 1996).

nental lithosphere of the South American Plate in a process called *subduction*. A similar convergent boundary exists today along the Oregon–Washington coast. Here, the oceanic lithosphere of the Juan de Fuca Plate is impinging against, and being thrust (subducted) beneath, the continental lithosphere of the North American Plate. A line of large, active volcanoes (including Mount Rainier, Mount St. Helens, Mount Hood, Mount Shasta, and Mount Lassen) parallels the coast and results from the immense forces attending this subduction. The volcanoes occur along a line above the locus where the top of the subducted oceanic plate lies about 60 miles beneath the land surface.

The process that produces the magma that is injected into the crust to feed these volcanoes is still not well known. The dewatering of the downgoing subducted oceanic slab—the contained water is driven off by intense heat—is a potent factor. Much of the water in the volcanic rocks and sediments of the downgoing plate is baked out of the slab as it passes deeper and deeper and becomes hotter and hotter downward into the Earth's mantle. The rise of this superheated water (which would vaporize if the pressure were not so great) and other volatile-rich material into the overlying plate lowers the melting point of surrounding rock and causes it to melt. This melted rock rises through the mantle part of the upper plate and into its crust. A part of this magma remains beneath the surface as magma chambers that have invaded the overlying rocks, but some eventually erupts to build volcanoes. Considerable research is under way to attempt to determine what part of the final lava is of mantle origin and what part is of crustal origin.

Many researchers believe that the Mesozoic granitic rocks of the Sierra Nevada were formed by a subduction process near the plate margin where an oceanic plate impinging from the west was subducted beneath the North American Plate. Most of the thick pile of volcanic rock that was produced in this process has been eroded away. But many of the now-cooled magma chambers formed during the subduction-related melting of mantle and crust have survived erosion. They are now exposed as the Sierra Nevada Batholith, where they represent a vast belt of individual magma chambers stacked side by side and extending the length of the Sierra Nevada.

The transmutation has been effected by the influence of subterranean heat acting under great pressure, and aided by thermal water or steam and other gases permeating the porous rocks, and giving rise to various chemical decompositions and new combinations. . . .

—CHARLES LYELL, 1838

METAMORPHIC ROCKS

Most of the bedrock in the highest Sierra is granite. Moving upward from depth as a hot liquid mass, it intruded and shouldered aside the older roof and wall rocks. Erosion has since cut deeply into the older rocks, exposing the long-since solidified granite, but small remnants of the older rocks remain (Fig. 5.1). These early rocks were originally made up of a variety of materials, mainly sedimentary and volcanic rocks, but they have since been considerably transformed, or metamorphosed. *Metamorphism* is the geologist's term for the process that causes marked changes—resulting from great heat and immense, prolonged pressure—in the original rocks. Even though most of the metamorphic rocks in the central Sierra have been uplifted and eroded away, those remaining tell an important story about the earliest history of the region. The story is difficult to read, however, because of the brutal processes that have affected these rocks since their deposition.

Because of their location near the convergent plate boundary of the North American and Pacific Plates, the rocks have been subjected to repeated deformations. They have been squeezed, thrust-faulted, and dismembered by the compressive forces of subduction, and transform-faulting has produced further fragmentation and offset. As hot granitic magma intruded, it pushed aside, deformed, baked, and recrystallized the rocks,

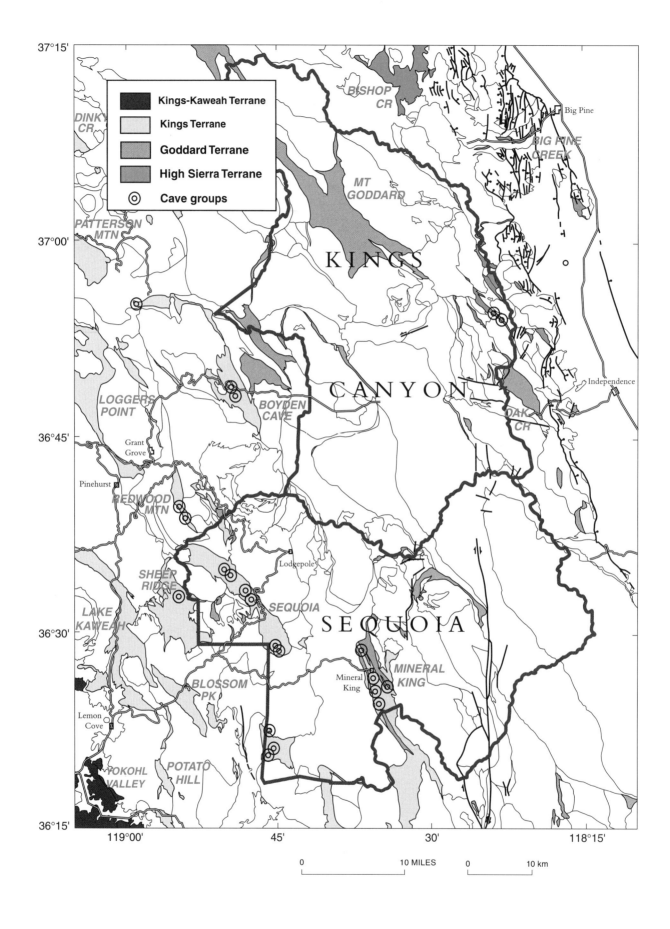

not only destroying much of the original structures in the rocks, but also leaving only a few recognizable fossils. The sequences and ages of the rock remnants, therefore, are poorly known.

Within the metamorphic rock, the originally horizontal sedimentary bedding has been tightly folded and sheared (Fig. 5.2), and is today often nearly vertical (Figs. 5.3, 5.4, 5.5). Faults that parallel the bedding are difficult to recognize, even though some of them are of regional significance. The remnants of layered rocks have been termed *roof pendants*, because in an earlier day they were visualized as flaglike sheets hanging down into the magma from the ceiling of the chamber containing the granitic melt. Dozens of these roof pendants occur within the two national parks. Most of them occur as screens, or *septa*, between individual granitic intrusives. The largest ones, generally more than 5 miles in length, include the Bishop Creek, Mount Goddard, Boyden Cave, Sequoia, Sheep Ridge, Lake Kaweah, and Mineral King Pendants (Fig. 5.1).

Because of their dark and variegated colors, the masses of metamorphic rock are generally easy to spot in the field. During metamorphism, much of the rock became dotted with pyrite, an iron sulfide mineral commonly called "fool's gold." The pyrite weathers readily to iron oxide, which stains the rocks a rusty brown, distinctly darker than the surrounding light-gray to nearly white granitic rock. Many of the place names emphasize such color contrasts. Black Kaweah Peak is underlain primarily by dark metamorphosed crystal-rich dacite. Black Divide and Black Giant Peak in the Mount Goddard Pendant are made up of dark-colored metamorphosed volcanic rocks. Rainbow Mountain in the Mineral King Pendant is underlain by variegated metamorphic rocks.

The metamorphic rocks that make up the roof pendants were originally sedimentary or volcanic rocks. The most abundant sedimentary rocks were shale, sandstone, and limestone. Recrystallization has converted the shale to slate, phyllite, and schist; the sandstone to quartzite; and the limestone to marble. Recrystallization commonly produced new and larger mineral crystals in the rocks. The clay-rich shales developed quartz and mica as their primary metamorphic minerals, but also feldspar, andalusite, sillimanite, cordierite, and garnet. The limy shales were transformed into layered rocks containing quartz and feldspar, along with epidote, garnet, and pyroxene. The sandstone was transformed to quartzite, a dense rock composed predominantly of quartz, with varying amounts of other minerals, depending on the content of the original clayey material. The limestone, composed originally of fine-grained calcite and dolomite, was simply recrystallized, the individual crystals of its minerals now increased in size. The resulting marble may have interlocking calcite and dolomite

5.1. Four belts of metamorphic masses (roof pendants) with distinctive rock assemblages have been grouped into four distinct terranes in the parks area. The major pendants are named, and the sites of cave systems that have formed in metamorphosed limestone (marble) are indicated.

5.2. An outcrop of metamorphosed marine sediments on the north side of State Route 180 east of Boyden Cave. The thin-bedded sandstone and limy shale have been recrystallized and intricately folded as a result of the heat and pressure associated with neighboring granitic plutons. The height of the outcrop is about 18 feet. U.S. National Park Service photograph.

grains up to as much as one-half inch across. The limestone was originally formed of organic shell debris or coral-reef material, but almost everywhere the recrystallization has destroyed all traces of ancient life. Outcrops of marble are not good places to find fossils.

The Four Terranes of Metamorphic Rock

The general character of the metamorphic roof pendants defines four major northwest-trending belts. The belts are subparallel to the elongation of the roof pendants and the individual granitic intrusions. This "grain" trends 25–45 degrees west of north and generally averages about N30°W (Fig. 5.1).

A given belt contains a general assemblage of rock units that are characteristic of that belt and differ from those in another belt. Each belt, therefore, has been designated as a *terrane*, which represents a fault-bounded geologic entity with a distinct assemblage of rock units, structure, and geologic history, an assemblage differing from that of adjacent rock masses. A terrane can be visualized as having once been all or part of a tectonic plate that may have traveled some distance before being accreted to the continent at its present location. From west to east the terranes in the parks region of the Sierra are (1) the Kings–Kaweah Terrane; (2) the Kings Terrane; (3) the Goddard Terrane; and (4) the High Sierra Terrane. (These terrane designations have been somewhat modified and simplified from those originally proposed by Nokleberg, 1983.)

The Kings–Kaweah Terrane is the westernmost belt of roof pendants in

5.3. Metasedimentary rocks at river level of the Middle Fork of the Kaweah River, at Buckeye Campground in the southern part of the Sequoia Roof Pendant. These phyllites and schists are so extensively deformed that the original horizontal layering is now nearly vertical. U.S. National Park Service photograph by R. Badaracco.

the foothills (in Fig. 5.1). A characteristic rock unit contained in it is *ophiolite*, a group of rock types originating in oceanic crust and mantle, a rock sequence that includes dark and heavy igneous rock like gabbro, peridotite (and its hydration product, serpentine), basaltic pillow lava, and deep-sea sediment. Pillow lava, a form of basaltic lava characterized by masses resembling pillows, typically forms where molten basaltic lava is quenched beneath water (Moore, 1975; Fig. 5.6). It is common today where fresh lava erupts on the seafloor. The deep-sea sediment in the Kings–Kaweah Terrane contains chert, a hard, very fine-grained, silica-rich sediment made up almost entirely of microcrystalline quartz. Chert forms by the very slow accumulation and sedimentation of silica-rich organic remains on the deep ocean floor.

Other sedimentary and volcanic rocks, in addition to the ophiolitic rocks, are associated with the Kings–Kaweah Terrane, and apparently represent materials originally deposited on oceanic crust that were added to the west side of the continent as a result of plate motion. The ocean-floor igneous rocks have been dated as Ordovician and Permian, but because some of the overlying rocks may be as young as Middle Jurassic (Saleeby and Busby, 1993), the lodging of plates against the continent must have occurred after those times.

The Kings Terrane occupies an extensive belt of roof pendants next to, and northeast of, the Kings–Kaweah Terrane. This belt, some 25 miles wide, is dominated by metamorphosed shale but also includes rocks that were originally rather pure sandstone and limestone. Metamorphism has converted these sediments to schist (and slate), quartzite, and marble, respectively (Figs. 5.4 and 5.5). Rare fossils found in the Kings Terrane pen-

5.4. *(Above)* Two nearly vertical folded light-colored marble layers in deformed metasedimentary rocks of the Sequoia Roof Pendant near Deer Ridge on the road between Hospital Rock and Giant Forest. U.S. National Park Service photograph by R. Badaracco.

5.5. *(Below)* View east showing metamorphic rocks of the Kings Sequence in the Boyden Cave Roof Pendant near Horseshoe Bend, on the South Fork of the Kings River. The bold, rugged, light-gray ridge on the right is a vertical layer of marble 1,500 feet thick upended from its original horizontal position. Boyden Cave has been carved in this layer. The sharp peak on the left skyline is the Obelisk, situated on the boundary of Kings Canyon National Park. U.S. National Park Service photograph.

dants range in age from Late Triassic to Early Jurassic. The similarity of these rocks with rocks to the north suggests that sediments of Precambrian and Paleozoic age may also be present in the Kings Terrane (Schweickert and Lahren, 1991).

Apparently, the Kings Terrane rocks were originally deposited in a marine basin not far from a continental landmass. The presence of marble layers, for example, indicates that the water was shallow in some places. The marble was formed from limestone layers built from the debris of coral reefs, which require shallow water for their growth. The quartz-rich sandstone in the sequence derived from exposed, older, quartz-rich continental rocks.

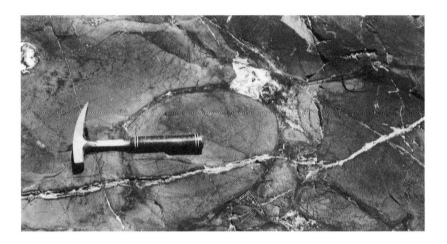

5.6. (*Above*) Basaltic pillow lava in the Kings-Kaweah Ophiolite belt of metamorphic rocks. Such ellipsoidal, pillow-like masses are commonly produced when molten basaltic lava is quenched in water, such as on the ocean floor (Moore, 1975). The outcrop is high (3,600 feet) on the south side of Bald Mountain, ½ mile north of State Route 180, 14 miles west of Wilsonia.

5.7. (*Opposite*) *Above.* Metamorphosed fragmental volcanic rock (andesite breccia at right) in sharp contact with granodiorite (at left) in a small metavolcanic pendant east of the Redwood Mountain Pendant, on Stony Creek. *Below.* Accretionary lapilli in air-fall rhyolite tuff near Summit Meadow in the Oak Creek metamorphic roof pendant. These lapilli, ¼ to ¾ inch in diameter, fall out of an ash-filled volcanic cloud as mud balls. They apparently form as moisture collects on an ash nucleus in the air and fine dust sticks to the nucleus and builds up the concentric layers in a fashion similar to the growth of a hailstone.

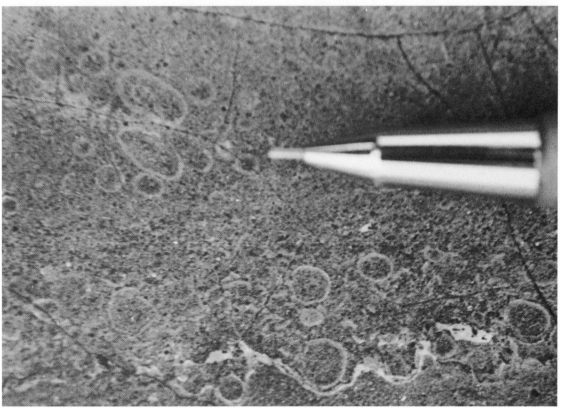

The Goddard Terrane, lying northeast of the Kings Terrane, is a rather broad belt, about 20 miles wide, of roof pendants made up principally of metamorphosed volcanic rocks. These pendants are dominated by rhyolitic tuff and breccia (Fig. 5.7) that had been emplaced during periods of explosive volcanism. They range in age from Triassic to Cretaceous and represent granitic melts that broke through to the surface and erupted, commonly with explosive violence. The magmas in these units are of continental origin.

This belt of roof pendants, composed primarily of metamorphosed volcanic rocks, crosses the Sierra Crest west of Independence (Fig. 5.1), crops out in the Alabama Hills west of Lone Pine, and continues southeast into the Inyo Mountains east of the Sierra Nevada. Four large masses of metavolcanic rocks in the Goddard Terrane occur in the area of the highest Sierra: the Mount Goddard, Oak Creek, Boyden Cave, and Mineral King roof pendants. Several smaller masses of such rocks are found, commonly as septa between individual granitic intrusives or plutons.

Uncommonly, *accretionary lapilli* are found in the fine-grained rhyolite tuffs. These are small spherical masses with concentric structure commonly ¼ to ½ inch in size (Fig. 5.7, below). They apparently form in the air in an ash-charged volcanic cloud, by accretion of ash around a core due to the condensation of moisture on the core, and then fall to the ground as mud balls (Moore and Peck, 1962). They have been found in the Mount Goddard, Oak Creek, and Boyden Cave roof pendants. Their appearance indicates that the volcanic vent and site of deposition were both above water when they formed.

In addition to the abundant rhyolite tuffs, lava of basaltic and andesitic composition occurs in these pendants. Some pendants contain older granitic rocks and fine-grained intrusive masses that probably represent shallow magma chambers formerly situated beneath volcanoes. Also included are sediments derived from the erosion of volcanic terrain. Extensive metamorphism has converted most of these rocks to layered sequences such as slate and schist.

The Mineral King Pendant contains about equal quantities of metamorphosed volcanic and sedimentary rocks, the two types interlayered. The oldest metavolcanic rocks here (apparently of late Triassic age) represent rhyolite ash flows that filled a caldera 5 miles in diameter and 1,600 feet deep, on the ocean floor. Rhyolitic volcanic episodes were repeated over nearly 30 million years, the youngest dated at 190 million years ago (Saleeby and Busby, 1993).

The High Sierra Terrane, the fourth belt of metamorphic pendants, lies east of the Goddard Terrane and includes, in the parks area, metasedimentary roof pendants of apparent Paleozoic age. These sedimentary rocks were originally primarily shale, but thin beds of limestone were also common. Metamorphism has converted them to schist and marble. Fossils available to the north of the parks area also indicate that these rocks are of Paleozoic age (Moore and Foster, 1980). Consequently, they represent rocks considerably older than the two central belts: the Kings Terrane and the Goddard Terrane. Limy sedimentary rocks of the Paleozoic metasedimentary belt in the Pine Creek Pendant, just north of the mapped area (Fig. 5.1), host one of the nation's largest tungsten mines.

Rare Fossils: Key to the Age of Metamorphic Rocks

The deformation and recrystallization of the ancient sedimentary deposits in the four belts have destroyed most traces of plants and animals originally contained within the sediments. Despite diligent search, fossils have been found at only a few places, these few in the largest roof pendants, where the recrystallization and deformation of the rocks is the least severe, because of a position most distant from a nearby fiery magma chamber.

Within the Kings–Kaweah Terrane, some limestone blocks contain tiny fossils of Permian age (Saleeby, 1978). These blocks are within the Yokohl Valley Pendant just off the extreme southwest corner of the parks map area (Fig. 5.1). The fossils, fusilinids, are a type of single-celled organism of the general group Foraminifera.

An important fossil locality in the parks region is that within the slaty rocks of the Boyden Cave Roof Pendant (of the Kings Terrane) along State Route 180, a few miles west of Kings Canyon National Park. In 1960, we found several fragments of the stems of crinoids (genus *Pentacrinus*) at this locality. A crinoid is a type of sea lily (animals, despite the name) most of which are fixed to the ocean floor. Those from the Boyden Cave area are of a type known elsewhere from Late Triassic to Early Jurassic rocks.

In 1971, as a result of a further search, we found a fossil ammonite within a hundred yards of the site where the crinoids were found, and in the same general rock layer (Fig. 5.8; Jones and Moore, 1973). An ammonite is a spiral shellfish (a mollusk, like a garden snail) related to an ancestral chambered nautilus. From an original shell diameter of about 4 inches, this fossil has been stretched by metamorphism to more than 6 inches in the maximum dimension. Despite the crushing and flattening that the shell impression has undergone, paleontologists have determined

that it is almost certainly of Early Jurassic age. Another ammonite, much smaller, was found in a boulder in the Kings River (Fig. 5.9).

Within the Mineral King Roof Pendant (Goddard Terrane), a sequence of interbedded sedimentary and fragmental volcanic rocks has yielded several types of fossils. These rocks were originally sand and mud, some limy, interbedded with volcanic tuff. Apparently, the rocks were deposited in a marine basin that collected debris from a nearby volcanic

5.8. *(Opposite)* A deformed early Jurassic fossil ammonite (reproduced here at natural size) from the Boyden Cave Roof Pendant. The two halves of the same fossil (*above*, exterior mold, and *below*, interior mold) were found about 100 feet apart by different collectors. This spiral shell is normally round; its oval, flattened shape here reflects the distortion that the enclosing rock has undergone (Jones and Moore, 1973).

5.9. *(Above)* Fossil ammonite in a boulder in the Kings River near Boyden Cave. An ammonite is a marine shellfish (mollusk) similar to a chambered nautilus.

center. The fossils, which include ammonites, range in age from late Triassic to Early Jurassic.

In the Big Pine Roof Pendant of the High Sierra Terrane of metamorphosed sedimentary rocks, trace fossils (the sponge-like archeocyathid) of apparent Cambrian age have been found. And in the eastern part of the Bishop Creek Roof Pendant (in the northeast corner of the Mount Goddard Quadrangle), Ordovician graptolites have been recovered (Moore and Foster, 1980). Graptolites were small, chitinous, colonial marine organisms restricted to the Paleozoic Era.

Marble: The Birthplace of Caves

In the canyon of the South Fork of the Kings River, about 3 miles upstream from the confluence of the South Fork and the Middle Fork, is a 1,500-foot-thick mass of marble, spectacularly exposed (Fig. 5.5). The sedimentary beds within this marble layer in the Boyden Cave Pendant are now vertical, having been deformed and rotated from their original horizontal position. Natural forces have carved several caves in this mass of marble, including Boyden Cave, which is open to the public. A second commercial cave in the parks area is Crystal Cave, which lies within a marble layer in the central part of the Sequoia Pendant. Many other caves occur in the parks area, primarily in the Kings Terrane roof pendants, within marble layers that are interlayered with other metamorphosed sedimentary rocks.

It is water running in fractures in carbonate rock that is responsible for cave formation (Palmer, 1991). Calcite ($CaCO_3$), the dominant mineral making up limestone and marble, is soluble in slightly acidic water. Rainwater is acid because of the natural atmospheric carbon dioxide dissolved in it, which produces carbonic acid. Rainwater that soaks into the ground becomes even more acid because of the presence of considerable carbon dioxide formed by the decomposition of organic material in the soil. Hence, where such waters percolate through fractures in rock that is dominantly calcite, the rock is dissolved, openings appear, and caves form. Roof pendants that contain layers of marble are therefore commonly riddled with caves, of all sizes (Fig. 5.1). Most of the marble lenses, and therefore the caves, occur in the Kings Terrane roof pendants. Not only is the marble most common in this metamorphic terrane, but the layers of marble are thickest there. Because of the near-vertical aspect of marble lenses, the cave systems are generally deep and elongate in plan (Fig. 5.10).

The caves generally form where surface water drains into fractures in marble, flows downward, and eventually exits the underground at a lower level, appearing at the surface as a spring in the nearest entrenched valley. As the water flows through the carbonate rock it dissolves the rock and enlarges the fractures to produce caves (Fig. 5.10). The upstream parts of most caves lie above the water table, that is, in the unsaturated part of the

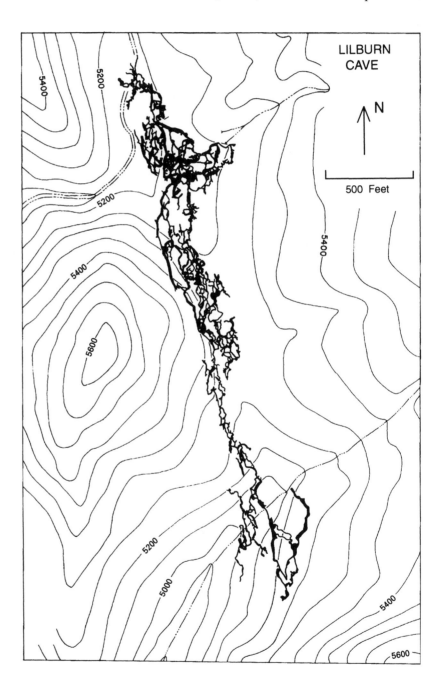

5.10. Map of Lilburn Cave in the Redwood Mountain Roof Pendant (Fig. 5.1). The cave system occurs within a steeply dipping layer of marble (Tinsley and others, 1981).

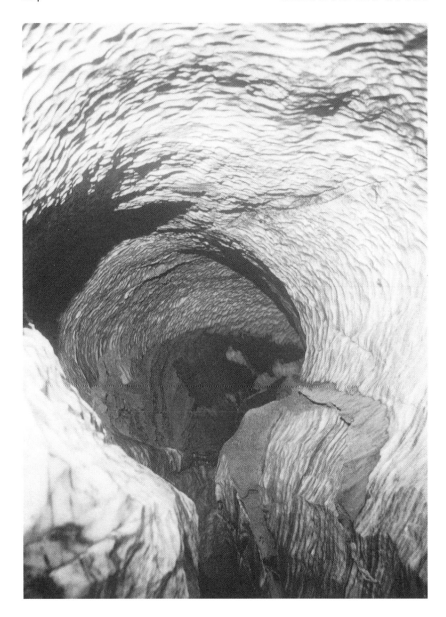

5.11. A solution passage about 10 feet in diameter in Lilburn Cave, showing the scalloped walls that have been produced by the corroding and dissolving action of rapidly flowing water. At times, water totally filled the tube and dissolved all parts of the solid marble walls. At another time, water half-filled the passage and deposited the sediment ledge on the right. Photograph by J. Tinsley, Cave Research Foundation

subterranean realm. Cave streams flow on steep gradients, and they commonly form canyonlike passages or even vertical shafts. The downstream parts of a cave system may extend below the water table, and there the passages are water-filled, especially in the wet season. The percolation of water through the saturated zone, however, continues to be an effective agent in dissolving the rock. This part of a cave system shows a low gradient, and the passages are often rounded or tubular in shape (Fig. 5.11). Because the water table in mountainous terrain can fluctuate widely in level, depending on the rate of precipitation, a single cave passage can alternately be subject to flooded conditions below the water table and dry conditions above the water table.

Cave formation requires thousands of years, but the torrential floods and surges of water from heavy rainfall and from the spring snowmelt are especially effective. During these periods, the muddy water, laden with sand, becomes a greater force in mechanically eroding—as well as chemically dissolving—the cave passages.

As soon as an open passage has formed, and remains above the water table for a considerable period, then calcite may be redeposited within the passage from water percolating down from the surface, if the water is saturated with $CaCO_3$. This may occur in circumstances where dripping water partly evaporates, or where the agitation of the water or its release from confinement causes carbon dioxide to be lost, thereby reducing the acidity of the water and rendering the calcite less soluble. When these factors cause the amount of calcium carbonate dissolved in the water to exceed its solubility, then calcite is deposited as an encrustation on the cave walls.

The slow deposition of calcite on the ceiling, walls, and floors of caverns produces the decorative cave formations that are a hallmark of this setting (Fig. 5.12). Cylindrical masses of calcite deposited from water dripping from the ceiling, in the fashion of icicles, are called *stalactites*, and those that grow upward, deposited from falling water splashing on the floor, are *stalagmites*. Water saturated in calcium carbonate flowing down the walls of a cave may leave sheetlike deposits known by cavers as flowstone or drapery.

The Origin and Assembly of the Metamorphic Terranes

Where and how the four terranes of metamorphic rocks now caught up in the Sierra Nevada Batholith may have originated are puzzling questions, ones that have excited endless debate among the scientists studying these enigmatic rocks (Saleeby and others, 1978; Nokleberg, 1983; Dickinson and others, 1996; Hamilton, 1978). Much of the evidence that might suggest answers is missing because of the extreme dismemberment, distortion, and metamorphism of the original rocks and the woefully incomplete age information available to us.

The evidence is particularly problematical for the westernmost zone, the Kings–Kaweah Terrane, where the sediments and volcanic rocks were deposited on oceanic crust, much of it in deep water some distance from the continent. Fragments or slivers of this oceanic crust were accreted to the

leading edge of the North American continent as the oceanic plate subducted beneath the continent (Fig. 4.6). By about 140 million years ago (the end of the Jurassic), this accretion of oceanic crust (Kings–Kaweah Terrane) in the western foothills region was complete, and the subduction zone jumped west to the region of the present Coast Ranges of California. But where did the slivers of oceanic crust come from and how were they attached? Some believe that the oceanic crust was generated at a distant spreading ridge (a diverging plate boundary), others at a small spreading ridge paired with, and just offshore from, the subduction zone associated with the continental margin, and still others as part of a separate offshore oceanic subduction zone (a converging plate boundary separate from that at the continental margin). Several workers, moreover, believe that the plate motion that brought the oceanic crustal fragments toward shore and "docked" them at the continent was oblique to the continental margin, such that the fragments may have traveled a long distance along the coast, perhaps even from equatorial regions, before docking.

The origin and age of the Kings Terrane pendants are also much debated. These continentally derived, relatively shallow-water marine sediments were deposited in offshore basins. Even though the meager definitive fossils are Mesozoic in age, some geologists believe that Paleozoic rocks are also included, because of the similarity of the rock types to some of the sequences of Paleozoic rocks east of the batholith. Transform faulting may have disrupted this sequence, and may have moved fragments or miniplates laterally for considerable distances.

Perhaps the least doubt attends the origin of the Goddard Terrane, which includes rocks from Triassic to Cretaceous age. That belt apparently represents a remnant of a volcanic arc. This volcanic arc was generated from an east-dipping subduction zone that may now be represented by major faults in the present Western Foothills Metamorphic Belt, currently exposed north of the parks area. As this arc developed, numerous plutons were generated, rose, and intruded the region to coalesce and produce the Sierra Nevada Batholith. During this stage, a major dike swarm intruded the wall rock and some of the earlier plutons. Widespread erup-

5.12. *(Broadside) Left.* Formations in the Dome Room in Crystal Cave, where stalactites growing down from the roof and stalagmites growing up from the floor have joined, producing columns. U.S. National Park Service photograph. *Right.* A drapery of flowstone decorating a wall of the Organ Room in Crystal Cave in the Sequoia Pendant. U.S. National Park Service photograph by S. Pusateri.

tions of volcanic rocks flowed onto the surface, then were themselves intruded from below by younger magma. But in the course of time, later regional uplift caused erosion to strip away all but small remnants of these volcanic rocks.

The older Paleozoic metasedimentary pendants of the High Sierra Terrane were formed in an environment near what was then the continental margin. They show alternations of shallow and deeper marine conditions that are much like the sedimentary beds of the same age adjacent to the east, in the Inyo and White Mountains.

In summary, the fragmentary evidence indicates that the two western terranes of metamorphic rocks may have traveled long distances as parts of tectonic plates from the west and south before they became attached to the North American continent in the Jurassic period. The two eastern metamorphic belts are currently in about their original site of deposition. The Sierra Nevada Batholith, predominantly of Jurassic and Cretaceous age, was intruded into, and welded together, this motley assemblage of rocks of different place of origin, different composition, and different early history, the whole marking a major transition zone, the western border zone of the continent at that time.

... which action has been termed "plutonic," as expressing in one word all conditions never exemplified at the surface. To this plutonic action the fusion of granite itself in the bowels of the earth as well as the development of the metamorphic texture in sedimentary strata may be attributed.

—CHARLES LYELL, 1833

GRANITIC ROCKS

Granitic rock is the most abundant rock in the Sierra Nevada. It underlies most of the range across a region 300 miles in length, more than 50 miles wide in places, and about 15,000 square miles in area (Fig. 6.1). This enormous mass of granite, nearly a tenth of the area of the state of California, is only a small part of the great belt of granitic rocks—all of about the same age—that occurs along the west side of North America along a span of 3,500 miles from Baja California to Alaska.

Plutons and the Sierra Nevada Batholith

Where a mass of granitic rock exceeds 40 square miles in aereal extent, geologists call it a *batholith*. Batholiths the world over are confined to the continents and owe their origin in large part to the melting of the continental lithosphere. The Sierra Nevada Batholith is one of several major masses of granitic rock, now largely exposed, that are disposed along the Pacific Coast of North America. Similar batholiths cover broad regions in British Columbia, Washington, Idaho, northernmost California, southern California, and Baja California (Fig. 6.2). All were largely emplaced in roughly the same time frame, during the Mesozoic Era. Most of these batholiths show a compositional zoning, such that the dominant rock in their western parts is tonalite (formerly called quartz diorite) and the dominant rock in their eastern parts is granodiorite. A line drawn through the several western batholiths, separating these two gradational groups in each case, roughly parallels the continental margin, except for an eastern embayment into Idaho. I call this the quartz diorite line (Moore, 1959; Fig. 6.2).

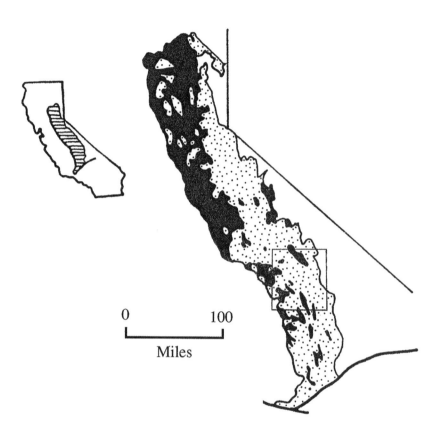

6.1. *(Above)* The Sierra Nevada Batholith (dotted area) in eastern California is 400 miles long and more than 50 miles wide in places. Metamorphic rocks are shown in black; note that the broad western metamorphic belt of the northern Sierra ends midway down the range (at the San Joaquin River) and is not present in the south. (The inset at left shows the main map's position in California; the box on the main map indicates the area of Fig. 1.3).

6.2. *(Opposite)* Major areas of Mesozoic plutonic rocks (batholiths) of the western United States. The quartz diorite line (the heavy dashed line) separates a western zone, in which tonalite (formerly classed as quartz diorite) is dominant, from an eastern zone, in which granodiorite and granite are dominant (Moore, 1959).

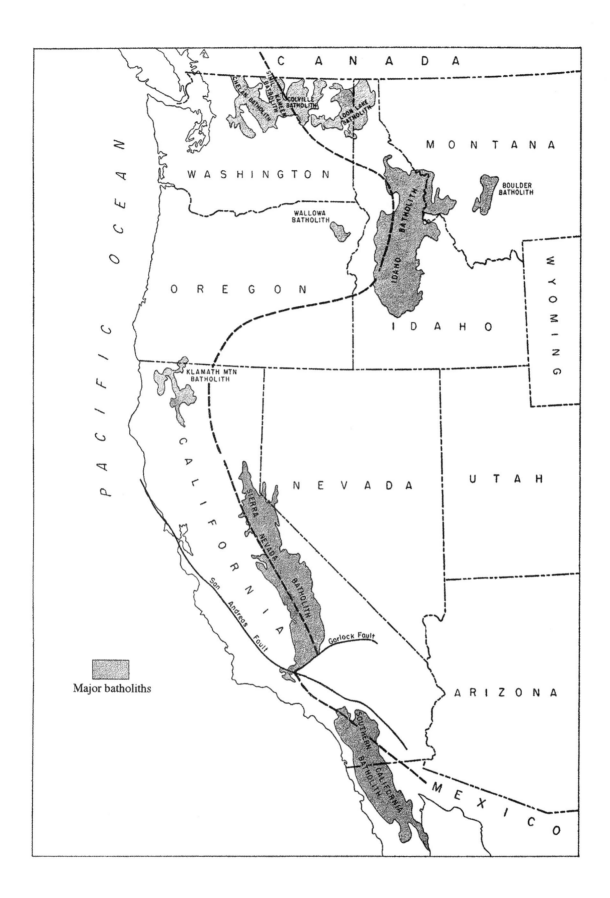

The Sierra Nevada Batholith, as well as the other Mesozoic batholiths of western North America, emerged from the interaction of the two tectonic plates that meet along the western continental margin. This band of batholiths defines a major convergent plate boundary that existed for much of the Mesozoic (Fig. 6.3). The batholiths apparently formed as a result of the subduction of oceanic lithosphere beneath the continental lithosphere.

The individual masses of granitic rock that together constitute a batholith, each generally representing a single event of intrusion and cooling beneath the Earth's surface, are termed *plutons*. Many plutons formed over a period of a few million years when rock material was heated at depth until molten, and then rose to a higher level. These masses of granitic melt were able to ascend because they were of lower density and greater mobility than the older, colder host rock into which they intruded. (Largely molten rock material below the Earth's surface is termed *magma*, and after it erupts on the surface it is called *lava*.)

Generally, the rock constituting a pluton is uniform in composition and texture and yields radiometric ages that are quite consistent from one part of the pluton to another, suggesting that each pluton was intruded at

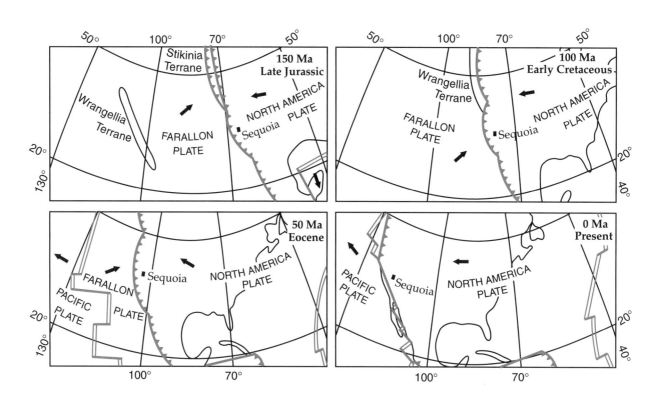

one time from a single batch of magma. The Sierra Nevada Batholith is composed of hundreds of plutons that were emplaced by the intrusion of magma over much of the Mesozoic Era, mainly during the Cretaceous Period, from 120 to 80 million years ago. The area embracing Sequoia and Kings Canyon National Parks is underlain by dozens of plutons (Fig. 6.4). The larger plutons are up to 500 square miles in areal extent, constituting large batholiths in their own right.

The individual plutons are commonly ovoid in plan and range greatly in size. A survey of 53 mapped plutons in the central Sierra Nevada indicates a median length of 7.5 miles and a maximum of more than 50 miles (Fig. 6.5). On average, they are somewhat more than twice as long as they are wide. The long dimensions of the plutons are nearly parallel with those of the overlying roof pendants of older rock, which average about 30° west of north (Fig. 6.4). This overall "structural grain" of the Sierra Nevada Batholith and its wall rock is believed to be a reflection of the subduction process, and parallels the positions of successive plate boundaries.

Generally, plutons in the Sierra Nevada range in average composition from 62 to 76 weight percent silica (SiO_2), their most abundant chemical constituent (Fig. 6.5). The largest ones range from 67 to 70 percent SiO_2, a span of just 3 percent, and the average composition across the entire batholith (at the parks area) is about 68.5 weight percent SiO_2.

The composition of individual granitic plutons is fairly uniform across their extent, yet neighboring ones may appear quite different. Each plu-

6.3. Approximate positions of the North American and neighboring tectonic plates during four periods spaced 50 million years apart, from 150 Ma (million years ago) to the present. The colored lines indicate plate boundaries of three sorts: toothed line, a subduction zone with the teeth on the side of the upper plate; double line, a spreading oceanic ridge; and single line, a transform fault, the plates moving past each other. At about 150 Ma the Stikinia Terrane collided with North America, causing a new subduction zone to form seaward. At 100 Ma the Wrangellia Terrane collided with North America, against a subduction zone that continued to produce the Cretaceous Sierra Nevada Batholith. At 50 Ma, the emplacement of Cretaceous batholiths along the west coast of North America had ceased, and the Pacific–Farallon Spreading Axis (double line) closely approached the coast. At the present, the Pacific Plate has intersected North America and created the San Andreas Fault. Maps after Ross and Scotese (1997), augmented by George Moore.

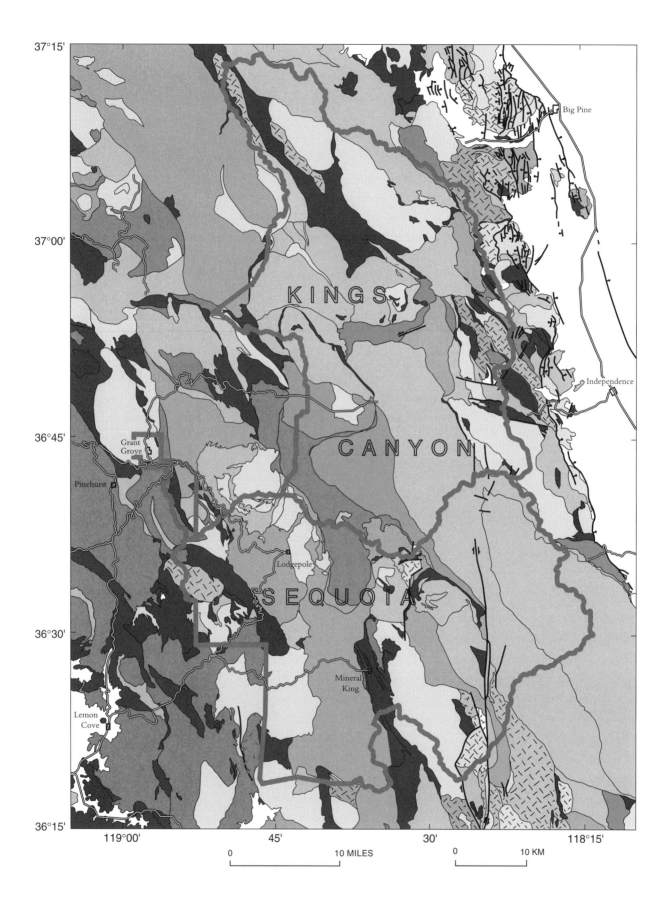

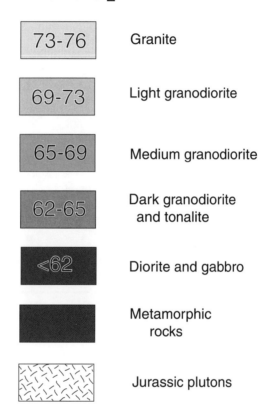

6.4. Bedrock geologic map of the Sequoia and Kings Canyon National Parks area. Granitic rocks are divided into four groups on the basis of silica (SiO_2) content, such that the rocks poorer in silica are darker in color. Note the consistent (although irregular) average trend of the mineral content of the plutons, becoming more silica-rich toward the east. Fine lines show the boundaries of all plutons, even where two of the same composition are in contact. All granitic rocks are Cretaceous except those designated as Jurassic. The post-Mesozoic sediments of Owens Valley are shown in white.

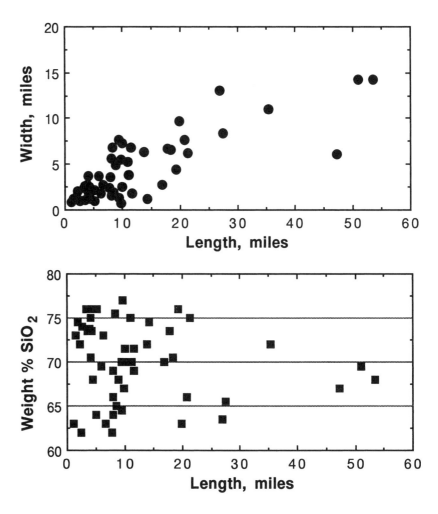

6.5. The nature of plutons, based on data from 53 mapped granitic intrusions in the central Sierra Nevada. *Above.* Plutons are generally two to three times longer than their width. *Below.* The largest plutons (at the right) contain intermediate levels of SiO_2 content, ranging from 67 to 69 weight percent.

ton commonly shows a similar texture, similar concentration and size of mafic (dark-rock) inclusions, similar spacing of joints, and similar overall aspect. This overall uniformity within plutons permits them to be mapped as discrete bodies that can be distinguished from their neighbors in the field. At some places, however, adjacent plutons made up of a very similar rock may be difficult or impossible to map separately.

In detail, the plutons do show variations in their overall makeup across their breadth, in some places irregularly, in others more systematically. The most common compositional change is for a pluton to be concentrically zoned, the outer shell darker-colored and poorer in SiO_2 than the center. This zoning is to be expected, for when a pluton moves upward as magma into the position where it solidifies and cools, the outer zone will cool more rapidly than the inner, because it is in contact with the cooler wall rocks. This cooling causes the minerals that crystallize at higher temperatures, thus before other minerals, to be the first to form in this outer zone, where they tend to accumulate against the walls, while the remaining liquid magma migrates inward. The major first-crystallizing minerals in a granitic melt are hornblende, biotite, and the more calcium-rich feldspar (plagioclase). The mix of these minerals that concentrates toward the pluton margins is darker and lower in SiO_2 than is the central body of the pluton.

In the zoned plutons we can study much of the sequence of liquid to crystal differentiation that otherwise would require examining—with less well-defined results—the compositional differences between many different plutons with different compositions. The Cartridge Pass Pluton (Fig. 6.6) in the northwest corner of the Mount Pinchot Quadrangle is a good example of a zoned pluton. This pluton, which has been extensively sampled and analyzed, is more dense (higher in specific gravity) and darker on its outer edge because of enrichment in hornblende and biotite there.

In some cases, the inner, hotter and more liquid part of the pluton will

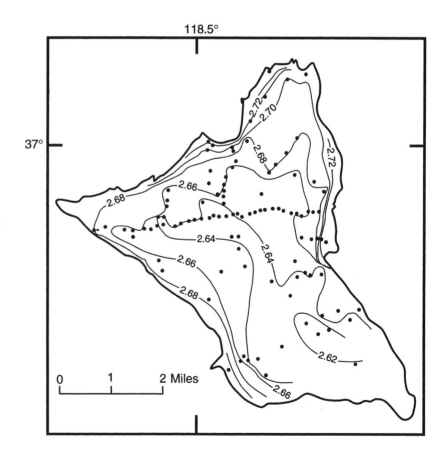

6.6. Outline map of the zoned Cartridge Pass Pluton 20 miles northwest of Independence (see Fig. 6.4), showing the measured specific gravity of samples of granitic rock by contour lines. Note how specific gravity increases toward the margin, indicating that the content of mafic minerals increases and silica content decreases toward the pluton margins.

surge upward and intrude its own walls, forming sharp contacts with the darker granitic rock of the earlier intrusive phase. Some plutonic complexes show several such mobilizations of the inner core, so that a sequence of concentric, nested plutons will form, each inner one more SiO_2-rich and younger than its outer neighbor.

The Whitney Intrusive Suite, covering more than 460 square miles and about 84 million years old, is such a nested complex (Fig. 6.7). This suite is one of the largest and youngest of the plutonic complexes in the Sierra Nevada. It is believed to have originally surged into position as a single near-molten granitic intrusion. Later, as the margins cooled and so-

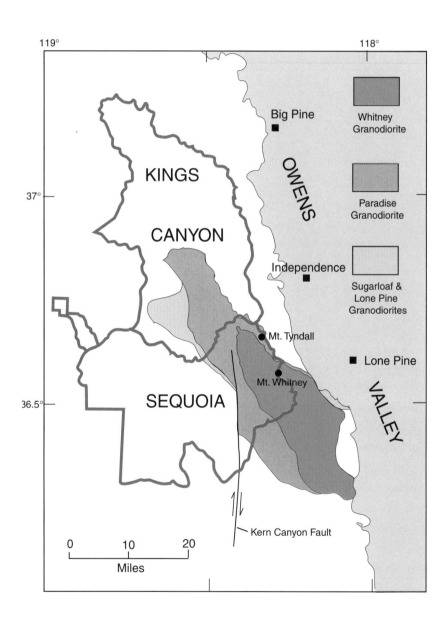

lidified, the Paradise Granodiorite rose upward and intruded its solidified margin, made of the Sugarloaf granodiorite on the west and the Lone Pine granodiorite on the east. Then, after further cooling, the inner, hotter part, the Whitney Granodiorite, rose again, intruding its own solidified margin, made up chiefly of the Paradise Granodiorite. The inner, most slowly cooled Whitney Granodiorite contains giant crystals of potassium feldspar, produced apparently because they had the longest period of cooling and crystallization.

Making Room for the Granite

The colossal Sierra Nevada Batholith is younger than its walls, and now occupies space that was formerly occupied by other rock sequences. Geologists have long pondered the problem of how this batholith—or any batholith—made room for itself. Where did the preexisting rock go that formerly occupied the site of the batholith? The fact that the batholith is composed of a mosaic of numerous, smaller plutons does not reduce the problem, because the aggregate of all the plutons requires an equal volume of space, as does the batholith.

It is clear that the granitic melt that moved upward and solidified to form the Sierra Nevada Batholith was created initially by the melting and assimilation of preexisting rock deeper down. Hence, when we transfer the magma upward, something must move downward (or laterally) to occupy the space vacated by the magma. In a gross sense, then, a regional downward movement of material has balanced the upward-moving magma. But the problem remains. If we consider the present-day surface of exposure, where did the rock go that formerly occupied the site of the batholith? Three mechanisms have been proposed for the em-

6.7. The nested Whitney Intrusive Suite (all three colored areas), one of the largest and youngest of the plutonic complexes in the Sierra Nevada, believed to have originally surged into its position as a single near-molten granodiorite intrusion. Later, as the margins cooled and solidified, the Paradise Granodiorite rose upward and intruded its solidified margin, made up in part by the Sugarloaf granodiorite on the west and the Lone Pine granodiorite on the east. Then, after further marginal cooling, the inner, hotter Whitney Granodiorite rose, intruding its own solidified margin, made up chiefly by the Paradise Granodiorite. The inner, most slowly cooled Whitney Granodiorite contains giant crystals of potassium feldspar (see lower photo, Fig. 6.11).

placement of plutons and displacement of older rock: stoping, granitization, and forceful intrusion.

Magmatic *stoping* is the process whereby intruding magma detaches and engulfs pieces of the wall rock, which then sink down into, or are assimilated within, the growing magma chamber. In this way the magma can quarry upward. It is evident that this process does take place, at least, on a small scale, because we see blocks of metamorphic and other wall rock included in the granitic plutons, especially near the edges. Large-scale stoping is probably not an important process, however, because plutons are generally found to be smoothly bordered, and long, thin, older metamorphic septa stand sheetlike between them. If stoping had occurred on a large scale, one would expect to find more fracture-dominated boundaries of wall rocks between plutons, and to find that the thin intervening septa of metamorphic rocks had been more disrupted or shattered.

Granitization is the process whereby preexisting rocks are so dramatically metamorphosed that recrystallization increases their grain size and percolating hot fluids modify their mineral composition, the rock thus being converted to a granitic rock. This mechanism implies that the granitic plutons did not crystallize from a magma that moved into place, but, rather, simply represent high-grade metamorphic rock that has been created from preexisting wall rock. The hallmarks of granitization are gradational (rather than abrupt) contacts between wall rock and pluton and the transgression of structural elements from the wall rock into the pluton.

In some small areas in the Sierra Nevada Batholith, contacts between granite and much darker igneous rock are indeed fuzzy, irregular, and gradational. Small masses of granite or aplite do occur in netlike masses in mafic rock, and appear to have been sweated out of the host rock. Such examples of granitization, however, are generally confined to small outcrops and do not occur on a regional scale. In most places the boundaries between two granitic plutons, or between a pluton and its metamorphic walls, are sharp, suggesting that they represent the original frozen contact between a solid wall and the liquid mass of the intruding plutonic magma. As we shall see, the nature of these contacts provides abundant information on the relative ages of the rock bodies on each side of the contact.

Forceful intrusion, our third option, is the process whereby plutons make room for themselves simply by pushing their walls aside and their roof upward. Within the area of the Sierra Nevada Batholith, there is ample evidence that country rock was dislocated, lifted, or shouldered aside by forceful intrusion. The existence of long, thin, curved screens, or septa, of metamorphic rock between plutons argues for a kind of emplacement where the pluton squeezed between layers in the wall rock and expanded to produce an ellipsoidal body of rock. The deformed wall rocks behaved plastically in some cases, allowing the intrusion relatively smooth passage, but in others they are extensively sheared. Another mechanism, related to forceful intrusion but distinct from it, is the notion that regional shear stresses would develop regimes of greater and lesser compressive stress. Magma waiting below for a route of ascent would seek out those zones with minimal compressional—or perhaps extensional—stress in which actual opening of the passage would be facilitated.

The need for a rigorous solution of the room problem has been eased in the light of plate tectonics theory, with its concept of the mobility of the lithospheric plates. The emplacement of the granitic plutons along continental margins, above the site where one plate is subducting beneath another, no longer compels us to think in terms of a rigid crust with a fixed area that cannot expand laterally. The movement of the plates permits—and no doubt causes—changes in the shape and size of the plates. Parts of the upper (continental) plate may be alternately under conditions of tension and compression, and the forceful intrusion of granitic plutons may be physically favored, leaving us with no need for recourse to the stoping or granitization mechanisms put forth prior to the advent of plate tectonics theory.

Structures of Granitic Rocks

Despite the fact that the overall aspect of granitic rocks is one of uniformity, small-scale structures can generally be found within them. Layering, dikes, and mafic inclusions are common structures, but other rarer structures are also present. These features prove to be valuable keys to an understanding of some of the processes that operated before, during, and after the emplacement of the host granitic plutons.

Layering. On first inspection, a mass of granitic rock appears to be homogeneous and to lack internal structure, if we discount the fractures and joints that may have cut it after cooling and solidification. It is true that some of the lighter-colored plutons are virtually structureless, and one can walk for a great distance staring at outcrops without observing a hint of any directional structure in the rock. In most places, however, a close look will reveal a planar structure throughout the body of the rock. This structure is commonly defined by the elongation of the dark, fist-sized inclusions that are so common in granitic rocks (Fig. 6.8). These inclusions are commonly pancake-shaped and aligned parallel to one another, thus

6.8. Elongate mafic inclusions in granodiorite; the inclusions show an imperfect preferred orientation.

defining a planar structure through the body of rock. In plutons with large, conspicuous, rod-shaped hornblende crystals, the crystals may show a preferred orientation that, again, defines a planar (or linear) structure.

This planar structure, or *layering*, when carefully mapped over the extent of a pluton, is usually found to dip steeply at the outsides of the pluton, parallel to the pluton's steeply dipping margins. Moreover, it is usually better developed near the margin. Apparently, the layering marks the flow planes of the magma as it pushed into a fissure, expanded, flowed upward, and eventually inflated to its present size.

The marginal zones of plutons commonly exhibit prominent dark, streaky layers that allow us to deduce something of the processes that operated during the growth of the pluton. These features, however, only define the very last stages of flow along the pluton's margins; all earlier flow structures will have been eliminated by the last pulses of flow. Some of these layers are decidedly graded—dark and sharp on one side and light-colored, fuzzy, and gradational on the other—and the grading may be systematically repeated among several layers (Fig. 6.9). Within each layer, the rock of the dark-colored side is finer-grained than the rock of the light-colored side. Commonly, the sharp darker side is toward the outer boundary of the pluton.

These graded layers are believed to have been formed by two processes, flow sorting and gravitational settling. The concept of *flow sorting* relies on the fact that when a liquid mass flows parallel to a solid wall, it flows most rapidly some distance from the wall and more slowly near the wall, due to the effect of frictional drag. This causes the larger quartz and feldspar crystals (which are lighter-colored) suspended in the flow to migrate away from the wall to the region of higher flow velocity—and less drag-induced shear—and the smaller biotite and hornblende crystals (which are darker) to migrate toward the wall. Hence, the sharp, dark side of the layers is on the old (first-formed because first-cooled) side of the layer that faces the pluton contact with the wall rock.

Some graded beds, whether horizontal or inclined, are apparently due to *gravitational settling*. In this mechanism, the heavier minerals (the smaller hornblende and biotite crystals) settle down during flow through the larger and lighter particles. Again, the sharp, dark side of a layer will be on its older side. In both of these mechanisms, repeated surges of flow after solidification of the older (and now cooled) layers deposit repeated but similar layers.

In some places, surge-like magma flow has apparently eroded channels in a sequence of previously deposited layers (Fig. 6.9). The younger lay-

6.9. Nearly vertical, curved layering in granitic rocks near Volcanic Lakes, the growth direction to the left (in the direction of the hammer handle). Note how the layers on the left define a channel that has eroded and cut earlier beds on the right. In places the dark beds are graded, such that each layer begins (on its right side) sharp and dark, and ends (on its left side) gradational and light-colored. A second channel occurs in the extreme upper left background.

ers are laid down in a channel that has eroded into the older layers and truncated them. We can determine the direction of growth of the sequence by observing in what direction the younger layers cut the older ones. This pattern confirms the same age sequence as was independently determined from graded layers, and in most cases indicates that the layering grew inward from the pluton walls.

Dikes. Dikes are tabular intrusions of magma that squeeze into and solidify within cracks in the wall rock. The kind of dikes that most commonly cut granitic rocks in the Sierra are light-colored, SiO$_2$-rich, gra-

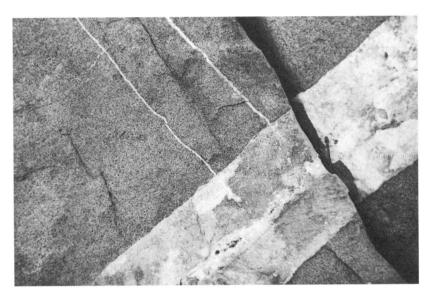

6.10. The appearance of dikes. *Above.* One thick and two thin aplite dikes cutting darker granodiorite. *Below.* Mafic dikes cutting granodiorite, with an angular inclusion of granodiorite in the thick dike.

nitic rock of a sort that remains molten to a relatively low temperature. These dikes are intruded when the outer walls of a pluton cool, solidify, contract, and crack. The still-molten material in the pluton core, now somewhat differentiated, is then free to squeeze into the marginal cracks. The source of the dike material is thus the pluton itself.

Fine-grained light-colored granitic dikes are called *aplite* dikes (Fig. 6.10, upper photo). Coarse-grained ones (commonly with a grain size of ½ inch or more) are *pegmatite* dikes. These two types of dikes are enriched in the low-temperature constituents of granite and depleted in the high-temperature constituents such as hornblende, biotite, and calcium plagioclase. They consist principally of quartz, potassium feldspar, and sodium-calcium feldspar, with minor biotite and/or white mica (muscovite) components, and hence are nearly white. In some places the pegmatites may contain rarer minerals such as tourmaline and garnet. This mineral makeup indicates that the dike rocks represent the final, low-temperature liquid remaining from the crystallization of a large body of magma. Under these conditions, the magma is likely to contain a large amount of volatile constituents (chiefly water), and it is the action of hot water and other volatile materials that favors crystallization of the large crystals in pegmatite.

Although the light-colored (felsic) dikes, whether of aplite or pegmatite, are the most common in granitic rocks, dikes of other compositions occur. Mafic (or dark-colored) dikes generally composed of dark granodiorite or diorite are, in fact, common in some places (Fig. 6.10, lower photo).

Mafic inclusions. Nearly all granitic rock bodies contain fragments of other rock types enclosed within them. These inclusions range in size, shape, and composition. Some are clearly recognizable as angular fragments of metamorphic rocks that have broken off and been incorporated within the liquid granitic magma as it moved into position and cooled. Commonly, they occur close to the edge of a granitic pluton, near the boundary with metamorphic rocks, and are fragments of slate, schist, or marble that are little changed from the rock within the main metamorphic mass from which they were derived. For them to have survived relatively unaltered, the cooling of the engulfing magma had to have been rapid.

Much more common than these recognizable inclusions of metamorphic wall rock, and obviously different, are rounded masses of dark rock termed *mafic inclusions*. These masses (Figs. 6.8 and 6.11, upper photo), also called *enclaves*, are darker and finer-grained than the enclosing granitic rock and are richer in mafic minerals. Relative to the enclosing rock, they show a concentration of hornblende, biotite, and plagioclase and

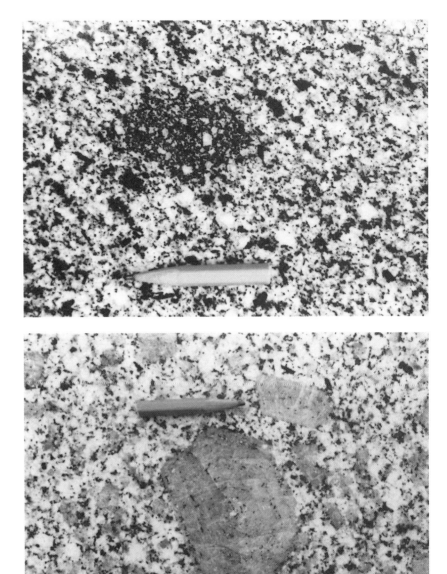

6.11. Coarse-grained granitic rock. The pencil stub is 2 inches long. *Above.* A small, ragged mafic inclusion in dark-colored biotite-hornblende granodiorite. *Below.* Large crystals (phenocrysts) of potassium feldspar in light-colored granodiorite.

a paucity of potassium feldspar and quartz. Geologists long ago recognized that these dark inclusions seem to be separate from, and of a different origin than, plainly recognizable fragments of wall rock.

The mafic inclusions, generally ranging from 1 inch to 1 foot across, are rather evenly spaced throughout the body of the rock, and in a given pluton their size and density of distribution are usually uniform and characteristic of the pluton. They may be pancake- or rod-shaped, all with a

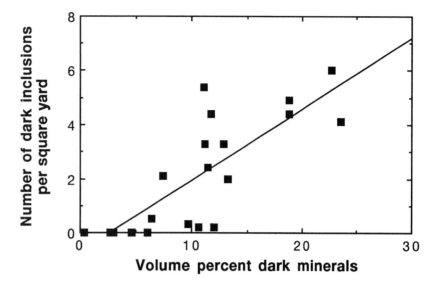

6.12. Average number of mafic inclusions occurring within one square yard of outcrop, for 20 separate plutons. Generally, inclusions are more abundant in darker granitic rocks.

similar, parallel orientation, and thus define a planar or linear structure in the rock mass. In the darker, more SiO_2-poor plutons, mafic inclusions tend to be more abundant, and each square yard of outcrop contains six or more. In the lighter-colored, SiO_2-rich plutons, they may be as scarce as one for every 4 square yards of outcrop, or may even disappear entirely (Fig. 6.12). The inclusions tend to be evenly distributed throughout a pluton, but in places swarms of dozens or hundreds of inclusions are to be found. In the zoned plutons where the granitic rock is darker and more silica-poor at the margin, the inclusions are commonly more abundant. Inclusions are nearly absent in coarse-grained, light-colored plutons, but diligent search will usually reveal a few small ones.

We know that the mafic inclusions contain the same minerals as the host granite, and they are remarkably evenly distributed over the extent of the pluton that contains them. These facts suggest that they resided in the host granitic magma (and solidified rock) at high temperature for a long time, and have attained mineralogic equilibrium with the host granite. Their origin, however, is problematic, because the recrystallization and chemical modification they have undergone leaves only a hint of their original character. To account for most of the inclusions, most specialists debate two origins: (1) that they are pieces of unmelted residue remaining from the melting of the lower crustal rocks that gave rise to the granitic magma, or (2) that they are blobs of hotter and more mafic magma—ultimately from the mantle—injected into a chamber filled with granitic melt that somehow became thoroughly distributed throughout the chamber. Probably each of these mechanisms has operated.

Comb layering and orbicular structure. A special type of layering within the granitic rocks (and also in diorites and gabbro) is termed *comb layering*, because it resembles a hair comb. Elongate crystals are arranged not parallel to the plane of the layering but at right angles to it, like the teeth of a comb or the bristles of a brush (Fig. 6.13). These elongate crystals are of the normal silicate minerals of granitic rocks, usually hornblende and plagioclase. Elongate forms of olivine, pyroxene, and potassium feldspar may also occur in this perpendicular orientation.

These layers of perpendicular crystals most often occur on the steep planar surfaces that bound the outer margins of a pluton, that is, along the boundary between the solid, outer part of the intrusion and the former liquid or plastic material of its inner, hotter part. The comb layering is commonly associated with the more normal types of layering in which elongate crystals are arranged parallel to, rather than perpendicular to, the plane of the layering.

Associated with comb layering in some places are spherical, walnut-sized to basketball-sized structures that have formed where comb layers enclose a rock core. The nucleus may be either a fragment of metamorphic or granitic rock, a larger than normal crystal of feldspar from a neighboring pluton, or an angular fragment of older comb layering. These spherical structures, called *orbicules*, show a characteristic radial crystal growth (Fig. 6.14). The comb layers wrap around the nucleus in such fashion that their right-angled crystals radiate outward from the nucleus.

Orbicules commonly occur in groupings in a near-vertical cylindrical mass usually only a few yards in diameter. In such a mass at Little Baldy Saddle, within the Giant Forest Granodiorite, the orbicules in the center are larger, those near the margin are smaller (Fig. 6.15).

Comb layering is beautifully exposed in the Volcanic Lakes region (Fig. 6.13), about 6 miles north of Cedar Grove on the north side of the Monarch Divide, near Volcanic Lakes (Moore and Lockwood, 1973). Here, the comb layers occur where steep pluton margins are indented with flutes or grooves. This relationship suggests that the layering formed when minerals crystallized at the pluton margin from hot, low-density, water-rich fluids that streamed upward in nearly vertical channels between the still pasty mush of the pluton and its walls (Fig. 6.16). Because these water-rich fluids are less dense and less viscous than the granitic melt, they flow rapidly up through the melt and are channeled against the solidified walls of the pluton. Crystals nourished by the fluid, and protected by it from the movement of the pasty magma, adhere to and grow out from the solid walls. Movement of the pluton causes the channels to change shape and size, thus affecting the flow rate—and the temperature—of the rising fluid. These changes can account for changes between the layers, the darker ones being precipitated during periods when the fluid temperature was higher, the lighter-colored layers when it was lower.

Orbicules apparently form where rock fragments fall into the stream of hot, rising, water-rich fluid that is nourishing the comb layers on the plu-

6.13. Comb layering at Volcanic Lakes. The hallmark of comb layering is that elongate crystals grow perpendicular to the layering. At the lower right, the horizontally arranged crystals in the wide dark layer are hornblende crystals that are perpendicular to the overall vertical light and dark layers.

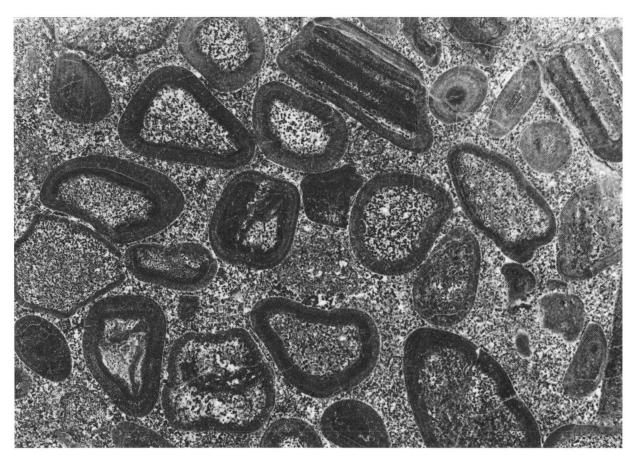

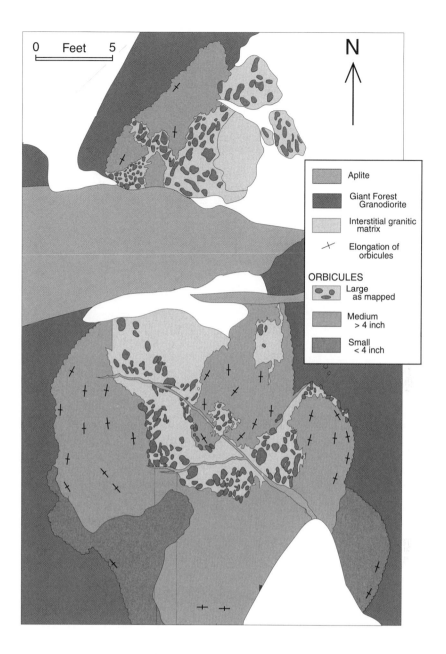

6.14. *(Opposite)* Orbicular granite. *Above.* Orbicular diorite from Deer Creek, 4 miles west of Wishon Reservoir (Moore and Lockwood, 1973). The orbicule at top center (9 inches long) has a nucleus of banded comb layering, as does an orb at upper right. Note that the thickness and layering of the orbicule rims (with radially arranged crystals) differ from one another, indicating that the individual nucleus fragments were added at different times to the fluid-filled channel where the outer layers were precipitated. *Below.* Orbicular body on the south wall of the Kern–Kaweah Canyon. The individual spherical orbicules (commonly with a nucleus of a large K-feldspar crystal) apparently were suspended in a hot, rising, water-rich fluid that precipitated mineral material on their outer surfaces and sorted them by size.

6.15. *(Above)* Orbicule-filled channel cutting the Giant Forest Granodiorite at the Little Baldy Saddle just west of Generals Highway. Note the concentric arrangement of the orbicules and a tendency for larger orbicules to gather in the center of the cylinder-like mass. White areas are covered and not exposed. Mapped by Kazuya Kubo, 1986.

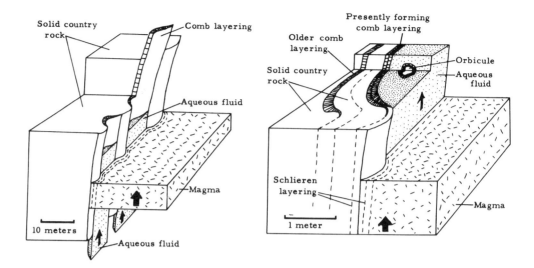

6.16. Diagram showing the mode of concentration of a low-viscosity, low-density fluid in grooves along the nearly vertical walls of a parent pluton. Crystals of comb layering attach to—and grow away from—the solid pluton walls. They are nourished by the rising low-density fluid and are protected from the viscously moving magma by the fluid. Orbicules grow when comb layers attach to rock fragments that are suspended in the rapidly rising low-density fluid (Moore and Lockwood, 1973).

ton walls. The rapidly rising fluid holds each fragment in suspension, allowing comb layers to coat the fragment (nucleus) while it is bobbing up and down in the stream (Fig. 6.16). Fragments that are small will be carried up and away by the stream, and those that are large will fall to the bottom of the channelway. This balance of flow velocity with size can account for the fact that commonly the orbicules are sorted by size. In some places the stream of fluid breaks away from the margin of the pluton and flows into a chimney-like structure in the wall rock, which may become filled with orbicules. As is the case in a blood vessel, the largest orbicules (or corpuscles) move toward the center of the pipe-like passage.

Field Criteria for the Relative Ages of Plutons

The individual intrusive masses or plutons that make up the entire Sierra Nevada Batholith were for the most part emplaced as magma across a short span of geologic time. Their outer boundaries or contacts at one time marked a boundary between hot molten rock and solid, cooler, and older wall rock. These contacts are clearly demarcated nearly everywhere, and the features along the contacts can tell which side was largely liquid and which side was solid at the time of a pluton's emplacement. This, then, enables us to determine which mass was the younger (liquid) and which the older (solid) at the time of emplacement. Generally, more than one type of feature occurs along the contact, and a comparison of several feature types can test the reliability of the relative-age determinations we make. Some types of features are consistently reliable, others are suggestive only if examined in isolation, and still others are missing in all but a few contacts and hence of limited use.

Among the most useful criteria is a progressive change in one of the masses toward the area of the contact, whereas the other mass remains unchanged on its side of the contact. It is the younger mass that shows the greatest change toward the contact, because the contact marks the outer wall of an actual magma chamber that began chilling and crystallizing after emplacement. Commonly, the younger unit becomes finer-grained or darker toward the contact; dark inclusions increase in abundance there, and they become flatter and more elongate.

The internal structures of the younger intrusive mass are increasingly better developed toward the contact. Dark inclusions are more nearly parallel there, and dark elongate minerals such as hornblende show a better parallel orientation there. At some contacts, the younger rock develops a layered concentration of alternating bands of light and dark minerals. And in some places the dark minerals or inclusions define a festoon or swirled structure parallel to or tangential to the contact.

Where an intrusive contact clearly cuts and truncates a structure within one of the plutons, that structure belongs to the older intrusive mass (Fig. 6.17). This may be a planar structure defined by the preferred orientation of dark minerals or elongate dark inclusions. The structure may also be a vein, dike, shear, or elongate inclusion, or a contact between two rock types.

6.17. The contact between the Cartridge Pass and Lamarck Granodiorite Plutons in the headwaters of the South Fork of the Kings River. The granodiorite of the Lamarck Pluton (upper right) has a layering defined by elongate mafic inclusions (parallel with the hammer handle) that is cut by the younger, finer-grained, and darker granodiorite of the Cartridge Pluton (at the left).

Where tongues or dikes of one rock type are intrusive into another, the intrusive rock is the younger. In most places, exposures do not show a dike that is clearly continuous with one unit and cuts another. More often, the dike crosses the contact and cuts both units. In some cases the rock type making up the dike is much more similar in composition and texture to the rock of one pluton than to that of the other. The pluton most resembling the dike rock is generally the younger.

Where a fragment of one rock is enclosed in another, the host rock is the younger, and the inclusion the older. Inclusions are not always positively distinguishable, but a distinctive composition or texture may enable us to correlate an inclusion with a rock unit on one side of the contact with more confidence than with one on the other.

Variations in the degree of shearing or chemical alteration can in some places be a guide to the relative age of plutons in contact. If one rock mass is sheared and recrystallized, and a mass in contact with it is not, the unaffected one is the younger.

Whether or not a prevailing structural element is present in a pluton can be an indication of the relative ages of two plutons in contact, even if the element is not continuous enough to be observed as truncated at the contact. The unit with the prevailing structural element, such as a swarm of dikes or a set of shears or veins, is the older. The dark-hued Independence Dike Swarm (taken up later in this chapter) is such an element. Plutons that are situated on the trend of the swarm, but contain no dikes, are considered to be younger than both the dikes and the plutons cut by the dikes.

Some granitic plutons occur in a nested, concentric sequence, with the outer mass enveloping the inner. Generally, the inner pluton is the younger.

These field criteria cannot determine the complete intrusive sequence among all the mapped plutons, however, because some plutons may not be in contact with one another, or, where they are, the contacts may not display the suitable criteria. Nonetheless the use of relative-age criteria is the first step in establishing an intrusive sequence. It can then be checked, elaborated, and quantified by radiometric dating.

The Age of the Sierra Nevada Batholith

A few localities have yielded fossil material that has helped to constrain the ages of the metamorphic rocks into which adjacent granitic plutons were intruded, and hence to provide maximum ages for the plutons. Until widespread use of radiometric dating became possible, our knowledge of the age of the batholith was dependent on this meager fossil evidence.

In the late 1950s, the potassium-argon system of dating igneous rocks was first applied to the granites of the Sierra. And in 1970 a major study utilizing many samples from localities scattered over the range allowed us to define the general features of the age of the batholith with some confidence (Evernden and Kistler, 1970).

Several dozen radiometric ages have been determined from samples collected within the parks area, from localities selected as representing key intrusions of granitic rocks. The ages represent the times at which the dated minerals in the granite crystallized—that is, when the host pluton cooled and solidified. Potassium-argon ages have been determined primarily on biotite and hornblende, minerals that are abundant in the Sierran granitic rocks. Biotite contains much more potassium and hence, through time, will produce more argon by radioactive decay, rendering analysis easier and more effective. The crystal structure of biotite, however, is such that it is more likely to lose the daughter isotope argon (a gas), yielding an apparent age that is too young. For this reason, the hornblende potassium-argon ages are considered more reliable.

Two minerals in granitic rocks, zircon ($ZrSiO_4$) and sphene ($CaTiSiO_5$), initially contained enough uranium to have produced (by now) measurable amounts of the daughter isotopes of lead and thorium, which can be used for age determinations. Most uranium-lead ages have been determined on zircon, which occurs sparingly as tiny crystals in most igneous rocks, but ages have also been determined on sphene, which is somewhat more abundant but has a lower uranium-lead content. The ages thus determined generally agree closely with one another and with the potassium-argon ages as well. The fact that the uranium-lead ages and the potassium-argon ages utilize analyses of different isotopes having different radioactive decay processes and different half-lives provides a good check on the reliability of the resulting ages when both methods are applied to a given sample, or a given pluton.

The radiometric dating that has been done demonstrates that most of the granitic rocks in the Sierra were emplaced in middle Cretaceous time, 120 to 80 million years ago. Two older series of granitic intrusions in the central Sierra Nevada Batholith, however, are of Triassic and Jurassic age. No Triassic granites occur within the parks area, but there are some in the region to the northeast, near Mono Lake, emplaced about 210 million years ago.

The Jurassic granites, which occur in scattered areas within and near the parks, yield ages generally of 155 to 175 million years and are commonly associated with roof pendants of metamorphic rocks. But their

6.18. Lead-uranium ages of granitic rocks across the Sierra Nevada in the parks area (Chen and Moore, 1982). The ages are projected onto a 75-mile line (line AB trending N 72°E, as shown in Fig. 6.22). The Cretaceous rocks (younger than 144 million years old) become younger toward the east (Chen and Moore, 1982).

measured ages are not internally consistent in places, perhaps because they themselves are metamorphic rocks, having been subjected to various episodes of heating from the intrusion of neighboring, younger Cretaceous plutons.

Samples from the more numerous Cretaceous plutons provide the most consistent ages, probably because they were least affected by later processes that might upset the radiometric ages, processes such as shearing, recrystallization, and the introduction of secondary minerals. Interestingly, the Cretaceous ages decrease toward the east across the batholith. Individual plutons were first intruded on the west side of the range, and later ones marched across the batholith at a rate of about 1.7 miles per million years (Fig. 6.18). A separate set of ages in the Yosemite region to the north shows a similar rate of movement of magmatism across the batholith (Chen and Moore, 1982).

Within the area of the parks map, the Cretaceous plutons in the western foothills are about 110 million years old, and the giant plutons of the Sierra Crest are 80 million years old. Hence, the period of Cretaceous granite-making in this part of the Sierra lasted for about 30 million years. Apparently, during this span of time, the relative positions and motion of the Pacific and North American plates remained remarkably steady, allowing a uniform pattern of change to produce the slow, more or less steady eastward migration of granite-making.

The ages of the plutons show a general relationship to the terranes of metamorphic rocks described in Chapter 5. Those surrounding the Kings–Kaweah Terrane in the western foothills are among the oldest dated Cretaceous rocks, and also include some still older Jurassic granitic plutons. Associated with the Kings Terrane are somewhat younger granitic plutons, mainly of Early Cretaceous age and commonly greater than 97 million years old. East of the Kings Terrane, in the central part of the batholith, is the Goddard Terrane, which includes the youngest wall rocks abutting the Sierra Nevada Batholith; it is associated with the youngest and largest granitic plutons of Late Cretaceous age (97 to 80 million years old). The High Sierra Terrane to the east, straddling the Sierra Nevada Crest, again is associated with older granitic rocks of Jurassic age, but scattered plutons of Early and Late Cretaceous age also occur in this area.

The Depth and Thickness of the Batholith

The hot and partly liquid granitic plutons of the High Sierra, which intruded upward from their site of melting deep in the Earth, were emplaced at considerable depth, where they cooled and solidified. They have subsequently been exposed at the surface by the erosional stripping away of the overlying rocks. The steep contacts of the plutons, the dense nature of the rock, the general lack of open cavities, the coarse grain size, and the rarity of transition into shallow volcanic rocks or structures all indicate that the plutons as we see them were originally emplaced and cooled at considerable depth. The depth of emplacement can be estimated by considerations of the pressure and temperature conditions required by the minerals that crystallized within the surrounding metamorphic rocks, and also by the minerals and structures of the granitic rocks.

Aluminum silicate (Al_2SiO_5) is a useful pressure indicator in metamorphic rocks, since it will crystallize as one of three different minerals—kyanite, andalusite, or sillimanite—depending on the ambient temperature and pressure. The absence of kyanite in zones of metamorphism around the granitic plutons indicates that the confining pressure was below 3,800 atmospheres (equivalent to the pressure of 8 miles of overlying rock). Hence, the properties of the aluminum silicate minerals indicate that the plutons cooled less than 8 miles below the surface.

The crystals of minerals of the granitic rocks themselves have been grown synthetically in the laboratory under varying conditions of pressure and temperature. The resulting data can be used to estimate the conditions of crystallization in natural rocks. Analysis of granitic rocks from several places indicates a pressure of 1,000 to 2,000 atmospheres, which is equivalent to 2.2 to 4.3 miles of cover (Bateman, 1992; Sisson and others, 1996). Hence, we can assume that approximately 2 to 4 miles of overlying rock has been eroded from the region since the emplacement of the granitic rocks.

The overall aspect of the plutonic rocks of the southernmost Sierra Nevada Batholith differs from that to the north in the parks area. Distinctly layered and darker plutonic rocks are more common there, and distinguishing igneous from metamorphic rocks is more difficult (Ross, 1989). The mineral makeup of these rocks, which includes greater proportions of the pyroxene minerals, also indicates crystallization at greater depth—perhaps as deep as 20 miles. Hence, the current depth of erosion (and exposure) is variable across the batholith and is distinctly deeper in the southernmost Sierra than in our area of interest at the location of the National Parks.

The present-day thickness of the batholith, that is, the depth from the present surface to the bottom of the granitic rocks, can be estimated by comparing the variations in the gravity field over the batholith with the specific gravity (density) of the surface rocks (Oliver and others, 1993). The strength of gravity was measured by sequentially setting up a gravimeter at 3,700 stations scattered over the area. The intensity of gravity, as measured in the field, depends on several factors, two of which are the elevation of the point of measurement and the average density of the column of rock below the instrument. After correcting for the elevation, the resulting gravity map reflects the variation in average rock density over the mapped area.

The specific gravity of 6,000 samples collected over the batholith between 36° 15' and 38° north latitude was carefully measured. These data were then plotted on a map to show the variation in the density of surface rocks over the area.

The next step was to prepare several calculated gravity maps, each of which assumed that the measured value of surface rock density extends down to a fixed depth. Six calculated gravity maps were computed for depths of 5, 8, 10, 12, and 15 kilometers (3.1, 5.0, 6.2, 7.5, and 9.3 miles). The texture of the 10-kilometer (6.2-mile) computed gravity map shows the best fit with the map of the measured gravity field. This means that the variation in density of the rocks measured at the surface seems to extend down to about 6 miles, either because the Sierra Nevada Batholith extends to that depth or because at that depth the compositions of the plutons become more similar and the density between them does not change horizontally.

Analyzing the Granitic Rocks

Many types of granitic rock have been distinguished and mapped in the batholith, even though much of the rock seems rather uniform in composition. To the casual eye, the rock looks about the same over broad areas, and, in fact, in the general region of Sequoia and Kings Canyon National Parks, 95 percent of the granitic rock ranges between 62 and 76 percent silica (SiO_2), its most abundant constituent (Fig. 6.5, lower diagram). To learn more about the batholith and how it was formed, we had to look more closely at its composition, at how that differs from place to place, and at how it varied through geologic time. The composition of hundreds of samples has been measured, in a variety of ways. These methods of analysis, in the order of increasing difficulty and expense, are: estimation of percent dark minerals, specific gravity measurements, modal analyses, and chemical analyses.

Volume percent dark minerals. Most of the granitic rocks contain the same minerals and have similar grain size and texture but differ in their proportions of dark and light minerals. In the field, the best way to quantify a given rock or pluton is to estimate the percentage of the dark minerals biotite and hornblende within it. This is usually done by making a simple visual estimate of the percentage of black minerals relative to white minerals. Such an estimate can be improved if one carries a few small standard rock specimens for which the percentages of dark minerals have been measured in advance (Fig. 6.19). A field sample can then be compared with the standards until its percentage is estimated at some point between the two standards that seem (on visual inspection) to bracket the sample, one being darker than the sample and the other lighter.

In the laboratory, when a series of samples is analyzed for both percentage of dark minerals and its chemical composition, we note an inverse relation between volume percent dark minerals and weight percent silica (SiO_2) (Fig. 6.20, upper diagram). At 60 percent SiO_2, the rock contains about 22 percent dark minerals; at 65 percent SiO_2, about 16 percent dark minerals; at 70 percent SiO_2, about 9 percent dark minerals; and at 75 percent SiO_2, about 2 percent dark minerals. Hence, the estimate of the percentage of dark minerals is also an estimate of silica content.

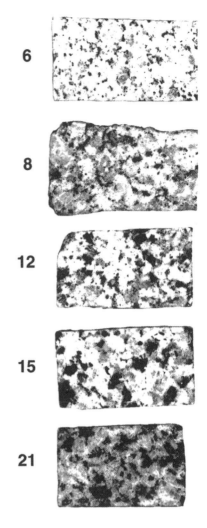

6.19. Slabs of granitic rock samples (actual size) arranged in sequence, the darker ones toward the bottom. The numbers indicate analyzed volume percent of dark minerals.

The rather broad scatter in the plot of the percentages of dark minerals and silica content in the figure indicates that this method is approximate at best. This is so because of several shortcomings. First, the distribution of dark minerals in a sample may not be uniform, with the result that the surface area where they are measured may reflect a concentration different from the bulk average that would be determined by the chemical analyses. Second, a fine-grained rock looks darker than a coarse-grained one, because more dark minerals can be seen through the smaller, thus more translucent, light minerals. Third, the percentage of dark minerals says nothing about the proportions of different light minerals; one rock may contain much more quartz relative to feldspar than another, and would therefore have higher SiO_2. But despite these shortcomings, the method provides a quick, quantitative way to characterize a rock, one that can be noted and compared with characterizations of rocks at other localities, and therefore has proved useful in mapping the different bodies of granitic rock.

Specific gravity. Another simple method of characterizing a granitic sample is to measure its density, or specific gravity. The specific gravity is the density of the rock as compared to that of water. The average specific gravity of the granitic rocks in the Sierra Nevada is about 2.7, meaning that they are 2.7 times denser (heavier) than water. Because by definition the density of water is 1 gram per cubic centimeter, the numerical values for the specific gravity and density of a material are the same when density is measured in grams per cubic centimeter. Specific gravity measurement requires that the sample be taken to a place where this simple measurement can be made.

Specific gravity is determined by comparing the weight of the specimen to the weight of an equal volume of water. An equal volume of water can be obtained by slowly immersing the specimen into a receptacle completely filled with water and catching the overflowing water, as

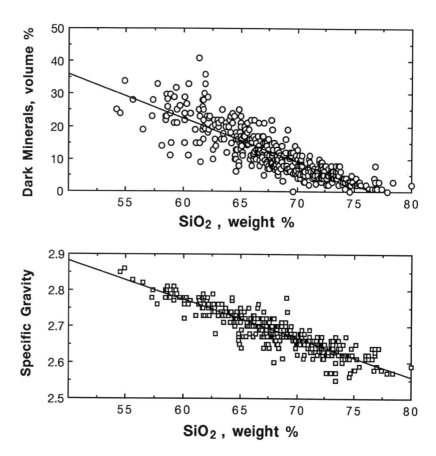

6.20. Plots of weight percent SiO$_2$ versus (*above*) volume percent of dark minerals and (*below*) specific gravity for 180 analyzed rock samples from the central Sierra Nevada. The average trend lines (determined by the method of least squares) allow one to make an estimate of SiO$_2$ when only the percent of dark minerals in a sample or the specific gravity of the sample is known. Rocks classified as granites (in the strict sense) fall at the right side of the two diagrams (samples greater than about 70 percent SiO$_2$) and are among the lightest, in both color and density, of all igneous rocks.

Archimedes did to determine if a king's crown was truly gold. A convenient way to do this is to use a beaker with a spout projecting from the side. The beaker is filled until it overflows at the spout, then, when the overflow has ceased, the weighed specimen is carefully lowered into the beaker and the water thus caused to flow from the spout is saved and weighed (or measured volumetrically). Specific gravity is calculated by dividing the weight of the sample by the weight of the displaced water.

Another method is to weigh the sample, then suspend it by a slender string and weigh it again when the sample is totally immersed in water. The buoyant effect of the water will cause the sample to lose weight when immersed. According to Archimedes' Principle, the amount of weight lost in the water is equal to the weight of the water displaced (which is of a volume equal to that of the specimen, unless the specimen floats). Hence, the specific gravity is equal to the weight of the sample divided by the loss of weight when it is immersed in water.

These methods require that the rock be compact, with no cracks or internal closed cavities. Otherwise, the measurement would determine the specific gravity of both rock and cavities, rather than that of the rock alone. If bubbles issue from cracks when the sample is immersed in the water, we must wait until all trapped air in the sample is released before a reliable reading can be obtained.

A comparison of specific gravity with analyzed chemical composition (SiO_2 content) of a group of samples shows a good inverse relationship. The data points for specific gravity plot closer to an average trend line (Fig. 6.20, lower diagram) than do the points for the percentage of dark minerals (upper diagram). Hence, specific gravity is a better estimate of chemical composition.

The specific gravity method is quick and inexpensive, unlike the more elaborate and expensive major-element chemical analyses. Generally, we determine specific gravity for every sample collected, and in some cases the analyses are done in the field so that samples need not be transported home. Samples can be collected along lines or grids laid across a pluton, to learn the details of change in composition across or within the pluton (Fig. 6.6). This specific gravity survey can then identify a few scattered specimens for later chemical analyses, which can then define in more detail the compositional variations of the sampled mass.

Modal analysis. Another method of analysis, one that requires rather simple laboratory equipment but neither a fully equipped chemical laboratory nor rigorous procedures, is *modal analysis*. This type of analysis deals not with the chemical elements of the rock but with the relative abundance of the minerals it contains. We saw in Chapter 4 how such data are necessary to properly classify granitic rocks.

Modal analyses are made on a flat surface of the rock specimen several square inches in area. A transparent film with a grid of about 1,000 points—each about the size of the periods on this page—is fixed on the flat face. Then under a low-power microscope, the mineral beneath each point is identified and tabulated. After 1,000 points are identified and tabulated, the areal percentage of each mineral on the analyzed area of the rock face can be calculated. The areal percent is equal to volume percent if the minerals are in fact scattered at random throughout the rock specimen.

In practice, all the common minerals in granite may not be readily distinguishable on a flat surface, especially when a judgment must be made on every spot where one of the points happens to fall. The two dark minerals biotite and hornblende are not distinguishable in every case, and consequently are lumped in the single category "dark minerals."

Likewise, the three light minerals, quartz, potassium feldspar, and plagioclase, are not always readily distinguishable. Quartz, however, can generally be separated from the other two because of its clear, glassy appearance, as opposed to their white, opaque aspect. Various stains can be employed, selectively dying each of the feldspars a distinctive color.

With such analyses, we can classify and name a given sample by noting in which field it plots in the standard classification scheme (Fig. 4.3). When plotted on the usual triangular diagram with corners representing quartz, plagioclase, and potassium feldspar (Fig. 4.3), the analyses of rocks from a series of localities are commonly found to lie on trend lines. Usually, these lines extend from the plagioclase corner to about the middle of the triangle.

A series of analyzed samples from a pluton can thus establish the compositional trends within that mass. The plagioclase-rich rocks fall in the categories of diorite and quartz diorite, and the rocks with about equal concentrations of the three major minerals are granite. In between is the granodiorite field. The chemically analyzed silica content of rocks that have also been analyzed for mineral content shows systematic trends when plotted on the triangular diagram. Establishing this relationship enables one to estimate the silica content from a determination of the mineral content (Fig. 4.3).

The melting temperatures of the various plutonic rocks differ. Extensive experimental study has shown that rocks plotting near the middle of the quartz–plagioclase–potassium-feldspar triangle (that is, in the granite field) melt at the lowest temperature. This minimum temperature for the melting of granite is about 650° C (about 1,200° F) when water is available under pressure. This temperature increases in every direction outward from the center of the triangle and is several hundred degrees higher for samples that plot outward toward the corners.

In addition to granite's low-temperature-melting character, it also has the lowest density of all common igneous rocks (Fig. 6.20), because it has a low content of heavy iron-bearing minerals—like biotite and hornblende—and a high percentage of potassium feldspar, the common plutonic mineral with the lowest density.

These facts together explain why granite is such a common rock in the upper crust. When we heat up any crustal rock material (which may plot anywhere in the compositional triangle), it will first melt a little, and at higher temperatures more and more, until eventually all will melt, and the final melt will have the same composition as the starting material. Regardless of the starting composition, however, the first drop of melt will be of the minimum-temperature-melting composition—the composition at the center of the triangle, thus within the granite field. Hence, during the heating of the crust, the first material melted will have a granite-like composition. If enough of it forms, it will coalesce and migrate upward as a magma, because of its melt-induced mobility, coupled with its particularly low density.

Chemical analysis. The most useful analysis of a granitic rock is a chemical analysis that reports the principal constituents of the rock as oxides. Such analyses are performed by first grinding all or part of the rock sample to a fine powder. The powder is then dissolved by the use of strong acids, or melted and then chilled to form a uniform glass wafer. In the former case, various chemical reagents are added to the solution to precipitate out insoluble compounds (those that can be filtered and weighed), each of them diagnostic of a chemical element. In the latter case, the glass is polished and bombarded by x-rays, and the nature and intensity of the emitted radiation is measured with a spectroscope to identify the different elements and their concentration within the rock.

Igneous rocks are largely composed of 14 chemical constituents, each a major element, but the 14 are generally expressed as oxides because of the oxidized nature of most Earth materials. The oxides that are those routinely analyzed are shown in Table 6.1.

When we examine the chemical analyses of the 14 major constituents of granitic rocks from the central Sierra Nevada Batholith, SiO_2 is clearly the most abundant, making up 60 to 75 percent of the rock (Table 6.2). Small wonder, then, that silicon is the second most abundant element on

TABLE 6.1
Common major elements of igneous rocks expressed as oxides
(Compare with Table 6.2)

	Formula	Chemical name	Common name
1	SiO_2	Silicon dioxide	Silica
2	Al_2O_3	Aluminum oxide	Alumina
3	Fe_2O_3	Iron oxide, ferric	—
4	FeO	Iron oxide, ferrous	—
5	MgO	Magnesium oxide	Magnesia
6	CaO	Calcium oxide	Lime
7	Na_2O	Sodium oxide	Soda
8	K_2O	Potassium oxide	—
9	H_2O+	Water released $> 100°$ C	Plus water
10	H_2O-	Water released $< 100°$ C	Minus water
11	TiO_2	Titanium oxide	Titania
12	P_2O_5	Phosphorous oxide	—
13	MnO	Manganese oxide	—
14	CO_2	Carbon dioxide	—

Earth, following only oxygen. The first seven or eight oxides on the list make up more than 95 percent of the Earth's crust, and the 14 constituents collectively make up more than 99 percent. The remaining 1 percent, however, includes all the other elements of the periodic table, although any or all may be present in vanishingly small concentrations in a given rock sample. Unlike specially formulated ultrapure synthetic compounds, rocks are "dirty."

The study of these analyses is a science in itself, a branch of geochemistry called *petrochemistry*. The overall balance and ratios of the oxides are an indicator of the reliability of the analysis, as is the comparison of analyses from sample to sample of related rocks. The nature of the analyses also indicates whether the rock has been chemically altered or weathered, and consequently whether it may be representative of the magma from which it originally crystallized.

The analyses of rocks in Table 6.2 are arranged in the order of increasing SiO_2 content. As SiO_2 increases, all other oxides decrease except for K_2O, which increases with the SiO_2. The order of increase in SiO_2 is also the order in which the melting or crystallization temperature of sam-

TABLE 6.2
Chemical analyses of major oxides of some common rocks
of the Sierra Nevada Batholith
(Compare with Table 6.1)

Oxide	1 Diorite	2 Tonalite	3 Granodiorite	4 Granodiorite	5 Granite
SiO_2	51.5	60.4	65.6	69.8	75.3
Al_2O_3	18.7	17.7	15.4	15.8	13.0
Fe_2O_3	4.80	0.82	1.26	1.25	0.79
FeO	4.20	4.48	3.06	0.94	0.53
MgO	4.90	3.36	1.73	0.65	0.24
CaO	8.80	6.55	4.02	2.46	1.15
Na_2O	3.60	3.36	3.53	3.87	3.11
K_2O	0.78	1.32	3.11	4.72	5.28
H_2O+	0.63	1.15	0.76	0.36	0.22
H_2O-	0.14	0.12	0.17	0.04	0.04
TiO_2	0.42	0.69	0.56	0.40	0.21
P_2O_5	0.16	0.17	0.12	0.16	<0.05
MnO	1.20	0.09	0.08	0.05	0.03
CO_2	0.0	0.17	0.10	0.09	<0.01
Totals	99.9	100.4	99.5	100.6	99.9

1. Diorite (Pyramid pluton), Moore, 1991, (G19) table 1, p. 15; Marion Peak Quad.
2. Tonalite of Blue Canyon, Krauskopf, 1984, (Md5) table 1, p. 14; Mariposa Quad.
3. Giant Forest Granodiorite, Sisson, 1992, (G-1) table 2, p. 15; Triple Divide Peak Quad.
4. Whitney Granodiorite, Moore, 1987, (W-14), table 1, p. 8; Mount Whitney Quad.
5. Granite of Tamarack Lake, Sisson, 1992, (T-3) table 3, p. 15; Triple Divide Peak Quad.

ples decreases, and it is commonly the order of intrusion of closely related plutons as well. The order of increasing SiO_2 is considered to be the order of *differentiation*, meaning the order in which the melt portion of a given magma will change its composition upon cooling and solidification. Cooling will cause silica-poor, high-temperature minerals to crystallize out, leaving a silica-rich, low-temperature granite melt. Again, this is why granite is such a common rock.

A common method of studying a group of chemical analyses is by the use of oxide-SiO$_2$ variation diagrams. In such diagrams the silica concentration (weight percent SiO$_2$) is plotted on the horizontal axis against the concentration of other oxides (Fig. 6.21). As noted above, an increase in silica, signaling a decrease in the melting point of the rock, is regarded as the direction of differentiation. This direction is toward granite in composition. Hence, in an oxide-SiO$_2$ diagram, differentiation occurs toward the right side of the diagram (toward an increase in SiO$_2$).

As SiO$_2$ increases in concentration, most of the other oxides decrease markedly (Fig. 6.21). Na$_2$O remains about equal across all levels of differentiation, and K$_2$O is the only oxide that shows a sharp rise. These changes occur because the magmas crystallize as they cool, and the first minerals to crystallize do not have the same chemical composition as the melt from which they grow. Hornblende, biotite, and calcium-rich plagioclase begin to crystallize and separate from the magmas early, at high temperatures. Hornblende and biotite are rich in FeO, Fe$_2$O$_3$, MgO, and Al$_2$O$_3$ (Table 4.2); hornblende is also rich in CaO, as is calcium-rich plagioclase. Together, the separation of these minerals impoverishes the remaining melt in FeO, Fe$_2$O$_3$, MgO, Al$_2$O$_3$, and CaO. Both hornblende and calcium plagioclase contain appreciable Na$_2$O, but not enough to seriously deplete the melt, and Na$_2$O thus remains relatively constant as the plutonic rocks form. Although biotite is rich in K$_2$O, the amount of biotite that forms is insufficient to reduce the K$_2$O abundance in the remaining melt. Instead, crystallization and separation of hornblende, biotite, and calcium-rich plagioclase enriches the melt in K$_2$O. The minor constituents TiO$_2$ and P$_2$O$_5$ are removed by the crystallization and separation of titanium-rich iron oxide minerals and sphene, and by the Ca-phosphate mineral apatite. All of these early-formed minerals are poorer in SiO$_2$ than the melt from which they form, and their crystallization enriches the melt in SiO$_2$. Finally, the melt becomes so enriched in SiO$_2$ and K$_2$O, and has cooled so much, that all the remaining liquid crystallizes as a mixture of quartz and potassium feldspar crystals.

This process of crystallization leads toward a shift in the melt composition (differentiation) that results in a series of igneous rocks that are progressively enriched in SiO$_2$ and K$_2$O, and progressively depleted in all other major constituents. This systematic change in rock composition reflects crystallization at continuously decreasing temperatures. In this way a series of related plutons—or a single pluton that is internally zoned—beginning with quartz-diorite, progressing through granodiorite, and ending with granite, may be produced by suitable tapping of melt from a single cooling parent magma.

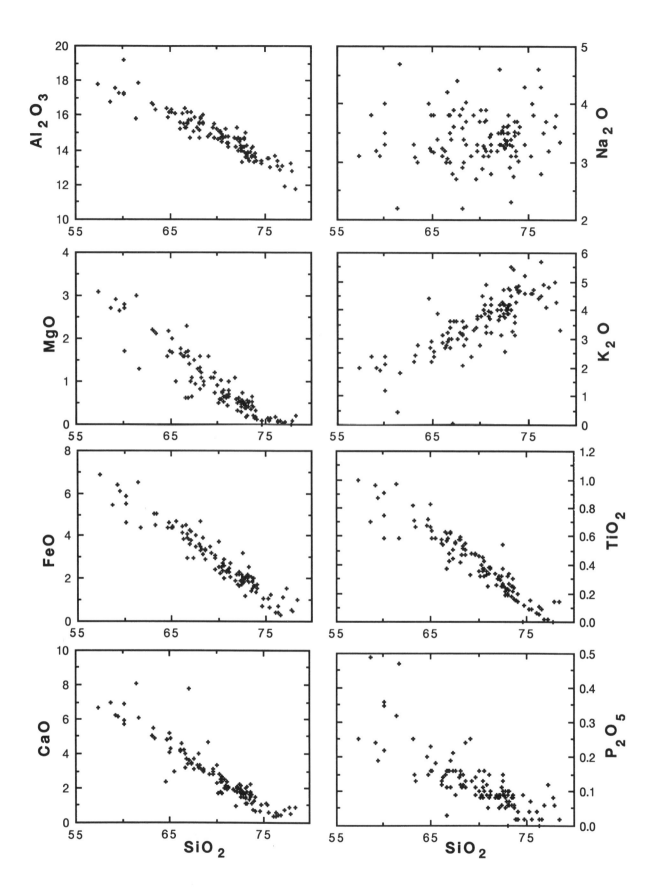

Age measurements show that as one pluton or group of related plutons was cooling and solidifying, new groups of plutons were forming in adjacent areas. This suggests that the supply of heat needed to sustain an active magmatic system shifted with time relative to the crust. Heat is carried to the roots of most actively forming plutons and batholiths by high-temperature magmas originating in the Earth's mantle. These magmas are SiO_2- and K_2O-poor but rich in other major constituents, particularly MgO, FeO, and CaO, and are generally referred to as "basaltic." When the supply of basaltic magma wanes, the supply of heat wanes with it, and the pluton or group of plutons cools and undergoes differentiation. Basaltic magma feeding into the roots of active plutons can also mix with the resident magma, this process temporarily retarding cooling and differentiation. But eventually, when the supply of basaltic magma is cut off or diverted to another place, cooling and differentiation will proceed, producing the SiO_2- and K_2O-rich plutons.

Changes in Composition across the Batholith

Broad-scale geologic mapping of the Sierra Nevada Batholith shows that tonalite (designated as quartz diorite in a former classification) is the dominant rock type in the western part of the batholith, and granodiorite the dominant type in the eastern part. The same trend is exhibited by other major batholiths along the western margin of North America, including the Coast Range Batholith of Canada, the Idaho Batholith, and the Southern California Batholith (Fig. 6.2).

Numerous chemical analyses (and specific gravity measurements) indicate that the most abundant rocks in the western central Sierra contain less than 66 percent SiO_2, whereas those of the eastern Sierra contain from 66 to 71 percent (Fig. 6.22). Plutonic rocks with more than 71 percent SiO_2, although less abundant than the other two groups, are scattered throughout the batholith.

The western part of the batholith, moreover, contains rocks that fall on a different trend line than do those of the eastern part of the batholith. The western rocks, therefore, have a different chemical pattern, through their entire compositional range, than do the eastern rocks. As expressed on the triangular diagram, analysis of the eastern rocks yields a trend line that is closer to the potassium feldspar corner (and therefore richer in potassium; Fig. 4.3). This higher potassium content is the hallmark of the plutonic rocks that occur eastward, more toward the continental interior.

We can ponder this change in chemistry across the batholith by com-

6.21. Diagrams plotting silica (SiO_2) against eight different common oxides for several dozen rock analyses from the central Sierra Nevada, within the parks area. Note that all oxides decrease with increasing SiO_2 except for K_2O, which increases strongly, and Na_2O, which shows no systematic change and much scatter. The direction of differentiation (cooling and increasing SiO_2) is toward the right.

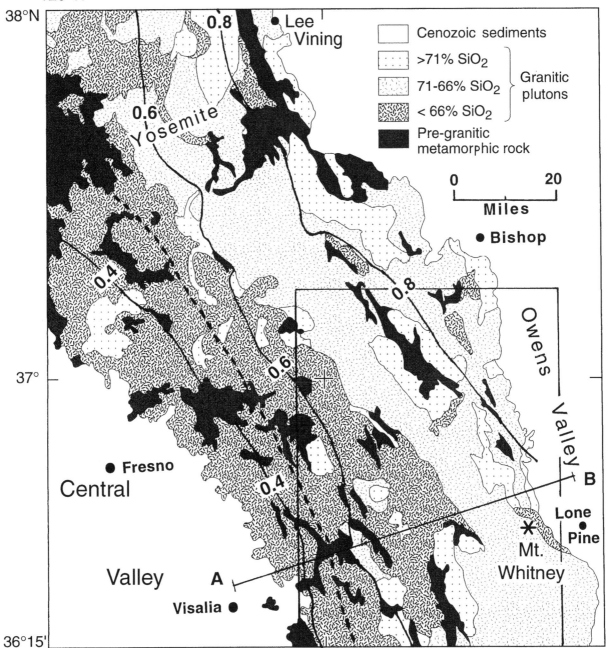

paring oxide-silica diagrams of analyzed samples from the western part of the batholith with those from the eastern part. Such diagrams indicate that the average trend lines for each element show a systematic shift. At a given silica content, all elemental oxides decrease toward the east except for K_2O and P_2O_5, which increase markedly, and Na_2O, total iron oxide (indicated by FeO*), and TiO_2, which show no change. The most extreme changes are those of the increase of K_2O and decrease of CaO toward the east (Fig. 6.23). These easterly changes are independent of the SiO_2 content (or the level of differentiation) because the changes occur at a fixed SiO_2 content.

To define this marked change in CaO and K_2O across the batholith geographically, we have determined the average weight percent ratio of K_2O/CaO at 65 percent SiO_2 for each of 23 15-minute quadrangles across the batholith (Fig. 6.24). This ratio is plotted at the center of each quadrangle and the data contoured on a map (Fig. 6.22) and plotted on a diagram (Fig. 6.24). These contours show a regular increase in the ratio toward the east, and a rather consistent relationship with the four metamorphic terranes (compare Fig. 5.1 with Fig. 6.22). The western King–Kaweah Terrane falls in the zone where the K_2O/CaO ratio at 65 percent SiO_2 of the granitic rocks is less than 0.4; the Kings Terrane falls in the 0.4 to about 0.65 zone; the Goddard Terrane falls in the 0.65 to 0.8 zone; and the High Sierra Terrane along the crest of the range falls in the 0.8 to 1.0 zone. The K_2O/CaO ratio at 65 percent SiO_2 thus more than doubles from west to east across the range (Fig. 6.24).

Many other chemical parameters can be used to quantify changes in composition across the Sierra Nevada Batholith, but I will mention only one other, because it has played a key role in understanding the origin of the batholith. The element strontium, which is present in minor quantities in all rocks, has several isotopes. The strontium isotope ^{87}Sr forms

6.22. Map of the central Sierra Nevada dividing granitic rocks into three general compositional groups: less than 66 weight percent SiO_2, between 66 and 71 percent SiO_2, and greater than 71 percent SiO_2. Solid lines indicate averaged K_2O/CaO ratio of granitic rocks normalized to 65 percent SiO_2 across the batholith (compare with Figs. 6.23 and 6.24). The dashed line is the position of the initial $^{87}Sr/^{86}Sr$ ratio of 0.706, which increases toward the east. These trends support the notion that the batholith formed on the western margin of the continental crust, and successively more continental crust was incorporated into the granitic magmas from west to east. The large box at the southeast indicates the area of Fig. 6.4.

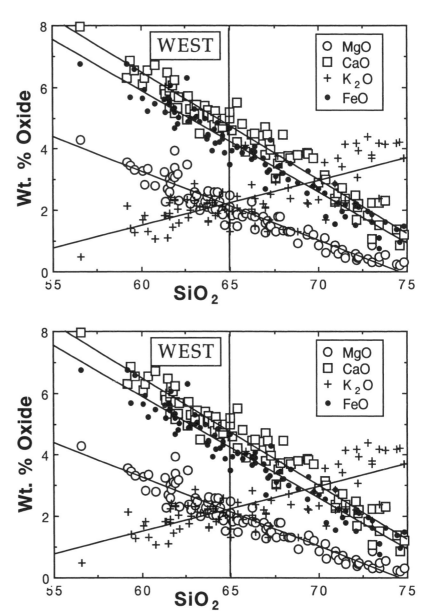

6.23. Diagrams plotting SiO_2 against four oxides in analyzed Sierra Nevada granitic rocks. *Above.* Rocks from the western foothills of the central Sierra. *Below.* Rocks from the eastern Sierra. The trend lines are displaced such that FeO★, CaO, and MgO are lower in the east and K_2O is higher in the east. This change is independent of the level of differentiation, because if we compare the oxides only at a fixed SiO_2 content, for example 65 percent, then the compositional shifts from west to east can be compared quantitatively.

from the radioactive decay of a rubidium isotope, ^{87}Rb (with a half-life of 47 billion years), in the same way that the argon isotope ^{40}Ar forms from the radioactive potassium isotope ^{40}K (half-life of 1.3 billion years; Table 4.1).

The strontium isotope ^{86}Sr is not produced by the radioactive decay of any other element, and hence the ratio $^{87}Sr/^{86}Sr$ reflects the long-term relative abundance of rubidium and strontium. Rubidium is an element similar to the element potassium, and both Rb and K are abundant in

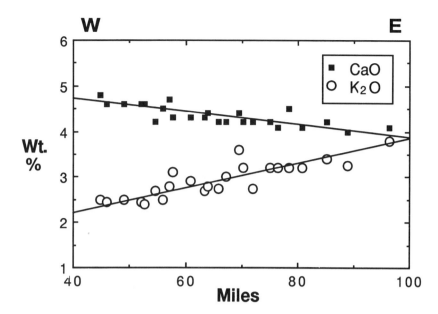

6.24. Average weight percent of CaO and K₂O, at 65 percent SiO₂, of rock samples within 24 15-minute quadrangles of the central Sierra Nevada, plotted against the distance across the batholith of the centers of the quadrangles. The average trend lines show the general decrease of calcium and the increase of potassium toward the east across the batholith among rocks at the same level of differentiation (same silica content; see Fig. 23).

most rocks of the Earth's continents and some parts of the underlying upper mantle (together termed the continental lithosphere). As a consequence of their relatively Rb-rich character, many old rocks of the continental lithosphere have high $^{87}Sr/^{86}Sr$ values of 0.705 to 0.725. In contrast, basaltic rocks of oceanic lithosphere have ratios of from 0.701 to 0.704. Ratios above 0.706 are commonly used as indicators for the presence of continental material in magma producing an igneous rock.

When rock of the crust or mantle is partly melted, its $^{87}Sr/^{86}Sr$ value is imparted to the magma. The magma may ascend long distances from its source, but its $^{87}Sr/^{86}Sr$ value provides a record of the types of rocks that it formed from. After the magma solidifies, additional ^{87}Sr forms slowly from ^{87}Rb in the rock, but the initial $^{87}Sr/^{86}Sr$ value of the magma at the time of crystallization can be recovered by making a small correction for this later production of ^{87}Sr. The initial $^{87}Sr/^{86}Sr$ value indicates whether a pluton or group of plutons formed mainly by the melting of old rocks of the continental lithosphere or from the melting of rocks of the oceanic crust and oceanic upper mantle.

The initial $^{87}Sr/^{86}Sr$ of granitic rocks increases systematically from west to east across the Sierra Nevada Batholith in a manner similar to the eastward increase in K₂O content and K₂O/CaO ratio. The initial $^{87}Sr/^{86}Sr$ value of 0.706, plotted on the map of the central Sierra Nevada (Fig. 6.22), is located about in the middle of the Kings Terrane of metamorphic rocks. This eastward increase in initial $^{87}Sr/^{86}Sr$ has been interpreted as resulting from an eastward increase in the proportion of old

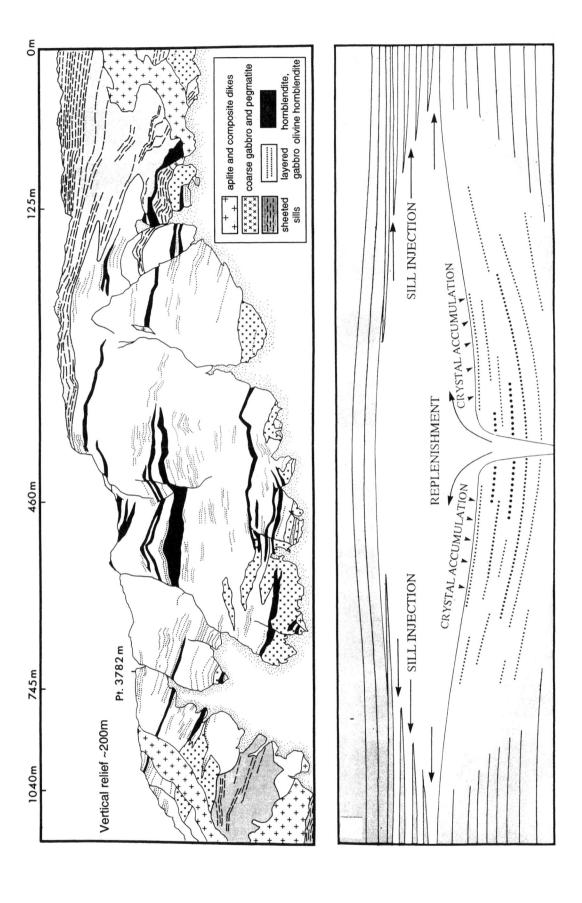

continental lithosphere in the source regions that melted to form the plutons. The mapped line marking an initial $^{87}Sr/^{86}Sr$ of 0.706 has been taken as marking the western edge of the old North American continent at the time the Sierra Nevada Batholith formed (Kistler, 1990).

Mafic Plutonic Rocks

Scattered through the Sierra Nevada Batholith are small masses of dark mafic plutonic rocks (Fig. 6.4). They are composed primarily of hornblende diorite and gabbro, and they generally occur in masses less than a few miles across, collectively comprising only a small part of the plutonic rocks. They are important, however, for they represent deep-seated mafic or basaltic magmas from the mantle that played a role in the formation of the batholith.

One of these mafic plutonic complexes near Onion Valley, west of Independence, has been studied in detail (Sisson and others, 1996). There, the complex is partly composed of a series of flat intrusive bodies (sills), stacked one upon another, that are composed of hornblende gabbro and diorite, as well as other, related rock types (Fig. 6.25). Within this sequence are dark, coarse-grained, hornblende-rich layers representing an accumulation of heavy crystals that sank to the bottom of these flat chambers while the magma was still molten.

The mafic complex at Onion Valley is the remnant of a long-lived chamber of mafic magma that was intruded into a series of granitic plutons already in place. This chamber, repeatedly replenished from below, solidified chiefly on top, allowing the early-formed crystals (mainly hornblende and plagioclase) to "rain down" to form a sludge on the bottom—a cumulate zone.

6.25. *(Broadside) Above.* Sketch showing mafic rocks exposed on a north-facing cliff 1 mile north of Kearsarge Pass in the Onion Valley Mafic Complex. Layers of heavy cumulate crystals (black pattern; rich in hornblende and olivine crystals) are overlain and underlain by numerous sills. *Below.* Diagram showing the processes that apparently produced the rock layers above. A mafic magma chamber (unpatterned) intruded into a swarm of sills that had solidified somewhat earlier (gray with lines). As the magma chamber cooled and solidified it was also replenished by repeated episodes of intrusion of new magma. Cooling caused heavy crystals to grow, sink to the bottom of the magma chamber, and accumulate as layers rich in hornblende and olivine (dotted lines) (Sisson and others, 1996).

One puzzling question is, how was the mafic magma, which is normally denser than granitic melt, able to rise buoyantly up into the sequence of granitic plutons? The answer is probably that the mafic melt was extremely water-rich, containing 4 to 6 percent water, and this water content reduced its density sufficiently that it floated up by buoyant forces. Perhaps other mafic magmas also lay at depth, but the dryer ones were too dense to ascend.

The chemical composition of these mafic rocks and that of the much more abundant granodiorite of the Sierra Nevada Batholith both indicate that the granodiorite formed by a mixing of the mantle-derived basaltic magma with silicic melts of granite composition. This silicic melt formed in part by the melting of old crust, the deep-sourced basaltic magma providing the heat.

The Independence Dike Swarm

6.26. *(Opposite)* Part of the eastern Sierra Nevada and desert ranges to the east, showing the extent of the Independence Dike Swarm, indicated by the many short dashes (modified from Smith, 1962). The box at the upper left is the area of the map of Fig. 6.4.

6.27. *(Above)* Dozens of dark-colored Independence Dikes cutting light granite south of Taboose Pass at the Sierra Crest.

Dark mafic dikes, scattered throughout the parks area, cut both plutonic and metamorphic rocks. Some seem to occur where masses of dark dioritic rocks were partly melted and remobilized by the intrusion of a large neighboring granitic pluton. A more extensive swarm of mafic dikes, however, which Cliff Hopson and I named the Independence Dike Swarm, occurs over a broad area, is generally steeply dipping, and trends toward the northwest. We found that it was convenient to divide granitic plutons in the Sierra into two groups: (1) older pre-dike plutons cut later by the dikes and (2) younger post-dike plutons that contained no dikes and in places cut off the dikes and their host plutons (Moore and Hopson, 1961).

This swarm of dark dikes crosses the area of the parks and can be traced south for a distance of more than 150 miles (Fig. 6.26). They are magnificently exposed in the high glaciated country, where the black dikes contrast starkly with the white granite (Figs. 6.27 and 6.28). South-

6.28. The Independence Dikes, some attaining widths of nearly 100 feet, on the peak south of Hell for Sure Pass over the divide between the North Fork of the Kings River and the South Fork of the San Joaquin River.

east of the parks area the swarm crosses Owens Valley near the town of Independence and appears in the Alabama Hills west of Lone Pine. It continues southeast across the Inyo Mountains, Argus Range, and Spangler Hills to butt against the Garlock Fault. Across the fault a matching swarm of dikes continues another 7 miles in the Granite Mountains, and clusters of similar dikes lie to the southwest in the Mojave Desert.

The groups of dikes on the two sides of the Garlock Fault are similar in composition, and both are about 12 miles wide, but they are offset 40 miles. George Smith (1962) proposed that these swarms were originally continuous, and that the offset was caused by 40 miles of movement along the Garlock Fault after emplacement of the dikes, a sense of movement termed *right-lateral strike-slip* (see Chapter 11).

The dikes are composed primarily of diorite and dark granodiorite, but in many places the swarm includes granite dikes. The dikes range greatly in thickness (Figs. 6.27 and 6.28), from less than 1 inch to more than 30 feet. In Woods Lake Basin, I measured the width of 238 dikes along a single line and found that the average thickness was between 1 and 2 feet (Moore, 1963).

Radiometric dating indicates that most of the dikes are of Late Jurassic age—about 148 million years old (Chen and Moore, 1979). We have used the dikes to sort out the older Jurassic plutons from the more abundant Cretaceous plutons.

Cretaceous mafic dikes, however, are found in several places within the parks area (Carl and others, 1998). The Cretaceous age is verified where the dikes cut plutons that have themselves been dated as Cretaceous. Commonly, these Cretaceous mafic dikes occur near bodies of mafic plutonic rock, and they may have originated from such masses. In some places, the Cretaceous dikes occur within a mass of Jurassic dikes, and there the only positive method of distinguishing them is by expensive and laborious radiometric dating. But in general, the Cretaceous dikes seem more local, and they contrast with the widespread distribution of the regional northwest-trending swarm of Jurassic dikes.

The Jurassic Independence Dike Swarm, which was emplaced during a short time span about 148 million years ago, points to some regional processes. The dikes were emplaced at the end of the period of intrusion of the Jurassic plutons. The change in the direction and nature of plate motion that ended this epoch of granite-making may have generated stresses that caused fracturing of the lithosphere and permitted the intrusion of mafic magmas from depth to produce the Independence Dikes.

Plate Tectonics and the Batholith

The global Mesozoic plate-tectonic setting bears directly on the evolution of western North America and the Sierra Nevada Batholith. At the beginning of the Cretaceous Period (144 million years ago), a giant Farallon Plate occupied the site of the present northeastern Pacific Ocean. The dominant process at the time was the subduction of the oceanic lithosphere of the Farallon Plate beneath the entire western margin of the North American Continental Plate (Fig. 6.3).

The integral processes of subduction and magma genesis cannot, of course, be directly observed, and are the subjects of heated debate, but the following is a model that is consistent with many of the features of the Sierra Nevada Batholith (Fig. 6.29). As subduction proceeded, fragments and slivers were scraped off the downgoing Farallon Plate to be accreted onto the front of the overriding North American Plate. Some of this material of oceanic affinity was thrust on top of the continental lithosphere and some was carried down to underplate it. As the greater breadth of the Farallon slab went down into the mantle and beneath the overlying plate, it was heated. The heating became critical at a depth of about 60 miles and caused water and other volatile constituents to be baked out of the lithosphere and to rise. These volatiles have the propensity to lower the melting point of silicate minerals, causing rock to melt.

As a result of this fluxing by volatiles, the uppermost rigid mantle above the downgoing oceanic slab—which was already sufficiently hot—began to melt, and the initial product of mantle melting was basaltic

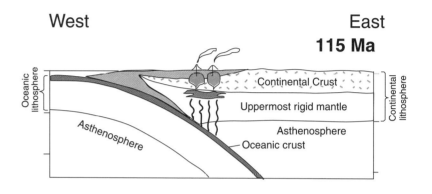

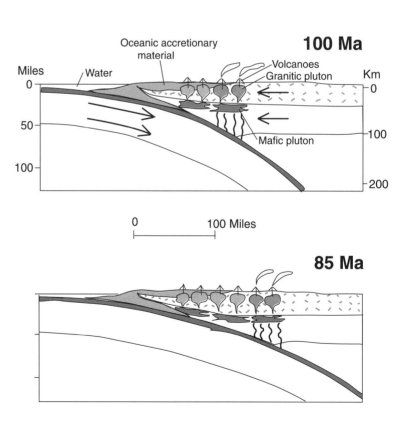

magma. This magma accumulated near the base of the continental crust, and formed pools that continued to rise and to melt the granitic rock of the continental crust, which has a considerably lower melting temperature than that of the basalt. A variety of magmas were produced by the processes of partial melting, cooling-induced differentiation, and mixing of basaltic and granitic magmas. Eventually, buoyant masses of granitic magma rose within the crust and formed magma chambers below the surface. Some of these chambers then vented to the surface and fed volcanoes, and eventually cooled below the surface to form the granitic plutons of the Sierra Nevada Batholith.

The presence of the accretionary wedges—partly composed of oceanic crustal slices—on the western edge of the North American Plate modified the continental crust, leaving it increasingly oceanic (basaltic in composition) toward the west. The granitic magmas that then formed by melting in the upper mantle and crust reflect this preexisting compositional difference. The granitic plutons on the west became more oceanic in composition than those on the east (Fig. 6.29). The quartz diorite line (Fig. 6.2), the generally more silicic character of the plutons toward the east (Figs. 6.4 and 6.22), the increasing K_2O/CaO ratio at 65% SiO_2, and of initial $^{87}Sr/^{86}Sr$ ratio (Fig. 6.22) all result from this preexisting chemical gradient across the edge of the lithosphere. Hence, the compositional change across the batholith was inherited from the wholesale melting and incorporation of parent rocks of differing composition.

The general age change of the Cretaceous plutons indicates that granite-making and intrusion migrated eastward across the batholith at a steady rate of about 1.7 miles per million years (Fig. 6.18). Some geolo-

6.29. Diagram showing three stages in the development of the Sierra Nevada Batholith, at 115, 100, and 85 million years ago (Ma), and the movement (subduction) of the oceanic lithosphere of the Farallon Plate under the continental lithosphere of the North American Plate. Magma chambers (color) form above the downgoing slab of oceanic lithosphere as water baked out of the dehydrating slab rises and promotes melting. Flattening of the subduction zone over time causes the locus of magma-making (and the growth of the batholith) to move toward the east as the depth of dehydration is maintained at about 60 miles. The thickening of the continental crust toward the east causes more of the batholith to be formed by the melting of continental material; this leads toward enrichment in silica and potassium (elements that are enriched in continental crust). Modified from Tobisch and others (1995) and Hamilton (1995).

gists believe that this migration can be attributed to a flattening of the subducting slab as it ages (Tobisch and others, 1995). Such flattening may have been caused by movement of the North American Plate to the west over the top of the subduction system.

As the subducting slab flattened, it moved farther inland—relative to the overlying plate—before it dove down into the mantle. This in turn moved the 60-mile-deep slab-dehydration zone (and thus the region of upper-plate melting) eastward, accounting for the plutons becoming systematically younger toward the east (Fig. 6.29).

The controlling process in western North America, and in California in particular, during the Cretaceous Period was the eastward sliding of oceanic lithosphere down beneath the continental lithosphere (subduction) along a deep-ocean trench like that off present-day Japan or Peru. Sediments and fragments of oceanic crust scraped off from the downgoing lithospheric slab accumulated along the landward side of the trench and formed a chain of small submarine mountains and perhaps some islands. These scraped-off rocks now form much of the Coast Range of California. Sand and mud eroded from the continent accumulated behind these submarine mountains in a narrow ocean basin where the Central Valley of California now lies. To the east of this sea was a chain of volcanoes that lined the edge of North America, in the fashion of the chain of active volcanoes that now forms the Cascade Range in Oregon and Washington. These volcanoes stood above and were fed by the actively forming Sierra Nevada Batholith. But as time passed, the volcanoes on the west ceased producing as their underlying plutons solidified, and new volcanoes formed farther to the east as the subducting plate flattened, causing the site of magma generation to migrate farther to the east. At about 80 million years ago, the subducting slab shallowed considerably and magmatism swept far to the east, leaving the area of the present-day Sierra Nevada and bringing an end to the development of the eastward-growing continuous batholith. Eventually, the youngest Sierra plutons—80 million years old—solidified and the Sierra Nevada Batholith was complete. Erosion initiated by later regional uplift stripped away the overlying rocks to expose the batholith.

Gold! We leapt from our benches. Gold! We sprang from our stools.
Gold! We wheeled in the furrow, fired with the faith of fools.
— *"The Trail of Ninety-Eight,"* ROBERT SERVICE (1874–1958)

MINERAL DEPOSITS

Prospectors poking about in the high country soon discovered that the gray to white granitic rock that underlies the great bulk of the southern Sierra Nevada is largely barren of mineral deposits. The series of gold-bearing quartz veins of the Mother Lode, in the northern Sierra, was the magnet that had drawn the Forty-Niners to California. The Mother Lode occurs in a broad belt of metamorphic rock in the Sierra foothills. This rock, older than the granitic rock, served as the walls of the chambers that hosted the hot intruding granitic magma. Metallic minerals, which were distilled from the cooling granite melts and mobilized from the wall rocks by magmatic heat, were deposited where the hot fluids that had transported them cooled, as the fluids streamed through fissures in the wall rock. Prospectors therefore concentrated their efforts on the dark reddish terrain identifying the metamorphic rocks, and on the sediments derived from the erosion of such rocks.

The Sequoia and Kings Canyon National Parks area contains no regional belt of metamorphic rock where large vein systems could have been emplaced. Previous Mother Lodes may have formed in the area, but if so, they have since eroded away as the range was uplifted. The smaller masses of metamorphic rock that do exist (the roof pendants), surrounded by granitic plutons, do host some mineral deposits. Most of

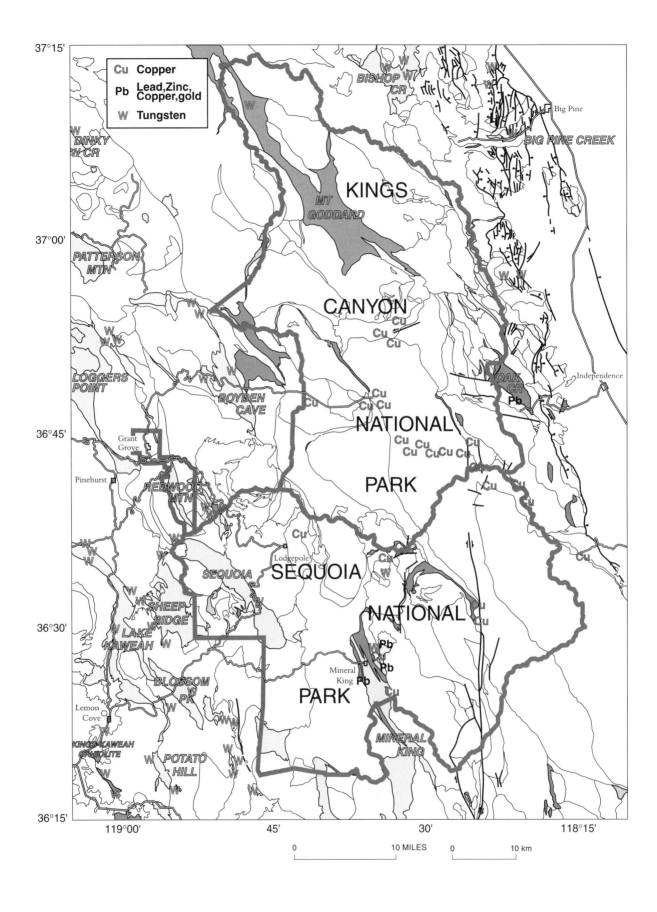

these were prospected, claimed, and explored by surface or underground workings very early. A few have indeed produced ore, but generally the deposits are small.

Several mines discovered east of the Sierra Nevada in the early 1860s stimulated trail building across the Sierra Nevada just south of the region of the highest Sierra. The first of these was the Coso silver mine, discovered southeast of Owens Lake in 1860 by Dr. Darwin French of Visalia. Soon other claims were staked in the Inyo Mountains, and a camp known as San Carlos was established northeast of Independence in 1862. The most productive of these mining camps was Cerro Gordo, east of Owens Lake, where deposits were discovered in 1865. This mine produced lead, silver, and zinc valued at $17 million. So that Visalia could compete with Los Angeles for trade with these camps, a trail from Visalia across the range, the Jordan Trail, was constructed in 1861. The next year a slightly shorter trans-Sierra toll trail, the Hockett Trail, was opened.

George Wheeler's 1871 map (Fig. 3.8) shows that numerous mining districts had been staked out east of the Sierra Nevada, but within the eastern Sierra, at that time, only the Fish Spring and Kearsarge Districts were known to hold deposits. By 1891, the map of the State Mineralogist (Irelan, 1891; Fig. 2.37) showed only eight known mineral deposits in the parks area. The most important were the deposits in the Kearsarge and Mineral King Districts.

The mineral deposits in the region (Fig. 7.1) consist of three general types: quartz veins containing gold, silver, lead, and zinc; mineralized shear zones containing copper and molybdenum; and calcium silicate deposits containing tungsten.

The Kearsarge District

Perhaps the first mineral deposits found in the parks region were the gold-and-silver deposits high on Kearsarge Peak on the east slope of the Sierra. They were discovered in the fall of 1864 by a party of prospectors, including Thomas Keough of Owens Valley, who built a trail up the east side of the Sierra to a pass then called Little Pine Pass, because it was at the head of Little Pine Creek (now called Independence Creek; Colby, 1918). Most of the group then crossed the pass, and while going west down into the canyon of the South Fork of the Kings River, they met the Brewer Party, which had about completed its monumental investigations in the Kings and Kern River drainages. The prospectors directed the survey party east over the Sierra pass they had just crossed.

7.1. Relationship of the three principal types of mineral deposits to granitic rocks (most of the uncolored area), metavolcanic rocks (dark color), and metasedimentary rocks (light color). Tungsten (Wolfram) deposits occur mainly at boundaries where granitic melt came into contact with limy sedimentary wall rocks.

7.2. Mill at the Kearsarge Mine on Independence Creek, as it appeared in 1875. Photograph by T. H. O'Sullivan (Wheeler, 1889).

The remaining prospectors, those who had not crossed the pass, explored the high country north of the pass and found gold-bearing quartz veins at about 10,000 feet. This gold discovery was made during the Civil War and the group were all Union men. They named the mine and district *Kearsarge* to honor the Union battleship that had just sunk the Confederate privateer *Alabama*. (The Alabama Hills, west of Lone Pine, had previously been named by Confederate sympathizers.) Later, the peak on which the mines were located, and the pass, took on the name Kearsarge.

The ore there is found in a series of northwest-dipping veins from a few inches to several feet thick that cut both granitic rock and metamorphosed volcanic rocks in the south end of the Oak Creek Pendant. These veins contain abundant pyrite, or fool's gold (FeS_2), and a gold- and silver-bearing lead sulfide mineral (galena, PbS). In the early days of the district, some of these veins yielded considerable quantities of ore, rich in both gold and silver.

A small settlement grew up along Independence Creek as the production from the mines increased, and three stamp mills soon were in operation, crushing the quartz-rich ore (Fig. 7.2). One 10-ton batch of ore is said to have been valued at $900 a ton. In 1867 a snow avalanche destroyed 11 cabins and killed one person on the slopes of Kearsarge Peak, but activity continued in the district until about 1870. By 1888, the town had disappeared, and the mines were abandoned, except for a single miner diligently working his claim (Goodyear, 1888).

The Mineral King Bust

The first person to enter the remote Mineral King area, in the headwaters of the East Fork of the Kaweah River, was apparently John O'Farrell (also known as Harry Parole), who stumbled on the valley in 1864. O'Farrell was a hunter who had been hired by the U.S. government to furnish meat for the crew building the Hockett Trail. This trail, passing over the Sierra, was built to supply Fort Independence, a U.S. Army camp established July 4, 1862, as well as the Coso Mines, where silver was discovered in 1860. Something of a prospector himself, he continued visiting the valley for many years, and noted evidence of mineralization in the abundant metamorphic rock, but did not file a claim until the fall of 1873.

The Mineral King gold-silver deposit occurs in one of the area's larger masses of metamorphic rocks, the Mineral King Roof Pendant. As this mining venture developed, it came to have a major impact on the history and development of the parks area, even though the district never produced a ton of ore for profit. The ore occurs primarily as poorly developed veins containing iron, zinc, lead, and arsenic sulfide, along with disseminated gold. The lead sulfide mineral (galena) contains some silver. The discovery of these deposits is credited to James Crabtree, who had settled in the White River area near Porterville and prospected in his free time. He made the first silver discovery at Mineral King in July 1873 and called his prospect the White Chief Mine, after a vision of an Indian who appeared in a dream and guided him to the place. He went back to Porterville, recruited five companions, and returned to the area to stake claims and form a mining district.

By the fall of 1873, 65 claims had been filed by 93 prospectors, including John O'Farrell. Ore from many of the diggings was sent out for assays, with reported values as high as $445 per ton (Porter, 1965). A rough trail was built to the mines up the canyon of the East Fork of the Kaweah, and many called for the construction of a wagon road. It is said that some miners produced bullion by smelting hand-picked ore by the "frying pan process," in which charcoal in a frying pan was covered by crushed ore. A blacksmith's bellows directed an air stream onto the burning charcoal, heating the ore until it was glowing, hot enough for the charcoal to reduce the ore and produce a mass of metallic silver.

In 1875 a stock company, The New England Tunnel and Smelting Company, acquired many of the mining claims, on a shared basis. Assay equipment was set up, and a furnace smelter was built to reduce the ores. The ores proved "rebellious," however, and for a variety of reasons, including a lack of high-grade ore, no bullion was produced. Nonetheless, in 1876 work began in earnest on the road up the Kaweah Canyon, and by March a bridge was completed across the East Fork of the Kaweah River. Financial troubles, however, beset the company, and it became necessary to assess the original claim owners. Finally, in September 1877, the company went bankrupt. During the following winter, a snow avalanche struck the company barracks with great force, causing extensive damage and injuring several miners. Amazingly, no one was killed.

In the fall of 1878 Thomas Fowler, a California State Senator from Visalia, bought the Empire Mine and formed the Empire Gold and Silver Mining Company (Fig. 7.3). The mining went forward, and a series of large natural caves were intercepted underground. In the summer of 1879, a 2-mile-long cable tram was built to carry ore down off Empire Mountain to the newly constructed mill 2,000 feet below.

The ore was first fed through rock crushers, then onto a stamp mill, where it was pulverized by fifteen 1,000-pound stamps, and then to shakers that separated the lighter, worthless material from the ore. From there it went through wooden troughs where freed gold and silver were passed over mercury, with which they combined, yielding an alloy with the mercury, called amalgam. The amalgam was put in buckskin bags, the excess mercury squeezed off, and the remaining material taken to a retort room, where it was heated to drive off the remaining mercury, leaving behind gold and silver sponge.

In the spring of 1879, Fowler organized the Mineral King Wagon and Toll Road Company to complete the remaining 25 miles of road from the bridge across the East Fork of the Kaweah River (known then as Toll Gate or Toll Bridge; Figs. 2.37 and 2.38) to Mineral King. The road was completed in August. Tolls were: one man, 50 cents; wagon and one span of animals, 2 dollars; pack animals, 25 cents each; loose horses, 10 cents each; sheep, hogs, and goats, 10 cents each; buggy drawn by one horse, 1 dollar and 50 cents; additional trail wagons, 50 cents each.

In early 1880 Fowler was rarely seen in Tulare County. In April his

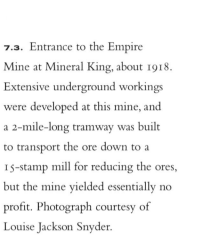

7.3. Entrance to the Empire Mine at Mineral King, about 1918. Extensive underground workings were developed at this mine, and a 2-mile-long tramway was built to transport the ore down to a 15-stamp mill for reducing the ores, but the mine yielded essentially no profit. Photograph courtesy of Louise Jackson Snyder.

companies were reorganized, and creditors were asked to register. During the spring of 1880, exceptionally heavy snow built up in the valley, and the roofs of many buildings collapsed. Finally, on April 18, a mammoth snowslide carried away the miners' boardinghouse at the Empire Mine. Again, there were no fatalities, but several people were injured.

The last run of the Empire Mill would be in July 1880, a run that produced bullion reportedly worth about $1,000, but even that was reduced when the bullion was found to have a high content of lead. Fowler continued to rebuild the mine buildings each spring, but his efforts became steadily feebler, and by 1882 little activity was to be found, anywhere in the district. Thomas Fowler died in April 1884.

Copper and Molybdenum Deposits

Many small shear zones in granitic rocks are mineralized by quartz veins that contain sulfides of iron, copper, and molybdenum. In the parks area, these are the only mineral deposits that are not associated with metamorphic rocks. At the surface, the copper minerals oxidize to form a prominent blue-green or turquoise stain, and in short order this staining led prospectors to many of these veins despite the old prospector's adage that "a penny's worth of copper can stain a mountainside." Some of the veins contained a small amount of gold.

In 1890 Joseph N. LeConte visited a copper prospect in Kings Canyon that was being actively mined at the time. The mine lies near where the creek now called Copper Creek joins the South Fork of the Kings River. "The ledge consisted of a fine vein of clean white quartz, thickly spotted with copper ore (bornite), and some silver and gold," LeConte wrote. He went on to say,

> The sunlight streaming up the canyon fell full on the ledge, reflecting from these metallic particles in splendid iridescent colors. The vein was about 12 feet wide, and must be very extensive, since the croppings could be distinctly seen on the inaccessible face of the opposite cañon wall ¾ of a mile away. They had driven a tunnel in 60 feet simply for the purpose of holding their claim. . . . Champion said that he had taken 80 lbs. of the rock down to the reduction works and got $80 a ton out of it.

This means he got $3.20 for mining the ore and packing 80 pounds of it out over a rugged trail for two days.

This prospect is one of several small occurrences of copper and iron sulfide (chalcopyrite; $CuFeS_2$) associated with molybdenum sulfide (molybdenite, MoS_2) in quartz-pyrite veins that extend over a broad area (Fig. 7.1). The molybdenite is a distinctive, soft, greenish-gray mineral with a metallic luster. These deposits occur in quartz veins that occupy late-stage fractures and shears cutting granitic rocks. The silica and sulfide minerals of the veins were apparently precipitated from hot fluids originating from the still molten—but crystallizing—core of the granitic plutonic mass. Weathering of copper in one of these small deposits has left a prominent copper stain high on the east-facing wall of Grand Sentinel. It can be seen from the road turnaround near Copper Creek.

The Tungsten Mines

Tungsten (also called wolfram; chemical symbol W) is a very heavy element associated with gold, silver, copper, and molybdenum. It is commonly found concentrated in zones of altered rock at the contact between granite intrusions and metamorphosed limy sedimentary rocks such as marble, and this environment is the key prospecting site. The heat and fluids of the granitic melt react with the calcium carbonate wall rock and fix the metallic elements in the host rock. The ore-mineral scheelite, a calcium tungstate ($CaWO_4$) that contains the tungsten, is usually found in a distinctive garnet-bearing metamorphic rock called *tactite*, formed by the interaction of limy rocks with the fluids expelled from the granitic melt. Key minerals in tactite include garnet, diopside, epidote, and idocrase. Small tungsten deposits (see Fig. 7.1) are found at many places, within and adjacent to the parks, where calcium carbonate metamorphic rocks in the roof pendants were invaded by granitic magmas. Most of the deposits are associated with the limy rocks of the Kings Terrane, some with the rocks of the High Sierra Terrane.

In 1916, a major deposit of this kind, rich in scheelite, was discovered near Pine Creek, west of Bishop and slightly north of Kings Canyon National Park. The mine occurs in the Bishop Creek Pendant (Fig. 7.1), in association with limy metamorphic rocks. This mine, which has been one of the world's largest tungsten producers, shut down in the early 1990s because of a declining market for tungsten. The mine still contains about 1.5 million tons of mineralized rock.

Scheelite is yellowish white or brown and difficult to distinguish from other common minerals. But it fluoresces with a distinctive blue color under ultraviolet light, and this property has proved to be an excellent prospecting tool. When the price of tungsten went up during the Second World War, many small tungsten deposits were prospected adjacent to limy roof pendants in the parks region, on both sides of the Sierra Crest.

The Garnet Dike Mine, 6 miles south of Patterson Mountain and about 1 mile north of the Kings River at an altitude of 3,000 feet, is a small tungsten mine that was active in the mid-1940s. The mine, like many of the Sierra tungsten mines, was never a large producer. Up to the end of 1945, 6,653 units of WO_3 (tungsten oxide) had been produced (one unit of WO_3 equals 20 pounds of WO_3) (Krauskopf, 1953).

The deposit is at the northern tip of a northwest-trending septum of marble about a mile long and one-fourth mile wide. The marble is surrounded and intruded by a hornblende-biotite granodiorite that in places is intimately mixed with metamorphic rocks. Interaction between the granodiorite and the marble has produced a thin layer of garnet-rich rock along the west and north margin of the marble. The ore in this layer consists mainly of the minerals garnet, pyroxene, epidote, quartz, and calcite.

Both the metamorphic and the granitic rocks are cut by a series of nearly horizontal dikes of aplite and pegmatite. The ore is concentrated both above and below these dikes. In places, large masses of ore follow the dikes away from the main granodiorite contact. The tungsten-bearing minerals are distributed as small grains throughout the ore, which is considered good if it contains more than 0.5 weight percent WO_3.

Several other small tungsten mines were active during the Second World War. The Tulare County Tungsten Mine, located near the south end of the Potato Hill Pendant in the southwest corner of the mapped area, was operated from 1942 through 1945. The mill processed about 40 tons of ore per day.

The Division Creek Mine, on the east side of the Sierra in Armstrong Canyon, lies near a boundary between metamorphic rock (mainly marble and tactite) and a complex of mafic plutonic rocks cut by granitic rocks. Tungsten production has been modest. Thirty tons of ore were shipped through 1941, and from 1952 to 1956 an additional 40 tons of ore were shipped.

Fires that shook me once, but now to silent ashes fall'n away

Cold upon the dead volcano sleeps the gleam of dying day.

—ALFRED TENNYSON (1809–1892)

CENOZOIC VOLCANIC ROCKS

From the time of the earliest explorers, young volcanic rocks were recognized in the Sequoia and Kings Canyon National Parks area. The basaltic cinder cones of the Big Pine Volcanic Field, at the foot of the Sierra in Owens Valley, are labeled as volcanoes on the map of the 1864 Brewer expedition prepared by Charles Hoffmann (1873). They do not appear, however, on the Goddard map of 1857. The 1883 map of the southern Sierra drawn by J. W. A. Wright (Fig. 2.38) shows the lava field and volcanoes (cinder cones) of Golden Trout Creek, called Volcano Creek on the map. This map also indicates "Basalt columns" where the eroded lava flows of Golden Trout Creek crop out on the east wall of Kern Canyon (Fig. 8.6).

Several distinct episodes of volcanic activity have deposited lava and ash in the parks region in the past few million years. Although the area covered by the volcanic rocks is small (Fig. 8.1), the deposits tell us something about the magma source, the age of the flows, and the nature of the conduits that brought the lava to the surface.

Volcanic vents erupted small basaltic lava flows along the base of the east escarpment of the Sierra Nevada. This lava apparently fed upward along the faults that were responsible for the uplift and tilting of the range. The group includes the Big Pine Volcanic Field in Owens Valley, east of Kings Canyon National Park, the Long Valley Volcanic Field east

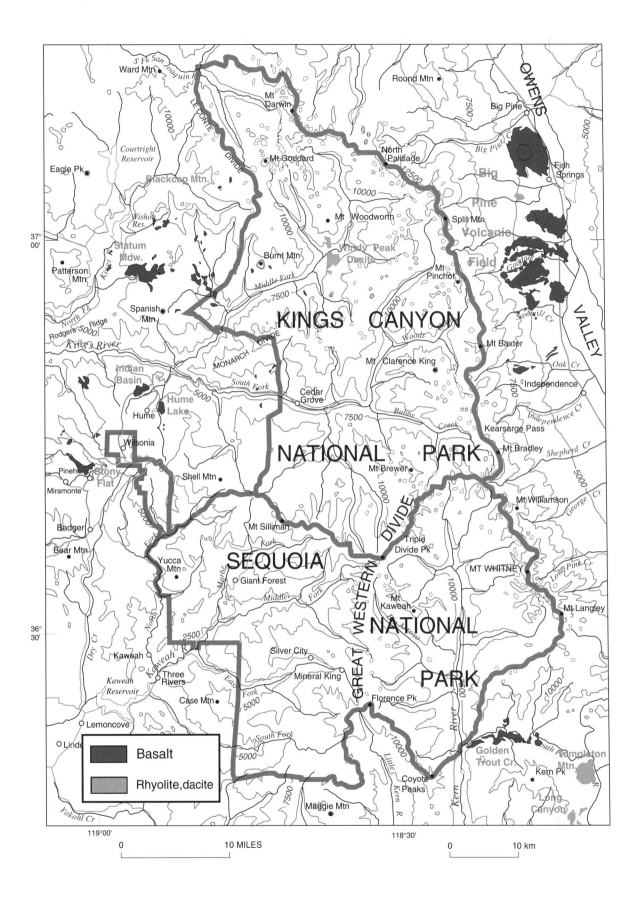

of Yosemite National Park, and the Coso Volcanic Field east of Sequoia National Park. The Big Pine Volcanic Field includes more than a dozen well-preserved cinder cones and associated lava flows. Small remnants of lava flows occur in two drainages of the east escarpment, those of Sawmill Canyon and Oak Creek. Scattered remnants of flows and vents also lie in the middle to upper reaches of the range west of the crest.

The Cenozoic volcanic rocks of the parks area erupted in three phases: (1) an early phase, chiefly of basalt about 4.5 to 2.5 million years ago, (2) a second phase chiefly of rhyolite, 2.5 to about 0.8 million years ago, and (3) a phase chiefly of basalt, from 0.8 million years ago to recent times. The composition and distribution of the lava suggest the following causes for these phases: first, a period of warping that caused widespread, but minor, cracking in the Sierra block, allowing small amounts of tapped primitive magmas from the upper mantle to rise to the surface; second, a period of compression (or at least reduced tension) that permitted ponding of magma within the lower crust and upper mantle, which then melted continental rocks, producing silicic-rich melts; and third, a period of renewed tension and block faulting that again tapped basaltic melts from great depth.

The Initial Volcanic Phase

The volcanic eruptions and intrusions produced by the first phase are now represented by small lava-flow remnants, dikes, and volcanic necks. They lie principally on the west slope of the range in the northern part of the parks area (Fig. 8.1).

Most of the first-phase volcanic rocks (4.5 to 2.5 million years old) are basaltic lava-flow remnants of small volume that are scattered throughout the upland areas (Fig. 8.2). Potassium-argon ages indicate that a flow remnant in the northwest corner of the Tehipite Dome Quadrangle and one near Hume Lake are both 3.3 to 3.4 million years old (Moore and Dodge, 1980). The Stony Flat Lava Flow 3 miles west of General Grant Grove is also 3.4 million years old.

Because of the exceptionally high potassium content of these basalts, biotite is a common mineral, and the rather rare mineral leucite, $KAl\text{-}(SiO_3)_2$, crystallizes at several localities (Moore and Dodge, 1980). The unusual composition of these basalts is best explained by very small degrees of partial melting in mantle rocks—a process favored by the exceptionally low heat flow beneath the present Sierra Nevada. The fact that the basalt melt could travel up through miles of granitic plutons with

8.1. Map of the parks region, showing areas of Cenozoic volcanic rocks. Dark color indicates areas of basalt; light color, areas of rhyolite and dacite.

8.2. The author examining Indian Basin olivine basalt lava flow (see Fig. 8.1) in a stream channel exposed in a roadcut on State Route 180, 10 miles east of the Big Stump Entrance station. The lava rests on about 6 feet of sediment composed of granitic sand and basaltic ash that in turn rests on weathered granodiorite.

very little contamination also points to a rather open conduit through cool crustal rocks.

A volcanic neck of intermediate composition (dacite) occurs near Windy Peak on the Middle Fork of the Kings River. This shallow intrusion, with its associated lava flows, is well exposed on the south wall of a major glacial canyon. It is about 1 mile in diameter, is composed mainly of hornblende-biotite dacite containing about 63 percent SiO_2, and was emplaced 4.5 million years ago (Moore and Dodge, 1980). For it to have become exposed as it is today on the wall of the canyon south of Simpson Meadow, the volcanic neck apparently underwent several thousand feet of erosion.

The Second Volcanic Phase

The second phase of volcanism in the parks area, 2.4 million years ago, produced the Templeton Mountain and Monache Mountain rhyolite domes (Bacon and Duffield, 1981), near the southeast corner of the mapped area (the Monache Dome is slightly outside of the area of Fig. 8.1). A similar, though smaller, rhyolite dome erupted 1 million years ago in the northern Big Pine Volcanic Field, a few miles southwest of Fish Springs. Chemical "fingerprinting" has established that obsidian contained in the Fish Springs rhyolite dome is the same as that found as artifacts in several high-elevation Indian campsites west of the Sierra Crest (Roper Wickstrom, 1993).

But the major volcanic event during this period was the 0.8-million-years-old Bishop Tuff, a widespread ash sheet originating from the Long Valley Caldera, north of Bishop. About 150 cubic miles of fragmental rhyolitic material erupted from this center, and winds blew some of it westward to the Pacific Ocean and some eastward as far as Kansas.

The Third Volcanic Phase

The third phase of volcanic activity began about 0.7 million years ago and has lasted essentially to the present. This phase, which saw basaltic lava flows erupting from fault-fissures along the east base of the range, created about 20 distinct vents in the Big Pine Volcanic Field (Fig. 8.3). Basaltic cinders erupted first and built cinder cones, and in some cases lava flows then issued from the cones (Fig. 8.4). The lava flowed east as much as several miles toward the center of Owens Valley. Natural levees flanking the flows formed when the hotter, more fluid central part of a flow drained out downslope (Fig. 8.4). Lava tubes also formed where the fluid lava drained out from beneath a solidified crust. Geochemical studies indicate that the magma source was deep, from the mantle part of the continental lithosphere (Ormerod and others, 1991).

Somewhat older lava flows are on the south side of the Big Pine Volcanic Field. These lava-flow remnants are exposed in Sawmill Canyon (Fig. 8.5) and Oak Creek, and some flow remnants can be found in Owens Valley, where they are partly covered by younger sand and gravel washed down from the mountains. The Sawmill Canyon lava is notable

8.3. Vertical aerial photograph showing the central and southern reaches of the Big Pine Volcanic Field (compare with Fig. 8.1). The cinder cones at left of center are aligned along range front faults; some fed basaltic lava that then flowed east toward central Owens Valley. Note the channels in the lava flows, and the lobate branching flow fronts with their concentric and transverse ridges, the ridges enhanced in appearance where they are partly covered by younger, light-colored alluvial material. Both U.S. Route 395 and the Los Angeles Aqueduct appear on the right. The image is 6 miles wide, and north is toward the top. U.S. Geological Survey aerial photograph taken in October 1948.

8.4. Lava-flow features in Big Pine Volcanic Field. *Above.* A partly collapsed basaltic lava tube in a flow fed from the cinder cone in the background. *Below.* Natural levees flanking the course of a basaltic lava flow.

8.5. Looking west at an erosional remnant of a basaltic lava flow (arrow) in the lower course of Sawmill Canyon, on the east side of the Sierra.

because it lies on top of a glacial moraine and is overlain in turn by a younger moraine. These two moraines are believed to be from the Tahoe and Tioga Glaciations, and the fact that they bracket the lava flow provides good constraints on its age.

Also included in the third volcanic phase are the basaltic lavas of the Golden Trout Creek Volcanic Field just east of Kern Canyon, near the southeastern border of Sequoia National Park. The lava covers only a small area, about 4 square miles. Mapping of the lavas, radiometric dating, and magnetic measurements indicate that three separate episodes of basaltic volcanism built small cinder cones and erupted lava flows.

The Little Whitney Cone was built during the first of these episodes 740,000 years ago. Ice from one or more glaciations overtopped the cone and eroded most of the lava or covered it with glacial deposits.

The South Fork Cone erupted during the second of the three episodes about 180,000 years ago, and fed the largest lava flow of the field, which moved at least 6 miles west down the canyon of Golden Trout Creek. The lava probably flowed down to the floor of the Kern Canyon, but now, much erosion having intervened, the lowest outcrop is 200 feet above the floor of the canyon. The main Kings Canyon glacier may have covered and eroded away the lower end of the lava flow. Well-developed columnar joints (Fig. 8.6) in the lowest exposures suggest that the eruption took place when the canyon was filled with glacier ice, and that ice or meltwater quenched the lava. The flow has been extensively glaciated over much of its extent, although ice did not override the cinder cone

CENOZOIC VOLCANIC ROCKS

vent. The Tunnel Cone, just north of the South Fork Cone, was probably active at about the same time.

The Groundhog Cone, which erupted during the third episode, is clearly the youngest in the area. The cone and the flow issuing from it can readily be distinguished from earlier lavas because they are not covered in any part by glacial morainal material and retain a very rough and clinkery surface (Fig. 8.7). The lava is also distinctive because the basalt does not contain the large conspicuous crystals (phenocrysts) found in the other, older lavas in the region. The flow has not been dated, but it is es-

8.6. The lower end of the South Fork Lava Flow in the Golden Trout Volcanic Field. The well-developed, fine columnar jointing suggests that the flow moved down a canyon filled with glacial ice and that meltwater quenched the lava. The view is toward the north, with the west wall of Kern Canyon visible in the background. U.S. Geological Survey photograph by Francois Matthes.

8.7. Blocky postglacial basaltic lava flow that erupted from Groundhog Cone, Golden Trout Creek Volcanic Field. U.S. National Park Service photograph by Howard Stagner.

timated to be 5,000 to 10,000 years old, making it one of the youngest lava flows in the Sierra Nevada. The Groundhog Lava poured 4 miles west down Golden Trout Creek, partly on top of the flow from the South Fork Cone.

The volcanic vents of the Golden Trout Creek area lie in topographically low areas, and the cones and flows have blocked and diverted streams. Eruption of the South Fork Cone dammed a previously west-flowing tributary of Golden Trout Creek, causing it to reverse direction, overtop a low divide several miles east, and join the South Fork of the Kern River. This diversion was further intensified when glacial moraines blocked the Tunnel Meadow drainage, which had previously flowed into Golden Trout Creek, and diverted it into the South Fork of the Kern River drainage. In the late 1800s, a tunnel was dug through the moraine to divert Tunnel Meadow water back into Golden Trout Creek for irrigation purposes farther downstream. Landowners on the South Fork of the Kern River objected, however, and the tunnel was blocked and the water returned to the South Fork.

The rhyolite of Long Canyon, 3 miles south of Templeton Mountain, may be related to the Golden Trout Volcanic Field. This curious young pumiceous rhyolite dome, with associated ash-flow lobes, yields a potassium-argon age between 35,000 and 335,000 years old and hence probably formed during the overall period of activity at the Golden Trout Volcanic Field.

Extensive travertine (calcium carbonate) hot-spring deposits lie adjacent to the South Fork and Groundhog Lava Flows where they are confined in the lower canyon of Golden Trout Creek. They were probably deposited from mineral-rich springs heated by the thick lava flows while they were still hot, after volcanic activity had ceased. The trail along Golden Trout Creek passes over a natural bridge at an altitude of 7,900 feet. The bridge was formed where mineral-spring water deposited travertine onto sand and gravel that was later undermined and eroded away, leaving the more resistant travertine arching above. This bridge, an important landmark in the region, was employed by the Hockett Trail (Fig. 2.38). John Muir, on his Kern Canyon trip in 1902, commented as follows on its origin: "The natural bridge across Volcano Creek [Golden Trout Creek] was not made by volcanic action, but by a hot spring. The spring is gone and lava is all around but there is no doubt about how the bridge was made. In the Yellowstone Park the hot springs there are making just such curious things."

Even horses and dogs gaze wonderingly at the strange brightness of the ground, and smell the polished spaces and place their feet cautiously on them when they come to them for the first time, as if afraid of sinking.

JOHN MUIR, ON GLACIAL POLISH, 1894

GLACIERS AND GLACIATION

When John Muir first visited Yosemite Valley, he, like most visitors, puzzled over how the valley had formed. After hiking over much of the surrounding country and documenting the geologic evidence, he became convinced that glaciation had shaped much of the landscape of the High Sierra, as well as Yosemite Valley. As we have seen, this view placed him in opposition with the catastrophic-collapse theory for the origin of Yosemite, which had been advocated by Professor Josiah Whitney. Whitney, as well as his young assistant Clarence King, had previously recognized the abundant evidence for glaciation in the Sierra Nevada, but because of the unique character of Yosemite Valley, Whitney laid forth the notion that it had been created by a special process of faulting and subsidence. The debate was complicated by the fact that several other factors were involved in the origin of the valley. Most workers agreed that stream erosion before and between the glacial episodes was important in excavating the valley, and that the jointing or "cleavage" of the granitic rock was an important element in the shaping of some of the valley's features.

The debate on the origin of Yosemite soon became polarized between Whitney's collapse model and Muir's glaciation model. To find more evidence to support his ideas, Muir decided to broaden his area of study. He had read the Whitney report of 1865, on William Brewer's expedition to

9.1. *(Above)* The confluence of the Middle (left) and South (right) Forks of the Kings River 1 mile upstream from the region where the canyon approaches 8,000 feet in depth. The V-shaped canyons shown are river-cut, but the steep granite walls midway up the Middle Fork mark the lower extent of sculpturing by massive glacial ice. Wren Peak is the high point between the rivers, and State Route 180 appears on the right. U.S. National Park Service photograph.

9.2. *(Opposite)* Map of the parks area, showing the extent of glacial ice during the culmination of the Tioga Glaciation, about 20,000 to 15,000 years ago. Note that virtually all the lakes occur within glaciated areas.

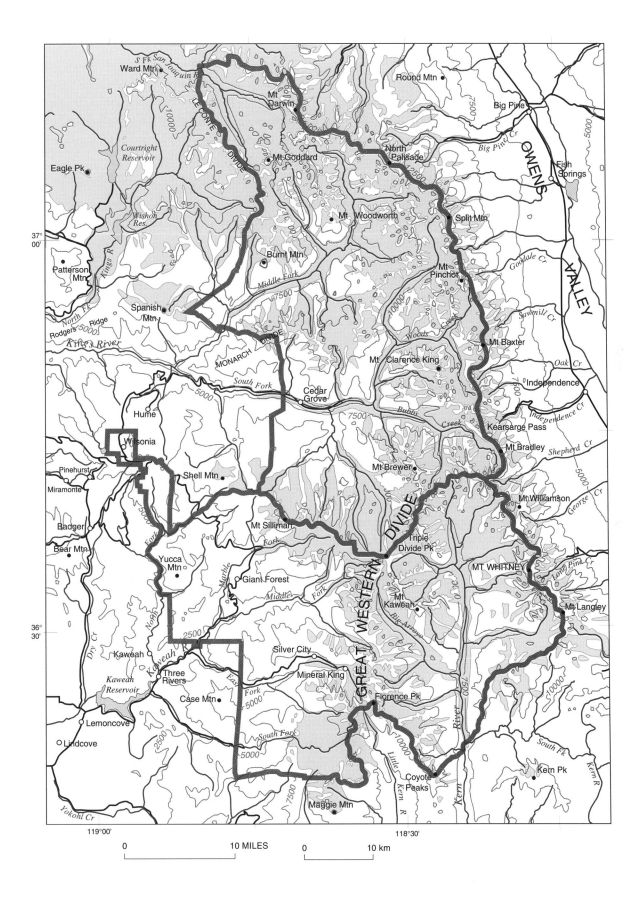

the Kings River region, and was no doubt excited by its description of the canyon of the South Fork, which "rivals and even surpasses the Yosemite in the altitude of its surrounding cliffs," and of the canyon of the Middle Fork, where "there is a precipitous descent to the north, into the canyon of the middle fork, which is perhaps even deeper than the one just described." Muir traveled south in the Sierra to the Kings River region to see for himself, and to marshal more evidence for his belief that the giant U-shaped valleys of the range were shaped by slowly flowing ice. He studied the great canyons of the South, Middle, and North Forks of the Kings River as well as the canyons and cirques of the alpine country to the east, and found what he was looking for. The compelling evidence left by glacier ice demonstrated to him the importance of glaciation in sculpting the landscape and carving these great canyons, which so resembled Yosemite.

Still, there is no doubt that water has been more important than ice in the overall erosion of the Sierra Nevada. Even during the Pleistocene Ice Age, from about 1.5 million years ago to 10,000 years ago, the ice cover in the highest parts of the range formed intermittently. During lengthy periods, similar in character to the climate of the present time, ice was virtually absent even on the highest peaks. What we find is that the lower slopes of the range are cut by enormous canyons (Fig. 9.1), deeper than those of the upper glaciated terrain, and among the deepest in the nation. These lower canyons, such as the lower Kings River Canyon, are distinctly V-shaped and thus carved by the erosive action of water.

But the nature of these canyons changes dramatically as we move upstream. They acquire a steep-walled aspect, bare rock walls tower above the valley floor, and the valley bottom broadens from a narrow slot, occupied largely by the river, to a wide strath with an "underfit" stream occupying only a small part of the broad valley floor. These changes are so striking and lead to such startling scenery that we should not be surprised to learn that the area of Kings Canyon and Sequoia National Parks is dominated by regions previously covered by ice (Fig. 9.2). This topographic change between unglaciated and glaciated terrain results from the difference in velocity of flowing water and moving ice. In a relatively steep canyon, water flows at a few miles per hour, ice at a few tens of feet per year. Since ice thus flows about a million times slower than water, the higher-elevation frozen part of a stream (the part locked up in a glacier) must have one million times the cross section of the melted part of the same stream. That is why the glacier fills up its canyon, and why it erodes in a fashion distinctly different from that of flowing water.

The Extent of Glaciation in the High Country

Individual glaciers are formed during cool periods, when the volume of snow that falls each year exceeds that which melts, causing snowbanks to linger through the summer, and perhaps many following summers. When the snowfall of successive winters is added to that already on the ground, the snow builds up ever thicker, becomes successively compressed at depth, recrystrallizes into ice, and eventually begins to flow slowly downhill, in response to its own enormous weight. The average rate of movement of glaciers generally ranges from 10 to 1,000 feet per year. Eventually, the down-flowing ice reaches an altitude where temperatures are higher and melting begins. Water then flows from the ice toe to form a meltwater stream. The toe of the glacier stands where the melting of the ice balances that delivered by iceflow.

More than half of the area of the parks was covered by glacier ice during the peak of the Ice Ages (Fig. 9.2), but the warmer weather of recent times has caused the ice to retreat, leaving only tiny glaciers in the northeast shadows of the high peaks. The first of these small glaciers in the Sierra was discovered by John Muir in 1871, on Black Mountain in the Yosemite region, and within a few years he had identified 65 separate glaciers in the Sierra Nevada. A few dozen of these are in the Sequoia and Kings Canyon parks area.

In the summer of 1872, Muir measured the flow rate of a small glacier on Mount McClure, east of Yosemite Valley near the range crest. He planted 5 stakes across the glacier in a straight line, as determined by "sighting across from bank to bank past a plumb line made of a stone and a black horsehair." After a period of 46 days he found that the stake nearest the center of the glacier, offset the farthest, had moved 47 inches, about one inch per day or 30 feet per year (Bade, 1924).

The small residual glaciers occur in the highest valleys, on the shady north to northeast-facing walls of the peaks, generally above 11,000 feet and generally somewhat above timberline (Fig. 9.3). In contrast, the lower limit of the giant glaciers that expanded during the most recent extent of the Ice Ages was at 4,000 feet on the west slope of the range and at 6,000 feet on the dryer east slope. But modern evidence does not support the notion advanced by Muir that the range was covered "from the summits to the sea with a mantle of ice . . . discharging fleets of icebergs into the sea" (Muir, 1894).

The movement of the ice even in the small existing glaciers can be demonstrated by the prominent crack (called the bergschrund) that occurs at the head of a glacier, where the moving ice tears away from a thin rim of stationary ice fixed to the mountain (Figs. 9.4 and 2.22). Commonly, in the lower parts of the glaciers, a prominent layering is revealed as the ice melts. These layers represent the annual layers of snow that were compressed by the weight of succeeding layers to form the glacier ice.

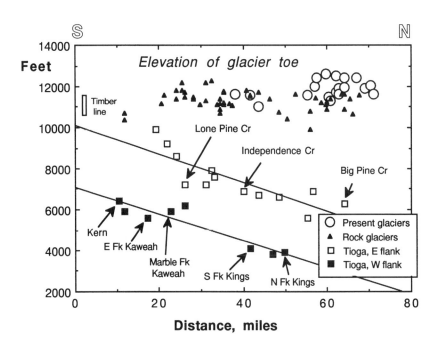

9.3. *(Above)* Altitude of the lower limits of glaciers in the parks region, south to the left, north to the right. The present-day glaciers lie mainly in the northern region and generally occur above 12,000 feet but do extend down to 11,000 feet. Active rock glaciers, generally cored by ice, lie at about the same or slightly lower altitudes. The lower limit of Tioga glaciers, the last major glaciation, was lower on the west slope of the range than on the eastern escarpment, probably because in the west there is a larger catchment area and greater snowfall.

9.4. *(Opposite)* Looking southwest across the Palisade Glacier and North Palisade (14,242 feet) on the Sierra Crest to the Black Divide and down the great canyon of the Middle Fork of the Kings River. The fine layering in the ice probably represents annual accumulations of snow, later compacted to ice. The main crack at the head of the glacier (the bergschrund) forms where ice is tearing away from the rock headwall. Small ridges of rock debris (moraines) beyond the toes of the glaciers mark the limits of a recent ice advance, probably during the Little Ice Age. U.S. Geological Survey aerial photograph by Austin Post.

Muir climbed down into the bergschrund of the Black Mountain Glacier, which gaped 12 to 14 feet wide at the surface. "Creeping along the edge of the schrund," he wrote (1894),

> holding on with benumbed fingers, I discovered clear sections where the bedded structure was beautifully revealed. The surface snow, though sprinkled with stones shot down from the cliffs, was in some places almost pure, gradually becoming crystalline and changing to whitish porous ice of different shades of colour, and this again changing at a depth of 20 or 30 ft. to blue ice, some of the ribbon-like bands of which were nearly pure, and blended with the paler bands in the most gradual and delicate manner imaginable.

The largest active glaciers in the parks area are in the high, rounded valley heads (called *cirques*) of the Evolution and Palisades region and on Mount Goddard. The largest, Palisade Glacier (Fig. 9.4) in the northeast cirque of North Palisade, is 0.9 mile long and 0.5 mile wide and descends from an altitude of 13,200 feet to 12,200 feet. A small glacier on the northwest face of Mount Goddard (Fig. 2.22) is 0.3 mile long by 0.2 mile wide and descends from 12,560 feet to 11,700 feet. The southernmost permanent ice patch in the Sierra is probably that on University Peak, just south of Kearsarge Pass.

In addition to the obvious glaciers with their exposed ice, many *rock glaciers* can be found at about the same, or slightly lower altitudes (Fig. 9.3). Rock glaciers are tongue-shaped masses of bouldery rock debris, generally less than one-half mile in length, that nestle in north- to northeast-facing cirques (Figs. 9.5, 9.6, 9.7, 2.17). They commonly show arcuate and rounded ridges in their lower reaches, ridges that are parallel to a lobate toe having a steep, unstable, rocky slope. Ongoing movement of the rock glacier is verified where the steep toe scarp, typically up to several hundred feet high, is mantled by large, fresh, angular boulders balanced at the angle of repose. Rock glaciers contain cores of clear ice or ice interstitial to the bouldery material, and they flow like ice glaciers at velocities commonly of 1 to 5 feet per year. They are apparently more abundant

9.5. View southwest across Kaweah Peaks Ridge, at the south end of the Great Western Divide. Rock glaciers with arcuate surface ridges and steep, unstable flow fronts impinge on glacial lakes. The actively moving rock glaciers contain a core of ice covered by rock rubble. Sawtooth Peak is the sharp dark peak in the middle background, 2 miles east of Mineral King, and the Kern Canyon trends south in the left background. U.S. Geological Survey aerial photograph by Austin Post.

than ice glaciers and occur to slightly lower altitudes, because of the insulating layer of rocky rubble that protects them from the Sun's heat.

Glacial deposits of rock debris left by ancient glaciers are far more widespread and situated at much lower altitudes than those of the currently active glaciers and rock glaciers. These deposits were laid down during the Pleistocene Ice Ages, when the glaciers were considerably larger and extended to lower elevations. These glacial deposits, called *till*, commonly form large ridges of unsorted clay, sand, and broken boulders. Because the ridges, called *moraines*, commonly occur on the sides of ma-

9.6. The front of a rock glacier in upper Bubbs Creek, in a cirque on the north side of Junction Peak. U.S. National Park Service photograph by Walter Huber.

9.7. Looking south across Lake Marjorie (11,150 feet) at the toe of a rock glacier seen from the John Muir Trail on the north approach to Pinchot Pass. The presently active movement of the ice-cored rock glacier is demonstrated by its steep face, delicately poised at the angle of repose. U.S. National Park Service photograph by Howard Stagner.

jor drainages, they show that these canyons were partly filled with ice (Figs. 9.8, 9.9). The most extensive moraines extend 2,000 to 7,000 feet below the active glaciers of today, proving that much colder weather prevailed during the Pleistocene (Fig. 9.3).

In many places, two sets of moraines, one inside the other (downslope and upvalley), give evidence of two major glacial advances—the Tahoe and Tioga Glaciations (Figs. 9.9, 9.10). The earlier Tahoe Glaciation was more extensive, and its deposits extend somewhat farther downslope than do those of the Tioga Glaciation. Mapping of the glacial deposits enables us to unravel the glacial history in a given canyon, to determine the former ice thickness, and to establish the down-canyon extent of the glacier. For example, the ice was about 880 feet deep at Lodgepole Campground during the Tahoe Glaciation (Fig. 9.10), and the glacier there at the time extended about 2 miles downstream from the campground, to an altitude of 6,000 feet.

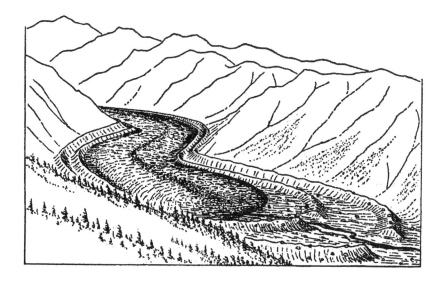

9.8. *(Above)* Moraines deposited beyond the lower end of a valley glacier. The piles of rock debris result either from a period when the ice front is stationary, so that melting dumps debris carried by the ice in a pile, or by a readvance of the ice, during which it bulldozes and pushes up a pile of debris. The two ridges, marking two successive shrinkages of the ice, are called *recessional* moraines. Clearly, any smaller, older advances of the ice leave no record, inasmuch as a later, larger advance destroys older moraines. Ridges on the sides (*lateral* moraines) are generally better preserved, because they are less likely to be eroded away by meltwater streams issuing from the ice. Sketch by Francois Matthes (1930).

9.9. *(Opposite)* Looking east across Roaring River Canyon to the Great Western Divide, with Mount Brewer in the complex of peaks near the center. The long, evenly graded pair of ridges on the far (east) side of the canyon are glacial moraines about 3 miles long. The farther, outer one (farther east) was deposited during the older Tahoe Glaciation, and the nearer, inner one (farther west) was deposited during the younger Tioga Glaciation. The Brewer party rode horses up the crest of this moraine system during their exploration in 1864. U.S. Geological Survey oblique aerial photograph taken November 7, 1956.

In addition to these two major glacial advances, other, older pre-Tahoe advances have been identified in areas to the north, several of which are best preserved on the east flank of the range near Mono Lake.

During the Tahoe and Tioga Glaciations an ice cap covered much of the higher parts of the range except for narrow ridges and isolated peaks that protruded above it (Fig. 9.2). This ice cap fed large trunk glaciers that flowed west down the major river valleys as much as 10 miles from the margin of the ice cap. These massive flow lobes, some more than 1,000 feet thick, carved the Yosemite-like canyons of the North, Middle (Fig. 2.27), and South (Figs. 2.28, 2.29 2.32) Forks of the Kings River, the Marble and East Forks of the Kaweah River, and the Kern River (Figs. 9.16, 11.7).

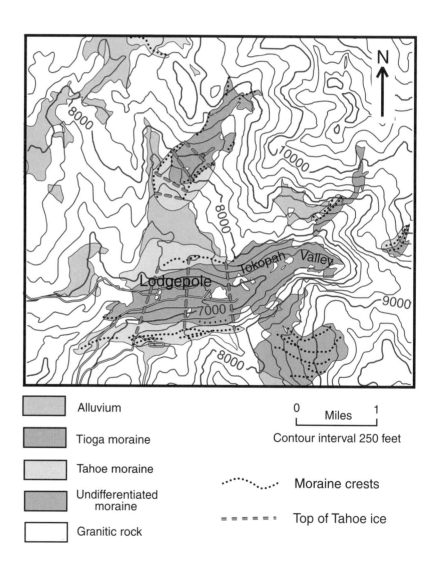

The ice in the canyons draining the west side of the range descended farther down than those draining the east side (Fig. 9.3). The mapping of moraines shows that the toes of the major Tioga glaciers, on average, descended about 3,000 feet lower on the west side of the range than on the east. This sizable difference results from the greater snowfall and larger high-elevation drainage basins on the west side, the two factors supplying the western glacial system with more snow and ice.

Glacial Processes and Their Products

The separation of ice from rock that produces the open crack of a bergschrund demonstrates that the ice is moving downslope, and that a live glacier rather than a stationary snowfield adorns the mountainside. Important glacial processes take place down to the base of the bergschrund. There the bedrock is plucked away as water freezes and expands in cracks, prying rock fragments loose to become frozen onto the moving ice and be carried downstream. In this way the glacier vigorously excavates the mountain slope nearly to the base of the ice in its upper reaches, and over time creates a steep-walled bowl-shaped valley head, or cirque (Figs. 9.11, 9.12).

These steep walls are subject to constant rockslides as well as, in winter, snow avalanches. As a result they are generally incised by steep gullies or avalanche chutes (Fig. 9.13). Rock debris falling from the surrounding mountain slopes onto the glacier, or into the bergschrund, is also carried downstream with the flowing ice. Increasingly embedded in the ice, the rocks provide effective tools for grinding away at the bedrock floor beneath the ice stream. Glacial polish on rock surfaces indicates where ice once flowed, and striations or scratches tell us the direction of the flow (Fig. 9.14).

9.10. Moraines at the lower course of glaciers that flowed west (from right to left) down the canyon of the Marble Fork of the Kaweah River in the Lodgepole area. The area covered by the Tahoe moraines (about 120,000 years old) and the Tioga moraines (about 15,000 years old) indicates the extent of the ice advances during these two glaciations. The colored dashed lines, representing contours at the top of the Tahoe glacial ice, indicate that during the Tahoe Glaciation the ice at Lodgepole Campground was between 750 and 1,000 feet thick. The western end of the main Tahoe glacier was about 2 miles west of the campground at an altitude of 6,000 feet.

Fractures, joints, and faults in the bedrock greatly aid the glacier in its work of deepening and widening its channel. Conversely, the highest walls and narrowest parts of a glacial canyon are usually made up of the most massive unfractured rock. Finer-grained rock commonly has more closely spaced joints, causing the rock to be fractured into blocks that are plucked more readily by glacier action. The remarkable excavation of the giant glacial canyon along the course of the Kern Canyon Fault (Figs. 9.15, 9.16) demonstrates how the fractured and weakened rock of the fault zone can control the location of these erosive processes.

The cirques feed ice downslope; the flows from neighboring cirques join one another and, thus joined, form major glaciers. Ice flowing downslope into canyons moves much more slowly than water because of its far greater viscosity. This meager pace causes the ice to fill up the canyon in time, and to erode the sides as well as the bottom of the gorge, thus producing the typical U-shaped cross section of glacial canyons. The moving ice also erodes off rock spurs and other irregularities that stand along its course, thus smoothing and straightening the canyon (Figs. 9.11, 9.12, 9.15). The straight U-shaped canyons carved by ice contrast with the curving V-shaped valleys and their interlocking spurs, cut by running water.

The marked contrast between the glaciated and unglaciated segments within a single canyon is nicely shown by the areas of transition within the three deepest canyons of the parks: the Middle Fork of the Kings River, the South Fork of the Kings River, and the Kern River. In each of the three canyons this transition occurs rather close to the park boundary,

9.11. Glacial modification of the landscape. *Top.* Before glaciation, showing rounded summits and V-shaped valleys resulting from normal weathering, mass wasting, and water erosion. *Middle.* During glaciation, showing previous valleys filled nearly brimful with slow-moving glacial streams. Plucking and grinding at the head and sides of glaciers has oversteepened the valley walls, producing steep, bowl-like valley heads (cirques) and straightened valley courses. *Bottom.* After glaciation and melting of the ice, showing sharp peaks (horns) and knife-edged ridges (arêtes) formed by the coalescence of glacially carved gorges. The U-shaped canyon of the main trunk glacier has cut deeper than the tributary glaciers, leaving hanging valleys on its side, from which waterfalls may cascade into the main canyon. Small glacially scoured lakes occupy the floors of the glacial canyons. Sketch by William Morris Davis (1906).

the glaciated zones all lying within the parks. The glacially scoured part of the canyon, with its broad meadowland floor, steep, smooth canyon walls, and plunging waterfalls, marks an abrupt transition into more scenic country that influenced those who established the park boundaries.

The steep residual bedrock between the glacial basins is carved into triangular peaks (*horns*) and sharp ridges (*arêtes*). The down-flowing

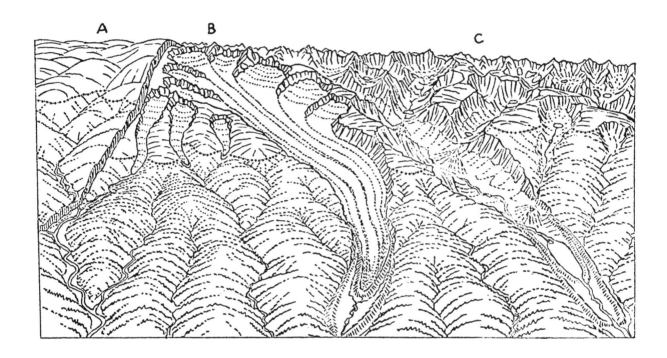

9.12. *(Above)* Three-stage diagram illustrating the changes wrought by glaciation on a fault-block mountain range like the Sierra Nevada. *A.* Before glaciation. *B.* During glaciation; note the longitudinal stripes of rock debris on the main trunk glacier, formed where rock debris on the sides of tributary glaciers coalesces where the tributaries join; such rock-debris ridges are termed *medial* moraines. *C.* After glaciation; note how the valley has been straightened and terminal and recessional moraines have been deposited at the end of the main trunk glacier. The position of timberline during each stage is indicated by a dot-dash line. Sketch by William Morris Davis (King, 1962).

9.13. *(Opposite)* Large avalanche chute on the canyon wall north of the High Sierra Trail east of Bearpaw Meadow. This chute has been carved in massive exfoliating granite. Others cut in more irregularly jointed granite have a ragged, commonly branching aspect. U.S. Geological Survey photograph by Francois Matthes.

9.14. *Above.* Glacial polish, with gouges and striations cut by fragments of rock embedded in the glacial ice that moved over a bedrock outcrop in the direction of the scratches. Headwaters of Kid Creek. U.S. National Park Service photograph by Wayne Acorn. *Below.* Glacier-polished surface strewn with boulders dropped in place when the ice melted. Upper reaches of Wallace Creek.

glacier ice gouges out areas of less-resistant rock and locally flows uphill at its base (even though the top of the glacier is flowing downhill), cutting many closed depressions into the bedrock. After the ice has melted, these depressions fill with water and produce the stair-step lakes that are a hallmark of past glaciation (Fig. 9.17).

The deepening and widening of the main glacial valley may leave tributary valleys at a higher level (Fig. 9.11). These *hanging valleys* are typical of glaciated terrain and account for the waterfalls and cascades often found where tributary streams flow or fall down the glacially steepened valley walls to join the main drainage (Fig. 9.18).

The regions of bedrock that glaciers have overridden commonly show areas of glacial polish (Fig. 9.14). This polish is best preserved in the highest parts of a glacial basin, where the glaciers have most recently melted away. It is produced by the abrasive action of the finer material embedded in the moving ice, and it is most striking on granitic rock. The glacial polish is commonly scored by grooves or striations of various sizes, engraved by particularly large, hard, or pointed rock fragments frozen into the moving ice. In some places hard rock fragments carried in the base of the glacier produce *chattermarks* by chipping the more brittle bedrock. The chattermarks are curved scars or cracks that cross the direction of ice flow and whose ends point in the direction toward which the ice traveled.

These features are found near the upper part of a glacier, where erosion—the removal of bedrock—is the dominant process. Other distinctive features are evident in the lower part of the glacier, where the deposition of the rock material carried in the glacier is the dominant process. Where the ice melts near the toe or foot of the glacier, admixed rock debris is dumped, forming piles of broken rock, sand, and clay (Fig. 9.19). Such *moraines* are loose piles of debris composed of angular rock fragments of widely varying size and shape. Commonly, the moraines occur as straight or curved ridges that may be several hundred feet in height (Figs. 9.8, 9.9, 9.13).

During cold periods, the toe of the glacier advances farther down valley, and the glacier becomes wider and thicker. These advances of the ice margin bulldoze the piles of broken rock down the valley and push much of them to the side, forming linear morainal ridges. During warm periods, the glacier front retreats and becomes thinner, and a new set of inner moraines is constructed (Fig. 9.8). These moraines can become important guides for geologists seeking to determine the location of past glaciers (Fig. 9.10).

About 100 lakes more than one-third mile long, as well as hundreds of

9.15. *(Opposite, top)* View to the northwest up the U-shaped glacial canyon of Goddard Creek, as seen from the head of Windy Creek. The roof of the batholith is shown on the peak in the middle background, where old, dark metamorphosed volcanic rock overlies light-colored granitic rock. The canyon walls are 4,000 to 5,000 feet high. U.S. National Park Service photograph by Howard Stagner.

9.16. *(Opposite, bottom)* Looking south from Rattlesnake Point, down the U-shaped glacial canyon of the Kern River. U.S. Geological Survey photograph by Francois Matthes.

9.17. *(Above)* Glacial stair steps in a canyon below Sawtooth Peak (12,343 feet), looking south from Black Rock Pass. Columbine, Cyclamen, and Spring Lakes (from top left to bottom right) occupy glacial cirques. U.S. National Park Service photograph by Howard Stagner.

9.18. Roaring River Falls, on the south wall of the South Fork of the Kings Canyon (Muir, 1887).

smaller lakes and ponds, are found within the parks, and virtually all are the result of glacial action. Probably most were formed when glaciers scooped out bedrock basins in regions of less-resistant rock (Fig. 9.17), but some were formed where glacial moraines dammed their downstream side (Fig. 9.20). Many of these former lake basins have since been filled by sediment carried down by the streams, and now survive as marshes and meadows.

The lake basins dammed by morainal ridges commonly formed at the toe of a glacier when glacial retreat permitted water to be impounded behind a recessional moraine. Likewise, lateral moraines, along the sides of a valley glacier, often dammed tributary streams to form lakes at the sides of the master stream.

Glaciers commonly deposited moraines during each of several succeeding periods of glaciation. Where a younger glacial episode is larger, and the ice extends farther downstream and higher on the canyon walls, any earlier moraines are overwhelmed, and the evidence of their former existence is destroyed. Where a younger episode is smaller, and the ice extends less far downstream, then the younger moraines are nestled within the older moraines standing lower down on the canyon walls. Hence, glaciation is generally recorded only in the last phases of the Ice Ages, where glaciations became successively smaller. Many different moraines may have been deposited during glaciations that were older and smaller than the later ones, but they will all have been destroyed and removed by a later, more extensive advance of the ice.

9.19. Detail of rock debris in a glacial moraine near Lodgepole. U.S. National Park Service photograph.

9.20. Glacial moraine impounding a small lake near Goat Mountain, east of Granite Pass. The glacier came from the left and pushed the rubble ahead, forming the moraine ridge that confines the lake water. U.S. Geological Survey photograph by Grove K. Gilbert.

The Causes and Timing of Glaciation

The Earth has experienced many past cycles of colder weather and increased cloudiness, all of which could have caused the polar ice caps to expand and glaciers to form in mountainous areas. The climate changes that brought on the glaciations of the Ice Ages were worldwide phenomena, and they began millions of years ago. In Antarctica, glaciation was well established 20 to 30 million years ago, and in Alaska and Greenland, 10 to 20 million years ago. Glaciers moved into Europe and North America 2.5 million years ago. The earliest well-recorded glacial evidence within the area of the parks is found in what has come to be called the Tahoe moraines, which are 150,000 to 125,000 years old. In other parts of the Sierra, especially to the north, older glacial moraines have been dated, but no good dates for pre-Tahoe glaciers have yet been established in the parks area.

Each cycle of cold weather followed by milder weather lasted about

100,000 years. The last major glaciation (representing only the cold part of its cycle) occurred from 20,000 to 15,000 years ago, and although most of the ice had retreated by 10,000 years ago, we are still emerging from this period today.

This modern warming trend has been interrupted by small episodes of cooling temperatures and advancing glaciers. Francois Matthes noticed tiny glacial moraines just below the existing glaciers in the Sierra Nevada (Fig. 9.4) and coined the term "Little Ice Age" as a name for the glacial stage that produced them. He had no means to date these glacial deposits, but estimated that they were produced 4,000 to 2,000 years ago.

Historic studies indicate that a global cool period occurred between the Middle Ages and the warmer period of the early part of the nineteenth century. Glaciers in many parts of the world expanded, grain would no longer ripen in Iceland, disastrous harvests hit much of Europe, and the areas under cultivation contracted. This period is now termed the Little Ice Age and is apparently responsible for producing the Sierra moraines that Matthes had described, named, and assumed to be somewhat older. The Little Ice Age, now considered to have prevailed from 1550 to 1850 (Grove, 1988), was no doubt responsible for some of the small high-altitude moraines that occur just below existing glaciers (Figs. 9.4, 2.26).

Just what brought on the Ice Ages—what it was that precipitated the global shift from warmer to cooler climate beginning tens of millions of years ago—has long been a matter of debate. The cooling may have been caused by a drop in the concentration of atmospheric carbon dioxide—a reverse of the so-called *greenhouse effect*—but there is no consensus on what caused this decline. One interesting theory holds that massive uplift brought on by the collision of two of the Earth's tectonic plates elevated the Himalayan Mountains. The rapid and sustained uplift of this largest of all mountain ranges caused weathering and erosion of continental rock on a giant scale, and the immense volume of pulverized calcium silicate rocks eroded from the range reacted with atmospheric carbon dioxide dissolved in rainwater. Enough calcium carbonate weathering products were produced to lock up the carbon in sediments and thereby reduce carbon dioxide in the atmosphere. Such a CO_2 reduction, and the resulting reverse greenhouse effect, is a plausible explanation for the shift to a cooler climate (Ruddiman and others, 1997).

There is more agreement, however, on the process that apparently forced the cycling of individual glacial and interglacial stages, once a general cooling was under way. An astronomical theory developed by the Yugoslavian mathematician Milutin Milankovitch in the 1920s accounts for the timing of the Pleistocene glacial-interglacial cycles. This notion holds that the cycles result from variations in the rotation of the Earth, both around its axis and around the Sun. Geophysicists recognize three orbital parameters that yield changes in the amount of solar energy received by the Earth; all had been identified and measured prior to the 1920s. The three are (1) variation in the shape of the Earth's elliptical orbit around the Sun, which changes through a 100,000-year cycle; (2) variation in the tilt of the Earth's axis of rotation relative to the plane of its orbit (now 23.5°), which changes several degrees over a period of 41,000 years; and (3) the progression of a wobble of a few degrees in the rotational axis, which has a period of 21,000 years. By combining these changes, and calculating their aggregate effect on the intensity of sunlight reaching various parts of the Earth at various points in time, Milankovitch was able to show that cool glacial stages occur on a cycle of about 100,000 years.

Support for the Milankovitch theory came from the analysis of small organisms that had been buried in mud on the ocean floor. The animals that produced these fossils, called foraminifers, lived at the ocean surface and had sunk to the bottom when they died, to be later recovered by deep-sea coring of the enclosing sediment. The concentration of the oxygen isotopes preserved in the shells of these organisms is controlled by the concentration of the isotopes in the ocean water in which they grow. Isotopic analysis of oxygen in the tiny shells thus gives us evidence of the ocean-water concentration at the time when the fossil foraminifers lived.

The element oxygen, an integral component of water (H_2O), exists in two isotopic forms, ^{16}O and ^{18}O. Water molecules made of the lighter ^{16}O are more likely to evaporate than are the heavier molecules made of ^{18}O. Hence, relative to the occurrence of ^{18}O, they have a higher concentration in clouds, rain, and snow, as well as in bodies of fresh water such as lakes and ice sheets, and a lower concentration in seawater. Therefore, when a relatively greater amount of ice lies on the land, as it does during a period of glaciation, the sea is relatively enriched in ^{18}O, and the foraminifera growing at such a time in the sea are also enriched in ^{18}O. Hence in a deep-sea core, those parts of the core that are most enriched in ^{18}O mark the time of an ice age. Inasmuch as sea level is at its lowest during glaciations (when more water formerly in the sea is tied up on land as glacial ice), then periods of low sea level mark times when foraminifers in the sea are most enriched in ^{18}O.

When the ages of the different layers within the cores, including those deposited during glacial periods, are measured by radiometric methods, we then have a dated sequence that indicates how much water was residing on the continents as glacial ice at various times (Fig. 9.21, upper). The last two periods of highest ^{18}O (equating to the lowest sea level and the most abundant glaciers) occurred at 150,000 to 125,000 years ago and 20,000 to 10,000 years ago. These are the ages, respectively, considered to represent the times of the Tahoe and Tioga Glaciations, as well as major concurrent glaciations in other mountain ranges on other continents.

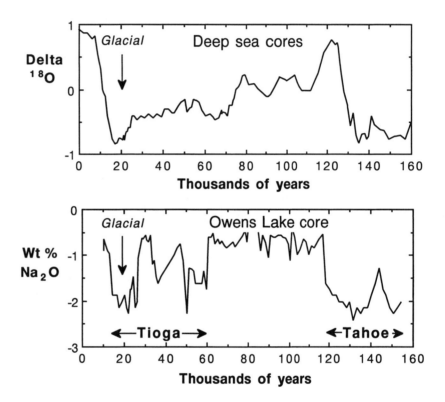

9.21. Approximate ages of the two principal glaciations (Tahoe and Tioga) in the parks area relative to the worldwide climate history (*above*) as indicated by the oxygen-isotopic composition of sea water (delta ^{18}O), determined from dated deep-sea sediment (Martinson and others, 1987); and the climate history of the Owens Valley region (*below*) as recorded by the abundance of glacial sediment (as measured by Na_2O content) within a drill core in Owens Lake (Bischoff and others, 1997). In both curves, downward deflections indicate glacial conditions. Note that the glacial advances for alpine glaciation (*bottom*) occur earlier than for continental glaciation (*top*). Time advances from right to left.

Another approach to dating the time of glacial periods within the southern Sierra Nevada was afforded by coring the sediment within the now dry Owens Lake Basin. The basin lies east of the Sierra Nevada, downstream from much of the east slope drainage of the Sierra. Several constituents in this core were chemically analyzed. The concentration of sodium oxide (Na_2O) proves to be a good climate indicator, because it is especially abundant in glacial silt carried into the Owens River and Lake Basin during a glacial period. High values of Na_2O, therefore, indicate a glacial period. A good correlation was found between the glacial periods as determined by oxygen isotope data from the world's ocean floors and the Na_2O data from Owens Lake sediment (Fig. 9.21). The periods of maximum Na_2O from Owens Basin somewhat preceded the periods of glaciation indicated by oxygen isotopes from the deep-sea cores (Fig. 9.21), probably because the smaller alpine glaciers of the Sierra Nevada responded more quickly to a return to glacial conditions than did the far more massive ice sheets of the main continental glaciers.

The nature of the geochemical peaks reflects the complex nature of glacial oscillations, and complete agreement on these ages has not been reached by glacial geologists. Some consider that moraines assigned to the Tahoe Glaciation in some places may actually belong to the early part of the Tioga glaciation, but to judge from the Owens Lake drill-core data, which is considered more representative of the local mountains, there were two general periods of glaciation in the last 160,000 years. The earlier occurred from 155,000 to 120,000 years ago, the younger from 60,000 to 15,000 years ago, and the two are assumed to be, respectively, the Tahoe and Tioga glaciations.

Any man looking out of any window sees a geological laboratory in constant and full-scale operation.

—HERBERT H. READ, 1943

LANDSLIDES

The processes of erosion are amply manifested in the Sierra Nevada. Running water and flowing ice have carved giant canyons and produced steep valley walls. The east flank of the range marks one of the country's highest scarps, and torrential streams have incised these slopes and others in the parks area, carrying pebbles and boulders downstream to be deposited in broad, coalescing alluvial fans at the base of the scarps.

The term *erosion* generally includes those processes that move earthy materials from one place to another by the action of some agency, usually running water, waves and currents, moving ice, or wind. In contrast, the dislodgement and downslope movement of soil and rock under the effect of gravity—that is, *not* carried by some other medium—is called *mass wasting*. This process includes creep, earthflows, rockslides, avalanches, and landslides. Angular blocks periodically break loose from steep slopes and cliffs in the wake of frost action and tumble downhill, where they pile up to form extensive rubbly heaps of talus. These talus piles comprise much of the loose surficial rock in the High Sierra.

Large landslides, however, are not common in the Sierra Nevada. The contrast between the landslide-prone California Coast Ranges and the paucity of landslides in the Sierra is striking. The cause of the disparity probably lies in the nature of the bedrock: the Sierra is largely underlain by hard, crystalline solid rock, either metamorphic rocks or the far more voluminous, more coarse-grained granitic rocks. The main part of the Sierra block, moreover, is almost completely devoid of Cenozoic faults. In this regard, it contrasts markedly with the rest of California, both to the east and west, where young faults are common. Earthquakes, like landslides, are uncommon within the Sierra block, though they are associated with the nearby Sierra Frontal Fault System on the east side of the range (Fig. 11.2). Solid bedrock and paucity of faults and earthquakes apparently explain the small number of landslides in the range.

Despite this general lack of landsliding, a few noteworthy slides, some of them historic, have modified the landscape. There is extensive evidence that a remarkable series of prehistoric landslides slid off the steep west wall of Moro Rock and moved south nearly 2 miles, down to the gorge of the Middle Fork of the Kaweah River. Granitic boulders as large as 30 feet, some of which are fragments of the exfoliated shells of the domelike Moro Rock, have fallen southwest down the steep slope toward the river. Some of them choke gulches cut in the Sequoia Roof Pendant, others spread out on the flat near Hospital Rock.

One can see these boulders in the gulches while driving up the Generals Highway from Hospital Rock to Deer Ridge. They are readily identifiable as landslide material by their granitic character, which contrasts sharply with the metamorphic bedrock on which they rest. Hospital Rock itself is one of these giant landslide boulders, one that has cracked and split open after coming to rest. Indians, who occupied camps in this region before the arrival of Europeans, painted pictographs on a flat wall of this crack. Other massive granitic boulders from this slide can be observed on the floor of the canyon at river level.

Landslides also create lakes. The two Kern Lakes lie in the bottom of Kern Canyon, near the southern limit of ancient glaciation in the canyon. Little Kern Lake, about one-fourth mile in diameter and nearly one-half mile downstream from Kern Lake itself, is the older of the two. It was dammed by a large rockslide, about three-fourths of a mile wide and 1 mile long, that came down from the east wall of the canyon. The slide descended more than 2,000 feet from a conspicuous scarp at 8,800 feet. The landslide dam is composed of giant blocks of granite chaotically piled into the canyon. Lawson (1904) estimated, on the basis of the size of trees growing on the debris, that the rockslide was less than a century old when he studied it in 1903. The lake is shallow today and surrounded by marsh, and it has no doubt been partly filled by sediment and vegetation.

Kern Lake, nearly one-half mile long (Fig. 10.1), is dammed by masses

LANDSLIDES 317

10.1. Looking north across big Kern Lake and up Kern Canyon, in 1903. The lake was apparently formed in the winter of 1867–68 by landslides; the dead timber seen standing in the lake is the remains of trees drowned by the formation of the lake. Tower Rock is on the right skyline. (Photograph from Lawson, 1904.)

of debris that choke the lower reaches of two streams entering the river from mountains on the east. Lawson (1904) talked to W. T. Grant, who had been in the canyon in the summer of 1867 and had noted that the upper lake did not exist at that time. He returned in 1868 and found the present Kern Lake, and he suggested to Lawson that it originated from landslides triggered by an earthquake in the spring of 1868 that had caused widespread rockfalls in the region. The winter of 1867–68 was an exceptionally heavy one, with high precipitation, causing torrents that mobilized and deposited the debris that dams the river below Kern Lake. These lakes are apparently the only ones in the entire parks area (excluding dammed reservoirs) that do not owe their origin to glaciation.

Many granite boulders lie scattered over the surface of the floodplain of the North Fork of the Tule River at an altitude between 3,000 and 3,600 feet. The boulders range from car-sized to house-sized. One colossal boulder measures 70 x 55 x 45 feet and is estimated to weigh about 10,000 tons. The boulders, contrasting sharply with the sand and gravel alluvium covering the valley floor, are out of place in this setting. They apparently were deposited by a major debris avalanche down the south flank of Dennison Peak (7,290 feet), which towers 3,500 feet above the valley on its north side.

The north side of the same mountain mass saw the largest historic landslide in the parks region, one that destroyed part of a grove of Sequoia trees in the canyon of the South Fork of the Kaweah River late in the night of December 20, 1867. The landslide came after 41 days of continuous rain and snow had saturated the ground. The slide began when a mass of rock and debris on the north side of 8,000-foot Dennison Ridge broke loose and swept some 2.5 miles down into the river canyon, 15 miles upstream from Three Rivers. It formed a dam across the canyon, reportedly 400 feet high, that impounded the river and caused a temporary lake to form behind it.

Joseph Palmer, a homesteader in the canyon several miles below the site of the landslide, reported (Fry, 1931),

> Just before midnight I was aroused by a heavy rumbling sound such as I had never heard before, and which lasted for an hour or more. Then a great calm set in, and even the roaring of the river ceased. On leaving my cabin in the morning, I found that despite the heavy rain the river was low. From this I knew that a great slide had blocked the canyon above and that later the dam would give way and cause a flood.

He watched and waited over the next several days, and finally, on December 23, the blockage burst and the river came to life:

> About 1:30 A.M. I was aroused by a tremendous thundering and rumbling sound which made my hair stand on end. I jumped out of bed, grabbed my clothing, and ran for safety up the mountain side some 200 yards from the river. In a few minutes the flood came along with a crest of water some 40 feet in depth that extended across the canyon, carrying with it broken up trees which were crashing end over end in every direction with terrific force and sound. The river remained high for several days, and all the while timber was going down and being swept clear out to the valley.

The bursting of the landslide dam let loose a great flood, and the impounded water spilled and smashed its way down the Kaweah River, carrying everything before it. Water and debris, including giant redwood logs, spread out on the floor of the Central Valley. The flood arrived at Visalia, 42 miles downstream from the landslide, late in the evening of the same day, flooding parts of the town with 5 feet of water and scattering logs and driftwood for miles around.

The avalanche itself had been caused when a thick layer of decomposed granite overlying granitic bedrock on Dennison Ridge, saturated by more than a month of continuous rain and burdened by snow in its upper reaches, became unstable and failed. The mass slid off the steep north slope of the ridge, carrying with it acres of timber and vegetation. About one-third of the Garfield Grove of Big Trees was swept away, and 350 million board feet of timber was destroyed.

It is useful to be assured that the heavings of the earth are not the work of angry deities. These phenomena have causes of their own.

—SENECA (4 B.C.–A.D. 65)

GEOLOGIC STRUCTURES

Geologists have learned a great deal about the distribution, composition, and age of different rock units, but to better understand the overall geologic history of a given region, we need to inquire into the nature of the geologic structures that affect these rock units. Structural geology deals with the general disposition, attitude, and arrangement of rock masses, regardless of scale. The largest structures define the makeup of the crustal plates and their movement. Intermediate-scale structures include batholiths, plutons, and the principal faults and joint patterns. We have already examined some of the smaller-scale structures that can be observed in a single outcrop, structures that provide evidence on the evolution of metamorphic, granitic, and volcanic rocks.

The Crustal Structure of the Sierra

The tectonic plates that collectively form the outer part of the Earth are up to 90 miles thick. They constitute a relatively rigid layer called the *lithosphere*, and they overlie a more plastic layer in the Earth's mantle called the *asthenosphere* (Fig. 4.5). The interaction of the tectonic plates as the lithosphere moves unevenly over the asthenosphere produces the major structures and events related to global plate tectonics.

The lithosphere is composed of an upper layer, the crust, that is 10 to 40 miles thick in continental areas and about 4 miles thick (plus 3 to 4 miles of water) in oceanic areas, and a lower layer, the lithospheric mantle, that is about 50 miles thick. Between the crust and the lithospheric mantle is the Mohorovicic Discontinuity (called the Moho for short), a boundary below which seismologists note a distinct increase in the speed of earthquake waves. The crust (which lies above the Moho) consists of an upper part and a lower part, the upper part in continental areas being largely granitic in composition, the lower part composed of heavier rocks such as diorite and gabbro.

Most mountain ranges are believed to be held up by a crustal "root," a zone of especially thick low-density crust that forces the range up by flotation in much the same way that the deep root of an iceberg floats the exposed part above sea level. Long ago such a root was proposed for the Sierra Nevada, and the early seismic studies suggested that the Moho dipped down to about 34 miles under the range, rather than the average depth of about 19 miles beneath the more typical continental crust.

In 1995, geologists and geophysicists conducted a major seismic experiment across the Sierra to measure the depth to the Moho in detail, so as to gain more understanding of the structure of the range, and perhaps the reason for its great height. This work used two long lines, an east-trending line, about 240 miles long, passing through the Mineral King region and across the Sierra near 36.5° N latitude, and a second, north-trending line, 200 miles long, east of the range in Owens Valley. Explosive charges were set off in drill holes at a total of 19 sites, and hundreds of seismometers that had been placed along the lines (as well as off to the side) recorded the arrival times and intensity of the minute earthquake waves generated by the explosions. Because the instant of each explosion (and thus the generation of waves) was known precisely, the arrival times and travel times of these waves could be determined with great accuracy—an accuracy unattainable in recording a natural earthquake. Each recording station received the initial wave and a series of later waves. The later signals identified waves that had been reflected or refracted along paths other than the fastest path. The subsequent analysis of the whole assemblage of earthquake waves enabled seismologists to study the crustal structure below the range. These data were then combined with other geophysical data, including surveys of the gravity field within the range, to enhance our understanding of the nature of the crust and mantle beneath the Sierra Nevada (Wernicke and others, 1996).

The seismic study showed that the Sierra root is not as deep as previously thought. Moho depths along the east-trending line were found to be 21 to 26 miles under the Sierra and 19 to 21 miles under the Basin and Range Province to the east. Along the axis of the Sierra the Moho deepens northward from about 18 miles near the southern end of the Sierra to about 27 miles east of Fresno. Another surprising finding is that the thickest crust is not under the crest of the range, but is displaced some 25 miles west, to the region of the western foothills. These features all argue against a crustal root holding up the highest Sierra.

What, then, holds up this stretch of the Sierra Crest? Part of the answer lies in the fact that the granitic crustal rocks in the Sierra are low in density. Those of the Sierra Crest region on the east side of the range are especially high in silica (Fig. 6.4), and therefore especially low in density and will "float" highest. But this density difference within the crust, as well as differences in the thickness of the crust, cannot account for most of the height of the range.

Further facts relevant to this puzzle came from gravity surveys that had been made previously over the range (see Chapter 6). The gravity evidence can place constraints on the nature of deeper structure (below the Moho) that cannot be clarified by the seismic experiment alone, the waves from small explosions being too feeble to penetrate that deep and return. The combined surveys support the notion that a north-trending rib of hot, low-density asthenospheric material can account for both the uplift and the westward tilt of the mountain block (Fig. 11.1). This interpretation suggests that the light-weight, plastic asthenosphere has welled up into the lithosphere, breaching the lithospheric mantle, and

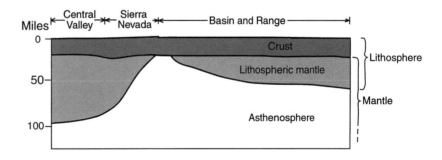

11.1. Lithospheric structure across the southern Sierra Nevada, based on seismic and other geophysical surveys. The rib of hot, lightweight asthenospheric mantle material under the Sierra is believed to hold up the range. No vertical exaggeration. Modified from Wernicke and others (1996).

now directly underlies the crust. The relatively low density of this asthenospheric rib, which underlies and parallels the Sierra Nevada Crest, is responsible for supporting the range at its present height. Most probably, the leakage of basaltic lava out of the fault zone bounding the Sierra Nevada Block on its east side had its origin in this asthenospheric welt.

We also know that east of the Sierra, during the Cenozoic, the crust of the Basin and Range Province in Nevada and Utah has undergone extensive regional uplift as well as extreme westerly stretching and extension. The hot mantle material that moved into the lower part of the Basin and Range lithosphere to accommodate the extension and uplift may have expanded to the west, thus providing the force needed to uplift the Sierra Nevada Block. (The Block is the total mountainous structure, consisting of the granitic batholith and all of the remnants of roof pendants, lava flows, and whatever else is integral to it.)

Faults in and near the Sierra Block

A planar crack or fracture through a mass of rock where the rock on one side has moved relative to that on the other is called a *fault*. A crack or fracture that has led to no slippage or movement is called a *joint*; the two sides of the joint may have separated from each other to a small extent, but one side has not slid notably past the other.

Faults range greatly in size and displacement. Small faults confined to a single outcrop may show displacements of less than an inch. Movement on the Lone Pine Fault in 1872 ruptured the ground surface for a distance of more than 60 miles, with lateral displacements of up to several tens of feet. Large fault movements are indeed the most common cause of large earthquakes. The fault slippage in 1872 produced, not a single quake, but a major swarm of earthquakes. These quakes, among California's largest, demolished the town of Lone Pine, 30 miles south of Independence (Figs. 11.2, 11.3).

There are three general types of faults, distinguished from one another by the nature of the offset, or displacement, along the fault surface (Fig. 11.4). *Normal* faults are those where the block on the upper side of an inclined fault moves down. *Reverse* faults (also called *thrust* faults) are those where the block on the upper side of an inclined fault moves farther up. *Strike-slip* faults are those where one side moves neither up nor down relative to the other, but horizontally. Strike-slip faults may be classified as *right-lateral* or *left-lateral*, depending on whether the block on the far side from a viewer has moved, respectively, to the right or the left (at a right-lateral fault, for example, the movement of the far side is to the right, whichever side of the fault one stands on).

When the details of fault offset are measured, many show a combination of vertical and horizontal movement, so that the resultant movement of one side may be, for example, both up and to the north relative to the other side. Such faults are termed *oblique-slip* faults.

The Sierra Block is remarkably free of young faults or seismic activity, despite the fact that faulting and earthquakes are common along its margins, especially on the east side (Fig. 11.2). In general, the range has behaved like a rigid block ever since the preexisting deformed sedimentary rocks and volcanic rocks were welded together by the intrusion and solidification of large masses of granitic melt in the late Mesozoic Era. The only sizable fault within the range in the parks area is the north-trending Kern Canyon Fault (Fig. 11.2), which has abetted the canyon-cutting propensities of river flow and glacial movement.

The faults within and adjacent to the Sierra Nevada can be divided into three main groups: (1) young faults along the east flank of the range, and those associated with Owens Valley, (2) the Kern Canyon Fault near the center of the range, and (3) older enigmatic faults that predate the emplacement of the batholith.

Young faults in and along Owens Valley. Several fresh major fault systems occur along the east base of the Sierra Block (Fig. 11.5). They include the fractures that accommodated the uplift of the range. Because the eastern margin of the range is also close to the east margin of the granitic batholith, the line of breakage along which the range has been uplifted is probably controlled by the transition of rock strength at the edge of the granitic mass.

Among the several major north-northwest-trending fault zones that have been active in late Tertiary and Quaternary time (see Fig. 4.1) at the eastern margin of the Sierra is the Sierra Nevada Frontal Fault Zone, which has seen the uplift of the Sierra Block on the west relative to the western side of Owens Valley and the Alabama Hills.

A parallel fault system, 7 miles to the east at the latitude of Mount Whitney, is the Owens Valley Fault Zone, a normal fault showing several thousand feet of displacement (Fig. 11.5). This fault separates the western part of Owens Valley, in which bedrock generally is within 1,000 feet of the land surface, from a structurally much deeper eastern trough, in which bedrock is buried by as much as 8,000 feet of young sediment. A segment of this fault zone is the Lone Pine Fault, which ruptured with disastrous results in 1872.

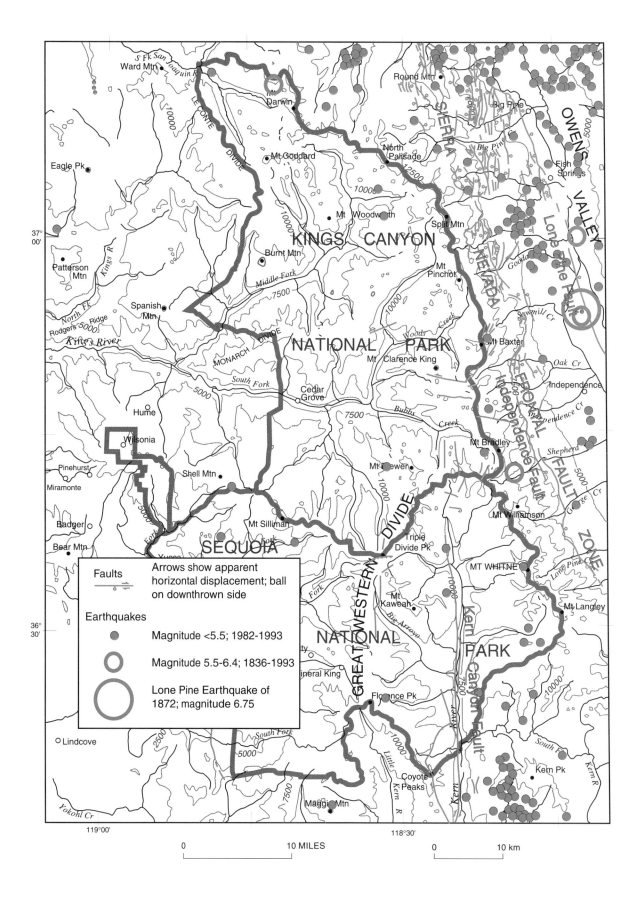

11.2. *(Opposite)* Major faults and earthquake epicenters within the parks area. Epicenters of quakes larger than magnitude 5.5 are shown for the 157-year period 1836–1993. The large circle indicates the epicenter of one of the two Lone Pine earthquakes of March 26, 1872. A larger one (magnitude 7.6) occurred 15 miles south, out of the map area, just north of the village of Lone Pine, which was destroyed with considerable loss of life. Epicenters smaller than magnitude 5.5 are shown for the 11-year period 1982–1993. Notice that few epicenters occur within the Sierra Block, and that most of those that do are associated with the Sierra Nevada Frontal Fault Zone and the Lone Pine Fault. Epicenters from Goter and others (1994).

11.3. *(Above)* View looking west at the Sierra Crest. The Sierra Nevada Frontal Fault lies at the base of the main east scarp of the Sierra, standing in the background. The Owens Valley Fault Zone, including the Lone Pine Fault that ruptured in 1872, lies between the town of Lone Pine (in the foreground) and the Alabama Hills (in the middleground), the latter representing bedrock exposed by uplift west of the Owens Valley Fault Zone. Peaks on the Sierra Crest are Mount Langley (L), Lone Pine Peak (LPP), Mount Whitney (W), Mount Russell (R), and Tunnabora Peak (T). U.S. National Park Service aerial photograph.

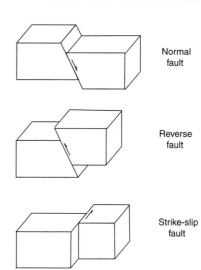

11.4. Block diagrams showing the three common types of faults: normal, reverse, and strike-slip; also called gravity, thrust, and transform.

A few miles east of the Owens Valley Fault Zone, bounding the east side of the deep structural trough of Owens Valley, is the Inyo-White Mountains Fault Zone, which separates Owens Valley from the uplifted Inyo Mountains Block of older basement rock to the east. Cumulative displacements on these fault systems have largely accommodated the uplift of the Sierra Block, the down-faulting of Owens Valley, and the uplift of the Inyo Mountains to the east, though in some places warping, rather than faulting, may also have operated. Presumably, the engine behind this regional uplift has been the upward rise of the asthenosphere rib discussed above.

The Sierra Nevada Frontal Fault Zone. This fault zone has been mapped as a discontinuous series of exposed fault traces that are commonly zigzag in plan (in map view) and steep in dip (the amount and direction of inclination), the dip commonly inclined down toward the east 60 to 80° from the horizontal. Most of these normal faults can be traced only a short distance. They occur at the base of the great eastern escarpment of the Sierra and are largely covered by the rock debris that has washed and tumbled from the mountains (Fig. 1.1). The Independence Fault (Fig. 11.2), west of the town of Independence, can be traced farther than most of the other individual faults. It forms a depression in the ridges of the lower part of the range front and can be traced for more than 20 miles. This fault is clearly quite young, for it cuts and offsets glacial moraines at the mouth of Independence Creek.

The lava flows of the Big Pine Volcanic Field are slightly offset by some of the faults. Other fault strands have produced relatively young fault scarps offsetting unconsolidated sand and gravel deposits. Those offsetting the Quaternary sediments commonly produce a youthful scarp 3 to 10 feet high, the east side down. In most places the individual faults are covered by alluvial material and show little surface expression, but in some places alignments of subtle changes in slope or vegetation characteristics, or the presence of springs, help define their position.

In general, the individual faults of the Sierra Nevada Frontal Fault Zone are normal faults exhibiting east-side-down displacement and little or no strike-slip movement. Total vertical displacement on the fault zone, apparently more than 2 miles, accounts for much of the uplift of the Sierra Nevada Block, relative to Owens Valley.

The small young volcanoes (basaltic cinder cones, with or without associated lava flows) that are aligned along the east base of the range (Fig. 11.6) were built where magma from depth rose up along strands of the frontal fault system. Two rows of these cones occur north of Goodale

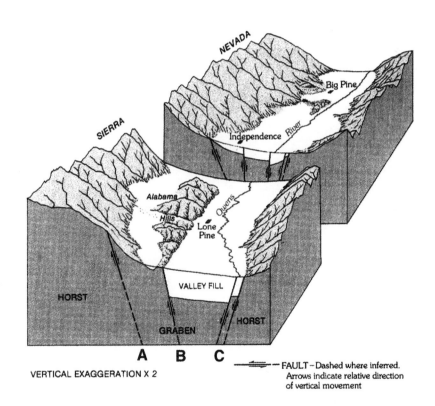

11.5. Block diagram of Owens Valley, showing the master fault zones that bound the valley and the Sierra. *A.* Sierra Nevada Frontal Fault Zone. *B.* Owens Valley Fault Zone (including Lone Pine Fault). *C.* Inyo-White Mountains Fault Zone. The Owens Valley Fault Zone is the only one that has a significant horizontal component of movement (right-lateral). Modified from Hollett and others (1991).

Creek. One 2-mile-long row of cones containing six small volcanic vents occurs directly at the base of the range front. A second subparallel row of three somewhat larger cones, and the lava flows fed from them, occurs ½ to 1 mile farther east, in Owens Valley. South of Goodale Creek, four more cones aligned along the base of the range apparently mark the position of a fault.

Certain segments of the Sierra Nevada Frontal Fault Zone appear as swarms of small faults that are displaced downward on the west side rather than the east. This curious mountain-down displacement occurs where the mountain front seems to be warped (rather than faulted up), and these small faults are providing an adjustment for the warp.

Many of the recent earthquake epicenters (Fig. 11.2) are distributed

along the Sierra Nevada Frontal Fault Zone. They demonstrate that the zone is still tectonically active, as is the Lone Pine Fault sector of the Owens Valley Fault.

The Owens Valley Fault Zone. This fault system extends north-northwestward for about 75 miles along much of the length of Owens Valley, from Owens Lake past Lone Pine, Independence, and Big Pine (Figs. 11.2, 11.5). The fault defines the western boundary of a deep, sediment-filled trough underlying eastern Owens Valley. West of the fault zone is a less depressed section of the valley that extends westward to the frontal fault zone of the Sierra Nevada Block.

It was a section of the Owens Valley Fault Zone that moved abruptly on March 26, 1872, to produce a swarm of earthquakes, including the great Lone Pine Earthquake. The trace of this movement along the Lone Pine Fault is marked by a nearly continuous zone of scarps, depressions, pressure ridges, and other linear features, some of which coincide with differences in groundwater level across the fault zone. The great Lone Pine earthquake (magnitude about 7.6) is among the most powerful historic earthquakes in California, exceeded only by those of 1857 at Fort Tejon and 1906 at San Francisco. Occurring on the same day as the great Lone Pine earthquake, and located about 15 miles north within the parks area (also on the Lone Pine Fault), was a quake of magnitude 6.75 (Fig. 11.2).

The 1872 earthquake resulted from rapid rupture along more than 60 miles of the fault, and was felt over much of California and adjacent states. Inasmuch as the quake occurred at 2:30 A.M., and most of the homes were of adobe construction, many of the casualties resulted from the collapse of the buildings on sleeping occupants. In the town of Lone Pine, 27 people died and many more were injured.

John Muir was in Yosemite Valley at the time, 110 miles northwest of Lone Pine. He was awakened in the predawn of March 26, 1872, immediately dressed, and raced outside to note that

11.6. Looking west at an 8-mile span of the steep east scarp of the Sierra Nevada. Cinder cones occur over volcanic vents that are aligned along strands of the Sierra Nevada Frontal Fault Zone (note in particular the six small aligned cinder cones near the number 3). Some of the vents have fed lavas that flow east (toward the viewer) to the floor of Owens Valley. Numbers refer to (1) mouth of Taboose Canyon, (2) Division Creek, (3) Armstrong Canyon, (4) Sawmill Pass, and (5) Taboose Pass. Photograph by Symons Aerial Survey.

the shocks were so violent and varied, and succeeded one another so closely, one had to balance in walking as if on the deck of a ship among waves, and it seemed impossible the high cliffs should escape being shattered. . . . Then suddenly, out of the strange silence and strange motion there came a tremendous roar. The Eagle Rock, a short distance up the valley, had given way, and I saw it falling in thousands of the great boulders I had been studying so long, pouring to the valley floor in a free curve luminous from friction, making a terribly sublime and beautiful spectacle—an arc of fire fifteen hundred feet span, as true in form and as steady as a rainbow, in the midst of the stupendous roaring rockstorm. The sound was inconceivably deep and broad and earnest, as if the whole earth, like a living creature, had at last found a voice, and were calling to her sister planets. It seemed to me that if all the thunder I ever heard were condensed into one roar it would not equal this rock at the birth of a mountain talus.

The main storm was soon over, and eager to see the newborn talus, I ran up the valley in the moonlight and climbed it before the huge blocks, after their wild fiery flight, had come to complete rest. They were slowly settling into their places, chafing, grating against one another, groaning and whispering; but no motion was visible except in a stream of small fragments pattering down the face of the cliff at the head of the talus. A cloud of dust particles, the smallest of the boulders, floated out across the whole breadth of the valley and formed a ceiling that lasted until after sunrise; and the air was loaded with the odor of crushed Douglas spruces, from a grove that had been mowed down and mashed like weeds (Muir, 1912).

Josiah Whitney, of the California Geological Survey, traveled to Lone Pine after the earthquake and wrote the first geologic description. He thought that the ground's cracking was a result of subsidence due to the extensive shaking, and did not appreciate the presence of the primary fault trace or the nature of the fault displacement. He also missed the evidence for horizontal as well as vertical movement along the fault trace.

John Muir, after his trip to Kings Canyon in 1873, crossed the Sierra by way of Kearsarge Pass and traveled south in Owens Valley to make an attempt on Mount Whitney. He passed through the town of Lone Pine near where the epicenter of the 1872 earthquake had been and noted the ground subsidence visible in the vicinity of the town.

It was not until 11 years later, in 1883, however, that G. K. Gilbert, then with the U.S. Geological Survey, visited the area and clearly described the nature of the faulting in his notebooks, as reported by Beanland and Clark (1994). He noted that in addition to vertical displacement, the terrain on the opposite side of the fault had moved to the right, indicating a right-lateral strike-slip displacement. The fault therefore had both dip-slip and strike-slip offset. Modern measurements of these features indicate that the average right-lateral displacement in 1872 was about 20 feet and the maximum was about 33 feet. The vertical displacement, down to the east, averaged about 3 feet, with a maximum of about 14 feet.

The Inyo-White Mountains Fault Zone accounts for the steep front and great length of the west scarp of the Inyo-White Mountains. This normal fault system (Fig. 11.4) is largely buried by alluvial material washed from the mountains, but like the Sierra Frontal Fault Zone it has served as a conduit for rising magma, which has vented and built young basaltic cinder cones at the west base of the Inyo Mountains, east of Big Pine.

Kern Canyon Fault. The north-trending Kern Canyon Fault is a major structure that dominates the landscape of the eastern part of Sequoia National Park, and at more than 80 miles is the longest fault within the southern Sierra Nevada Block. A zone of weakened, crushed rock along the fault has guided drainage and favored streams in cutting a giant canyon, the Kern Canyon, which was occupied repeatedly along its northern 20-mile segment by glacial ice during the Ice Ages (Figs. 9.2, 11.7). Major valley glaciers, fed from ice fields on both the main Sierra Crest at the east and the Great Western Divide at the west, turned and flowed south down this watercourse. The glaciers widened and straightened the rivercourse, producing the remarkably linear canyon of the upper Kern, which is several thousand feet deep and 25 miles long. This giant canyon, flanked by highlands on each side, has produced the double-crested aspect of the southern Sierra, a configuration not known until the 1864 exploration by the Brewer party.

Within the parks the trace of the fault is relatively straight or smoothly curving. Secondary subparallel faults and shears are common within a zone up to one-half mile wide on each side of the master fault. Most of the fault offset occurs across a region less than 300 feet wide in which the rock is extensively crushed and altered. Within this region, thin shear zones (less than an inch thick) are made up of totally pulverized rock that has been commonly recrystallized and replaced by small tourmaline crystals giving a blue cast to the sheared rock. The tourmaline, a silicate mineral containing both water and boron, was apparently precipitated by hot aqueous fluids that rose from depth, channeled by the crushed rock of the fault zone.

The west side of the fault has moved north as much as 8 miles relative to the east side, indicating that the fault displacement or offset is right-lateral, strike-slip. Displacements have been determined by measuring the

offset of near-vertical contacts of granitic plutons that have been cut and offset by the fault. The offset decreases systematically northward, and the fault cannot be traced north of the Kings–Kern Divide, where no offset can be demonstrated. Several small right-lateral faults near the divide may mark branches of the main fault where it splays and disappears. The decrease in displacement northward on the fault may be accommodated by this branching of the fault, and also by the possibility that the batholith was still hot enough to respond somewhat plastically when it was transected by the fault.

The Kern Canyon Fault is known to be quite ancient; no modern earthquake epicenters are concentrated along the fault trace (Fig. 11.2). A lava flow that has been radiometrically dated at 3.5 million years covers the fault south of the parks area. We know that the fault movement is older than the lava, because the lava is not fractured or displaced by the fault. Other evidence suggests that the fault is only slightly younger than the 85-million-year-old Whitney Intrusive Suite (the group of plutons that includes Mount Whitney), which it cuts and offsets (Fig. 6.7).

As old as it is, the Kern Canyon Fault still serves as a conduit for fluids rising from depth. A hot spring, a favorite of backpackers, occurs on the fault trace at the canyon bottom near Chagoopa Creek and Rock Creek, and a soda spring, popular during the cocktail hour, issues from the fault zone 10 miles farther south near the mouth of Golden Trout Creek. This mineral-soda spring, an important landmark and rest stop on the Hockett Trail, was first noted on the Wright map of 1883, near where the trail crossed the Kern River to ascend the canyon of Golden Trout Creek (then called Volcano Creek; Fig. 2.38). Additional evidence of the rise of volatiles along the fault is the presence of the secondary mineral tourmaline.

11.7. Aerial view looking 20 miles south down the remarkably straight Kern Canyon. Red Spur is the ridge in the right middleground, the Chagoopa Plateau behind it. Canyons on the left are, from the bottom, Tyndall Creek, Wallace Creek, Whitney Creek, and Rock Creek. Long, dark morainal ridges, surrounded by bare granite, can be seen flanking Rock Creek below the eastern (left) skyline. Kern Canyon was formed when stream erosion cut a valley in the weakened rock along a major right-lateral fault. Later a glacier, fed from the highlands on both sides, flowed down the valley and widened it, thus producing the present U-shaped valley. U.S. Geological Survey oblique aerial photograph taken January 1973.

Faults older than the batholith. Several important fault systems must have been active within the parks area in the past, as deduced from the distribution of older rock units as well as from the mapping of chemical gradients within the granitic rocks. These faults were active before the intrusion of the broad expanses of Cretaceous granitic rocks, and hence their location, offset, and importance are speculative.

On the west side of the batholith the various faults that were active during the subduction process no doubt provided the fundamental boundaries between the various belts of metamorphic roof pendants. The Boyden Cave Roof Pendant contains rocks belonging to both the Kings Sequence of metasedimentary rocks and the Mesozoic metavolcanic rocks. The zone separating these two rock sequences is considered to be an important fault (Nokleberg, 1983).

Extensive geochemical analyses of granitic samples collected over the expanse of the batholith show that the edge of the continental crust, as defined by strontium isotope content (Fig. 6.22), takes a wandering path both north and south of the mapped area. These variations suggest that a series of ancient faults have offset blocks of continental crust between 185 and 177 million years ago, prior to the development of the Sierra Nevada batholith (Kistler, 1990).

The Nature of Joints

Joints are cracks or fractures in rock along which little or no movement has taken place, except for the slight separation of one fracture wall from the other. Jointed rock is common in the mountain range and is especially evident in the High Sierra, where vast areas of bare rock are exposed, unobscured by soil or vegetation. Joints are most obvious in the granitic rock; because such rock contains little internal structure, there is no prominent layering, such as bedding or schistosity, controlling the alignment of rock fractures. As a result, joints are commonly rather evenly spaced, with three sets intersecting each other roughly at right angles so that the rock mass is broken into blocks. The resulting blocky aspect is characteristic of granitic outcrops (Fig. 11.8), and is an integral part of the landscape in granitic terrain. Major intersecting joints are beautifully displayed in aerial photographs of glaciated unforested regions (Fig. 11.9).

The spacing of the joints is commonly controlled by the grain size of the rock. A coarse-grained granite with a grain size of about ¼ inch may show joints spaced several feet or tens of feet apart, whereas a fine-grained

11.8. The highest part of the Sierra Nevada. Looking north to the summit of Mount Whitney across Third Needle, Day Needle, and Keeler Needle (compare with Fig. 2.10). Many of the needles and buttresses on the steep east face are shaped by two intersecting, nearly vertical joint sets. The stone cabin just visible on the Whitney summit was built in 1909 as living quarters for scientists using this high-altitude site for observations. U.S. National Park Service photograph by Bill Jones.

granitic rock with a grain size of $1/16$ inch may have an average joint spacing of a foot or less. This relation is nicely shown where a fine-grained light-colored dike cuts a coarse-grained granite. Commonly, the fine-grained rock is cut by many more closely spaced joints than is the host granite.

The high interior of the mountain range is cut by many nearly parallel joints, the larger of which can be traced on aerial photographs for as much as a mile or so (Fig. 11.9). These joints appear in the photos because erosion, by both water and ice, has commonly removed fractured rock along individual joints, or closely spaced joints, producing linear clefts and ravines or notches in cliff faces. Similar joints no doubt occur in the lower parts of the range, but they are not well exposed because of extensive forest, brush, and soil cover. Joints smaller than those visible in the aerial photographs, spaced only a few feet or yards apart (Figs. 11.8, 11.11), are ubiquitous.

The small, closely spaced fractures also prove to be faults in some places

because they show a slight offset parallel to the fracture. These small offsets, commonly of only an inch or so, are best seen where the joint-faults cut other, older structures in the rock, such as light-colored aplite dikes (Fig. 11.12) or dark inclusions. Examination of many of these small faults shows that those trending easterly are offset in a left-lateral sense, whereas those trending northerly are offset in a right-lateral sense. Where offsets can be determined on the larger lineaments seen in aerial photographs, they show similar offsets, as do the small faults.

The geometry of offset of these two joint sets, one trending easterly, the other northerly, suggests that they formed together in response to a regional stress (Lockwood and Moore, 1979). A compressive stress acting roughly at right angles to the northwest-trending grain of the batholith, that is, a compressive stress directed northeasterly-southwesterly, could account for the two sets of joint-faults.

These joint-fault sets are younger than the intrusion of the granitic plutons, because they generally cut across both the plutons and the boundaries between the plutons. They are usually mineralized with relatively high-temperature minerals, however, indicating that they were active and open at considerable depth when the rock was still hot, yet brittle. We may conclude, therefore, that they opened only shortly after the emplacement and solidification of the batholith.

In many places, a group of thin quartz veins invades these joints. The veins commonly contain sulfide minerals that incorporate copper, molybdenum, lead, zinc, and gold. In a few places these concentrations have been prospected, but no real mines have been developed.

In addition to these sets of nearly vertical joints, many joints in the granitic rocks are nearly horizontal and form roughly parallel to the ground surface (Fig. 11.13). In granitic masses that are relatively free of vertical joints and dominated by horizontal joints, the rock fracture produces a series of concentric sheets or shells somewhat resembling the lay-

11.9. Regional sets of joints, one predominating set trending roughly easterly, in the Mount Pinchot 15-Minute Quadrangle. The area is between the upper South Fork of the Kings River (upper left) and Woods Creek (bottom). The black oval areas are lakes: Bench Lake is at top center, and Lake Marjorie is at the upper right. North is to the top, and the vertical dimension is 7 miles. With respect to the map of Fig. 11.10, the area lies between the "Canyon" of Kings Canyon National Park and the "Clarence" of Mt. Clarence King. U.S. Geological Survey aerial photograph taken October 7, 1948.

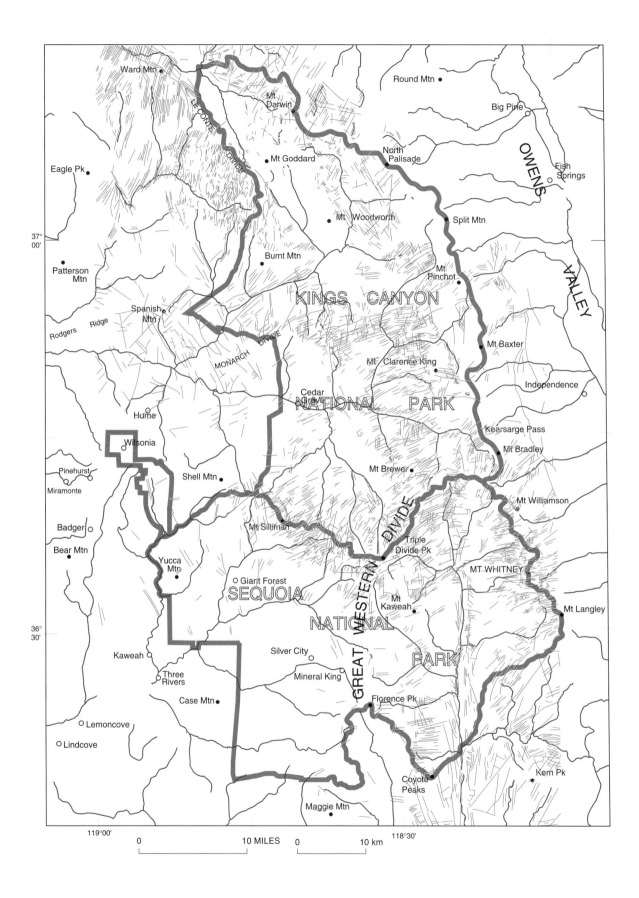

ered skin of an onion (Figs. 11.14, 11.15). Such *exfoliation* joints are spaced from a few feet to 30 feet apart.

Members of the Brewer party of 1864 encountered these joints in the area between Big Meadows and Mount Brewer, and were among the first to note that such joints are approximately parallel to the ground surface and not only arch over domes but bend down beneath canyons (Fig. 11.13). They supposed that the jointing "must have been produced by the contraction of the material while cooling or solidifying, and . . . in many

11.10. *(Opposite)* Map of the parks area showing traces of joints and small faults as they appear where exposed on aerial photographs above timberline (compare with Fig. 11.9).

11.11. Linear inclined rock face near Sugarloaf Creek bounded by joint surface. The bedrock on the right side of the joint has been removed by glacial action. U.S. Geological Survey photograph by Grove K. Gilbert.

11.12. Aplite dikes cutting granodiorite and offset by small faults. *Above.* Four-inch dike offset by seven small left-lateral strike-slip faults. *Below.* One-inch dike cut by several small faults, both right- and left-lateral.

places, we see something of the original shape of the surface, as it was when the granitic mass assumed its present position" (Whitney, 1865).

Further study of this problem 40 years later led Grove K. Gilbert (1904) to conclude that the exfoliation joints were not a result of cooling but, rather, formed by expansion of the rock mass as the confining pressure was reduced by erosional removal of the overlying material. We now know that erosion of up to several miles has occurred since the emplacement of the granitic rocks, and it is therefore unlikely that the original pluton roof shapes would be mirrored by today's exfoliation joints.

The exfoliation joints are common over much of the Sierra Nevada

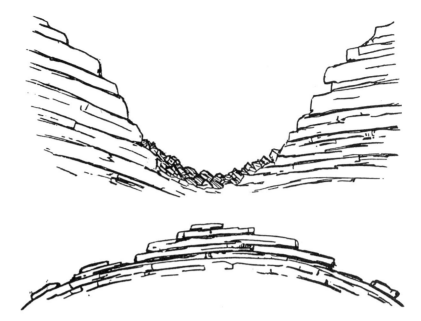

11.13. Exfoliating sheets of granite that maintain a rough parallelism with the ground surface, both beneath canyons (*above*) and on hills or domes (*below*) (Whitney 1865).

11.14. Nearly horizontal joints on an unglaciated remnant on a ridge crest near Mount Maddox. U.S. Geological Survey photograph by Francois Matthes.

and can be seen in favorable natural exposures and quarries even in the foothills of the range. Generally, however, two sets of vertical joints are also present, so that the rock fractures on two or three intersecting planes and is broken into blocks (Fig. 11.16). The bold aspect of cubic or blocky shapes is characteristic of granitic terrain. These blocks are commonly pried apart by the freezing and expansion of ice, are carried by glaciers and tumble downslope in rockslides, and thus become a common component of moraines and of talus piles at the base of slopes.

In places where a mass of granitic rock is not cut by sets of nearly vertical joints, the exfoliation joints become dominant. There they shape the spectacular domes that are a common feature of the Sierra Nevada at midlevel altitudes. The curved exfoliation sheets are well shown on Moro Rock, near General Grant Grove (Fig. 11.17), and on Ball Dome, near Silliman Crest (Fig. 11.18). The largest and most magnificent dome in

11.15. Exfoliating sheets on a steep 1,500-foot slope northwest of Elizabeth Pass, at the headwaters of the Roaring River and the Middle Fork of the Kaweah River. Intersecting vertical joints are visible on the glaciated surfaces in left background. U.S. Geological Survey photograph by Francois Matthes.

11.16. Tumbled, frost-riven blocks formed by sets of vertical joints cutting nearly horizontal exfoliated slabs, in the White Chief area south of Mineral King. U.S. National Park Service photograph by P. J. Wyckoff.

the parks area is the 3,600-foot-high Tehipite Dome (Fig. 11.19), rising above Tehipite Valley on the Middle Fork of the Kings River. John Muir described the dome as "a gigantic round-topped tower, slender as compared with its height, and sublimely simple and massive in structure. It is not set upon, but against, the general masonry of the wall, standing well forward, and rising free from the open sunny floor of the valley, attached to the general mass of the wall rocks only at the back. This is one of the most striking and wonderful rocks in the Sierra" (Muir, 1891).

Uplift and Erosion

Because the Sierra Nevada has undergone uplift for a long period of time, it has been eroded to a considerable depth. Most of this eroded material was deposited in a subsiding sedimentary basin to the west, now partly occupied by the Central Valley of California. These sediments indicate that erosion began to unroof the batholith in Cretaceous time.

The depth of emplacement of the batholith has been estimated by reference to the known conditions of pressure and temperature required for the original crystallization of certain metamorphic and igneous minerals collected at the surface. We have seen, in Chapter 6, that such evidence indicates that the currently exposed part of the batholith cooled at a depth of 2 to 4 miles and perhaps as deep as 9 miles. As uplift proceeded since the Cretaceous Period, therefore, a layer of roof rocks of at least this thickness had to have been eroded to expose the granite to its present level.

11.17. Moro Rock, showing nearly horizontal exfoliating shells near the summit curving over to nearly vertical on the side. U.S National Park Service photograph by W. Schoeb.

11.18. Ball Dome, near Silliman Crest, exhibiting several exfoliating shells. U.S. Geological Survey photograph by Grove K. Gilbert.

One hint to the timing of the uplift of the Sierra came from a study of fossil leaves in western Nevada. Every plant species, and especially a group of plants living together, grew within a rather restricted climate range, and this climate is related to altitude under certain climate and latitude regimes. According to this study (Wolfe and others, 1997), the region stood as a plateau, 3,000 to 5,000 feet higher than present ground elevations, 15 to 16 million years ago. The Sierra Nevada may have been the west flank of this plateau. What this scheme suggests is that the plateau has subsequently subsided, irregularly, to produce the present-day corrugated aspect of the Basin and Range Province.

11.19. Tehipite Dome, which soars 3,500 feet above Tehipite Valley on the Middle Fork of the Kings River. U.S. National Park Service photograph by W. Huber.

Part of the story of the uplift and tilting of the Sierra Nevada Block can be learned by examining the Central Valley deposits and those within the drainage basin of the upper San Joaquin River, which drains the west slope of the Sierra just north of the area of the parks. Eocene sediments in the Central Valley indicate that 50 million years ago the Sierra Nevada was a modest, low-lying range. Formerly, the San Joaquin River headed to the east of the Sierra Crest, and as recently as 3.2 million years ago it flowed west across what is now the range crest at Deadman Pass (28 miles north of the parks area).

Within the San Joaquin River Drainage Basin are the remnants of a lava flow that erupted 9 to 10 million years ago, as dated by the potassium-argon method. Today, the river has greatly eroded the flow and cut down below its base, so that only isolated patches of lava remain. Enough remains, however, that the altitude of the top and bottom of the original flow can be determined along the river valley for a distance of 15 miles on the west slope of the range. These remnants indicate that the ancient river valley in which the lava flowed had a wide, shallow cross section unlike the incised river valley of today. But the present longitudinal profile of the ancient valley is steeper than that of present streams. This indicates that the ancient lava-filled valley—and the range with it—has been tilted west, steepening the fossil profile.

Reconstruction of the ancient lava-filled valley allows one to estimate the history of uplift since eruption of the lava flow (Huber, 1981). Since 9 to 10 million years ago, about 7,000 feet of uplift has occurred at the range crest. About 3,100 feet of this uplift occurred during the last 3 million years, indicating that the rate of uplift has accelerated with time.

Taken together, this fragmentary evidence indicates that since the end of the emplacement of the Sierra Nevada Batholith 85 million years ago, the overall uplift of the range has been at least 2 to 4 miles and perhaps as much as 9 miles total, but that in general the range was leveled by erosion as it grew. At 15 to 16 million years ago, however, some of the terrain in Nevada east of the Sierra may have been standing at elevations of 10,000 to 12,000 feet, suggesting that parts of the Sierra may have been elevated at that time. Since 9 to 10 million years ago, the range crest has been uplifted about 7,000 feet, and since 3 million years ago, about 3,100 feet. Perhaps this estimated rate of the latest uplift, about one foot in a thousand years, is as fast as, or faster than, those uplift rates that prevailed in the past.

In completing one discovery we never fail to get an imperfect knowledge of others of which we could have no idea before, so that we cannot solve one doubt without creating several new ones.

—JOSEPH PRIESTLEY, 1775

AFTERWORD

Our insatiable quest to move beyond the frontiers of human understanding has been an unremitting force throughout our history, propelling the exploration and mapping that have prefaced the nation's growth to greatness. Various regions of the Earth had been known, more or less, for millennia, but the shape of the globe itself was not appreciated until the time of Columbus, and it was several centuries later before even the oceans and continents were outlined. The delineation of such features as river courses and mountain ranges followed, and was paralleled by the development of new mapping tools. These instruments and techniques facilitated the development of a global geographic reference system in which all such features could be located with some precision. North America did not see much of this first-generation mapping phase completed until the middle of the nineteenth century.

The search for transportation routes, especially for immigrant travel, railroads, and military concerns, stimulated finer-scale mapping of the broader unknown areas between the rivers and within the mountain ranges. Detailed mapping of both settled and promising areas was sorely needed, to provide a basis for property division and ownership. It was during this phase that efforts were made to begin the exploration of the highest Sierra, a region that had remained a backwater in exploratory programs because reconnaissance had shown no useful travel routes across it, no promising mineral deposits, and, other than the grazing of sheep, no pressing need to challenge its forbidding heights. The early settlers, a hardy lot who had already crossed the continent from the east, or had approached the mountains via ports along the Pacific shore, often after months at sea, simply avoided and circumvented the high country.

Other than Alaska, the region of the parks was one of the last areas in the United States to be systematically investigated and mapped, and the record of those efforts is thus more extensive and more complete than that in other regions on the continent that had been explored in much earlier times.

In the century since the time of the original exploratory incursions and the discovery of the highest Sierra, mapping and scientific studies have continued, trending constantly toward more and more detailed investigations. The scale of the mapping by the U.S. Geological Survey has systematically increased from the first 30-minute quadrangles (2 miles to the inch, completed about 1910), through the 15-minute quadrangle series (1 mile to the inch, completed in the 1950s), to the 7.5-minute quadrangles (0.377 mile to the inch, completed in the 1980s). What are the next 30 years likely to bring?

The more detailed maps have enabled all manner of fresh new observation and measurement, whether for more precise delineation of the land and its creatures or greater understanding of the origins, internal processes, and structures of the mountain mass. Improvements in the design of existing instruments, and the development of wholly new instruments and techniques, have facilitated geological investigations not dreamed of in the early decades. The exponential growth of science and instrumentation—more has been done in ten years than in the previous thirty—will no doubt continue to engender advances beyond our ken.

Several factors seem to have underlain the audacity and success of the early explorers. Strength of character, formal training, and personal connections loom large. Those who first probed the highest Sierra exhibited attributes often seen in those whose achievements have led the way in other branches of human endeavor. Most were driven men, possessed in the pursuit of their objectives, men who reveled in their accomplishments, and who infused their companions and others with their own energy and vision. Most were charismatic. They were leaders prepared to push forward even on an uncertain course, where the chance of failure was great and even simple survival was often in doubt. Work in the high country required strong will, robust health, and wilderness skills—horsemanship, camping experience, familiarity with weapons, and ingenuity in the face of the unexpected—in abundance.

But a second factor these investigators brought to their task was formal academic training. Most were trained professionals who were academically inclined by nature, and most had acquired invaluable instruction and experience under the guidance of outstanding mentors.

John Fremont, who first systematically mapped the eastern and western sides of the Sierra, and attempted crossings in several places, attended Charleston College, received intensive experience in navigation aboard a U.S. naval vessel, and assisted the renowned French geographer Joseph Nicollet for several seasons in the northern plains region.

William Blake graduated from the scientific school at Yale College and signed on with the Corps of Topographical Engineers in 1853 for a survey of railroad routes in California headed by the very capable Lieutenant Robert Williamson.

Josiah Whitney, the creator and director of the California Geological Survey, was born into a wealthy family, graduated from Yale College, and then spent several years studying at mining institutes in Europe. He came to California only after spending several years studying ore deposits in the eastern states and Michigan.

Clarence King and James Gardiner were young students fresh from the Yale Scientific School when they joined Whitney's newly formed California Geological Survey. They received instruction in the field under the dynamic group that included Charles Hoffmann, a young German topographer, William Brewer, a biologist and natural scientist who was also a Yale graduate, and Whitney himself.

Joseph N. LeConte was born into an academic family, his father the first geology professor at the University of California. The younger LeConte graduated from the University and continued his earlier mapping and exploration while serving at California as a professor of engineering himself.

John Muir grew up on a small farm in Wisconsin and developed both a robust nature and a close rapport with the natural world. He attended the University of Wisconsin and later devoted a solitary thousand-mile walk through the central U.S. to the study of botany. In California and especially in Yosemite Valley, he benefited from the many visits of both eminent naturalists and men of letters, luminaries who were drawn to the valley by its great renown.

Several of the explorers honed their literary skills by extensive letter writing, for centuries the sole means of long-distance communication and an exercise considered an art form in its day. Both Josiah Whitney and William Brewer maintained a lifelong correspondence with their brothers. Brewer later used these letters, in conjunction with his diaries, to assemble his book *Up and Down California*. John Muir—in addition to piling up sketch-filled notebooks—corresponded extensively with numerous people, including his bother and sister and Mrs. Ezra Carr, a fellow botanist whom he first met at the University of Wisconsin, but who later moved with her husband to California. In later life he continued an extensive correspondence, largely with his wife and daughters.

The success of these early explorers was also due, in most cases, to support from government agencies or other formal institutions. Support of this kind provided the funding indispensable to such undertakings. The right agency or connection could also see that investigation results found their way into print, and without publication these studies would be unknown and useless. Many also had the backing of powerful patrons, who could be counted on to support their bids for more ambitious projects.

Fremont was an officer in the U.S. Army Corps of Engineers, and had the funding and authority of the U.S. Government behind him. His expeditions to the west were clearly promoted by his father-in-law, Senator Thomas H. Benson of Missouri, who had a personal interest in western expansion, believing that California and the west should become a part of the union. From earliest times, exploration has been the first step—and seemingly the implied justification—in taking possession of a little-known territory.

The Williamson railroad route surveys were also funded by the Corps of Engineers, and the Brewer party received its support from the California Geological Survey, which had special appropriations—and a mandate—for the publication of its results.

Of all the explorers, John Muir was the conspicuous exception. At least in the early phases of his studies, he was a loner, with virtually no support and no powerful friends (and a powerful enemy in Josiah Whitney). But by virtue of his phenomenal ability to explore vast areas on foot, usually alone, and his keen observational and analytic skills, he was eventually recognized as an authority on the High Sierra. His exceptional writing skills enabled him to publish popular articles in both newspapers and scientific journals, and later in widely circulated national magazines. His publications included insightful essays on plant and animal life—as well as geological topics—and soon began exciting the interest and support of powerful conservation proponents.

Joseph N. LeConte moved rather easily into his mapping program in the Sierra because of his father's experience in the Sierra and position as professor at the University of California—a position much like that the younger LeConte would soon take. Being a charter member of the Sierra Club, Little Joe had a ready vehicle for the publication of his results.

But no explorer ever entered an "unknown territory" with no knowledge of it whatever. He at least knew it was there, and that was justification enough for exploring it. Commonly, bits and pieces were to be learned from the native people, or from travelers or trappers, and he would make good use of them. Fremont had the foresight to enlist the aid of Joseph Walker, a mountain man, and the famed Kit Carson as scouts. The Brewer party learned much from the party of prospectors from Independence they encountered in the high country, concerning the obscure trail up Bubbs Creek and over Kearsarge Pass. And from the time of John Muir onward, the great repositories of knowledge about travel routes in the Sierra were the sheepherders.

Beyond motives of adventure and achievement, the purpose of a journey of exploration was to gain more and better information, and perhaps, along the way, to discover something new. Before an expedition was undertaken, every effort was made to assemble all that was known about the area in question. The Brewer party of 1864 made use of Goddard's map of 1855, and in 1873 John Muir carried the Hoffmann map that had been published earlier that year. Exploring the extant written information before undertaking fieldwork is just as important today. This means research, research that can often be greatly aided by computerized filing and retrieval systems in the voluminous holdings of a modern library.

Future investigations in the highest Sierra will have the same objective as that of the early explorers, to collect ever more refined information. A well-formulated, well-executed map conveys the most information in the most economical form, and since antiquity maps have been the most treasured products of exploration and discovery.

The most salient recent advance in mapping techniques has been the development of the global positioning system (GPS), which permits precise location of one's position—anywhere on the planet—by the use of signals constantly transmitted from an array of Earth-orbiting satellites. It is now possible for a geologist on foot, using a self-contained device about the size of a deck of cards, to locate himself or herself at a single position, in just a few minutes, to a precision of a few tens of feet in the global coordinate system. This is a precision comparable to that obtained in the 1870s by Clarence King employing the most sophisticated field astronomical devices, telegraphed time, and weeks of labor by highly trained associates.

The GPS system will become increasingly important for future studies of all kinds, to ensure that the locations of field-collected samples or data are precisely established. This will then enable the testing of geographical correlations between diverse data sets. In its most sophisticated form, the global positioning system is now capable of locating field points to within a fraction of an inch, a capability that will no doubt improve and become more widely available in the years to come. The GPS is valuable not only in the construction of accurate maps and the location of field stations, but also in measuring the rate of tectonic processes. It will be especially useful in measuring, over time, major movements of regional crustal units. Such measurements will furnish, for example, information on the rate of uplift of the Sierra Nevada Block relative to that of Owens Valley.

Future geologic studies will generally follow the same pattern, that of mapping in increasing detail the extent and character of the various geologic rock units and the features and structures within them. The wave of new technologies will be pertinent, as well, for studies advancing the geophysical and geochemical sciences. Extended arrays of seismometers will no doubt be laid out in the highest Sierra, where very few are now in place. They will help not only in defining the deep geologic structures marked by precise location of small earthquakes, but also in recording distant earthquakes, and by so doing improve our knowledge of the continental crust.

Further investigations of magnetic properties in the region of the Sierra are especially needed, including an aeromagnetic survey over the full extent of the parks region. The results could be compared with extant field measurements of the magnetic susceptibility of plutonic rocks, a comparison that can lead to a better understanding of factors controlling the oxidation of iron-rich minerals in the granitic rocks.

Paleomagnetic studies can also be employed to record the Earth's magnetic field by determining the alignment of the iron-bearing minerals that form during the time that igneous rocks cooled, and from the data on that fossil magnetism thus amassed, study the movement of tectonic plates since that time. Finally, the magnetization of a rock body can define a structure within the rock resulting from the alignment of mineral elements in the late stages of crystallization.

The unraveling of the geologic history of a region has become ever more reliant on age determinations. What we know of the age of the Earth and Moon, and of the ancient travels of the continents, derives from these determinations of the age of rock units. But we will never have enough radiometric measurements of critical rock units, because every new age we determine raises new questions about the detailed sequence of local and regional rock history.

In a time when only a few excursions had been made into a new land, it was reasonable for one person to expect to become knowledgeable about all that was known about the area, but that is clearly impossible today. Modern computerized geographical information systems (GIS) have been created to manage the huge bodies of data now available on almost any given area. Future studies will use such systems not only to test correlations among all of the data sets collected within the area of interest, but also to compare these data with the data collected worldwide in similar environments—clearly a task only a computer can handle.

Much of the work done in times past by exceptionally gifted explorers and pioneering scientists is increasingly being relegated to systems of remote sensing and analysis, systems having limits of discrimination and refinement that seem almost magical, even to those who employ them. Astronomers now watch the growth and deformation of entire hypothetical galaxies, across many billions of years, in minutes, on their laboratory monitors. Aeronautical engineers know precisely the flight characteristics of a new aircraft before it leaves the ground. Such new systems are frontiers for future exploration that could not have been dreamed of by the pioneers of the highest Sierra.

| APPENDIX | GEOLOGIC ROAD AND TRAIL GUIDES

Only a few roads penetrate very far into the highest Sierra, and none crosses the crest of the range along the 175 miles between Tioga Pass, in the Yosemite region, and Walker Pass, 70 miles south of Mount Whitney. Nonetheless, many interesting geologic features, some of them quite spectacular, can be studied along the existing roads that mount the western slope of the range. Because of its dearth of vegetation, the highest country is overwhelming in its sheer majesty, truly a geologist's Valhalla of continuous outcrop, but it can be visited only by trail walking. This Appendix presents mile-by-mile geologic guides for the following roads and trails, which comprise the principal routes of travel on the west slope up to the crest of the range, but not over the crest or down the east escarpment:

Kings Canyon Highway, State Route 180, from Fresno past Cedar Grove to road's end in the Kings Canyon (85 road miles, Fig. A.1).

State Route 198 and the Generals Highway (as designated past the park entrance), from Visalia past Ash Mountain, Giant Forest, and Lodgepole to the connection with State Route 180 near Grant Grove (81 road miles, Fig. A.1).

The Mineral King Road, from State Route 198 to Mineral King (25 road miles, Fig. A.2).

The John Muir Trail, that part within Sequoia and Kings Canyon

National Parks, from Mount Whitney to Paiute Creek (102 trail miles, Fig. A.2).

The High Sierra Trail, from Crescent Meadows, near Giant Forest, to the trail's connection with the John Muir Trail on Wallace Creek (49 trail miles, Fig. A.2).

Mileage for field-trip stops, points of interest, and reference points for the road log is given in the left-hand column as distance in miles from the beginning of each of the field-trip routes. The second column records the mile notations of selected Department of Highway roadside mile markers, which are small white numbered signs mounted on metal posts. The trails do not have mile markers. Field-trip stops on the roads shown in Fig. A.1 indicate stops on Route 180 by numbers and those on Route 198 by letters.

Analyses of the mineral content of granitic rocks at various localities are often listed in the road log. The analyses are listed in volume percent for each mineral, with the following abbreviations: Pl, plagioclase; Kf,

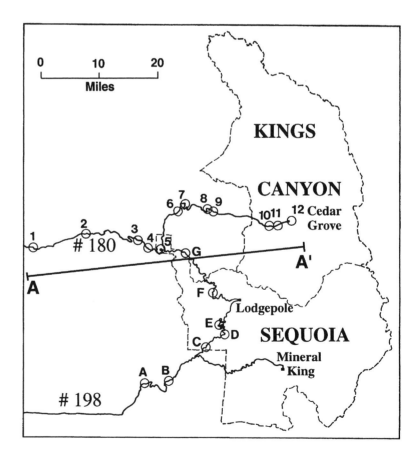

A.1. Map of the field trip road routes and stops. The twelve stops on Highway 180 are designated with numbers; the seven stops on Highway 198 (and Generals Highway) are designated by letters. The geochemical trends of Fig. A.4 are plotted on line A-A'.

A.2. *(Opposite)* The field trip routes within the vicinity of the national parks.

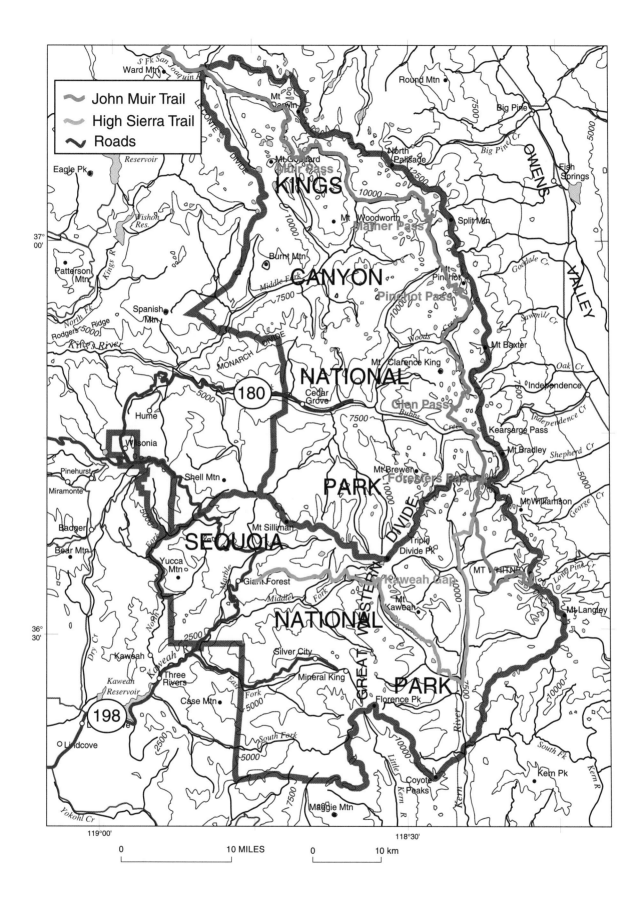

potassium feldspar; Q, quartz; and Mf, dark minerals. Note that all dark-colored minerals (augite, hornblende, biotite, iron-titanium oxides, sphene) are grouped together.

The geologic guides to the roads in Sequoia and Kings Canyon National Parks are taken from a U.S. Geological Survey Open-File Report (Moore, Nokleberg, and Sisson, 1994), with modifications.

Kings Canyon Highway, State Route 180

The Kings Canyon Highway, State Route 180, offers an 85-mile-long geologic traverse nearly across the southern Sierra Nevada (Fig. A.1). Only the highest part of the range is not crossed. Proceeding 13 miles east from Fresno, the route crosses to the south side of the Kings River, and then climbs 40 miles up the west flank of the range to Grant Grove, at 6,500 feet. For the next 15 miles, the route drops into the Kings Canyon near the confluence of the Middle and South Forks of the Kings River (where the canyon exceeds 8,000 feet in depth) and ends about 20 miles farther east at the head of the Yosemite-like canyon of the South Fork of the Kings River. The road's end lies only about 15 trail miles short of the separate road head on the east side of the Sierra, west of the town of Independence; between the two stands Kearsage Pass. The eastern part of the Kings Canyon Highway route is shown on Fig. A.3, which is an oblique physiographic map of the central Sierra Nevada, viewed from the southwest (Alpha, 1977).

This route follows rather closely parts of the trail of the first scientific exploration into this part of the Sierra Nevada, the 1864 Brewer Survey of the California Geological Survey (Fig. 2.2). During the 1864 expedition, the party explored part of the southern Sierra Nevada and discovered and named many of the loftier peaks in the range, including Mounts Brewer, Goddard, Gardiner, Cotter, Clarence King, Williamson, Tyndall, and Whitney.

The field-trip route provides a look at three of the four belts of metamorphic roof pendants in this part of the Sierra: ophiolitic rocks of the Kings–Kaweah Terrane on the west, metasedimentary rocks of the Kings Terrane, and Mesozoic metavolcanic rocks of the Goddard Terrane. Only the eastern High Sierra Terrane of Paleozoic metasedimentary rocks is not crossed.

Compositional trends across the Sierra Nevada Batholith are well displayed along the field-trip route (see Figs. A.1 and A.4). The transition from dark granodiorite and tonalite on the west to lighter-colored grano-

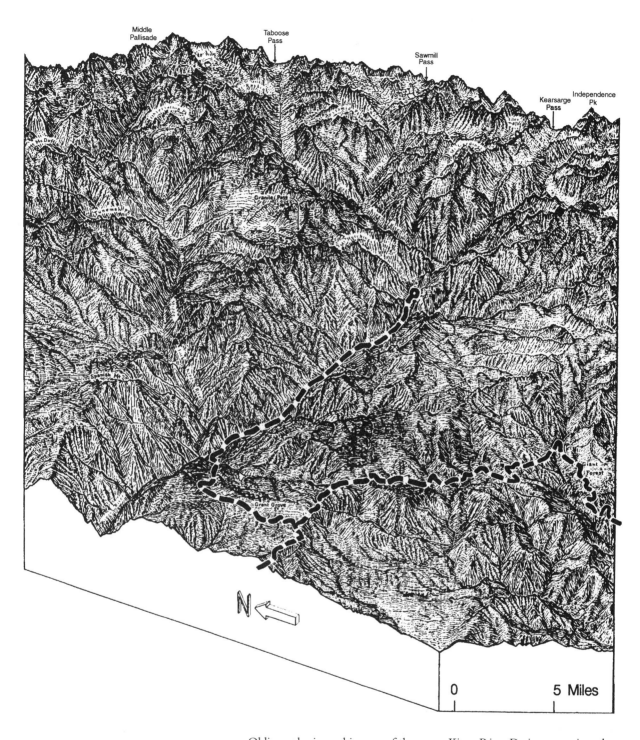

A.3. Oblique physiographic map of the upper Kings River Drainage, as viewed from 30 degrees above the horizontal from the southwest. State Route 180 extends east up Kings Canyon in the center, and Generals Highway contours around to the right (south). The deepest part of Kings Canyon, as measured down from the summit of Spanish Mountain, is in the left foreground (Alpha, 1977).

diorite and granite on the east is notable. Moreover, toward the east, the volume percent of potassium feldspar increases, the initial strontium 87/86 ratio increases, and the emplacement ages of the granitic rocks decrease (Fig. A.4).

The Kings Canyon Highway provides an opportunity to examine the results of glacial erosion in the South Fork of the Kings River Canyon, and to observe small remnants of late Cenozoic potassium-rich basaltic lava flows.

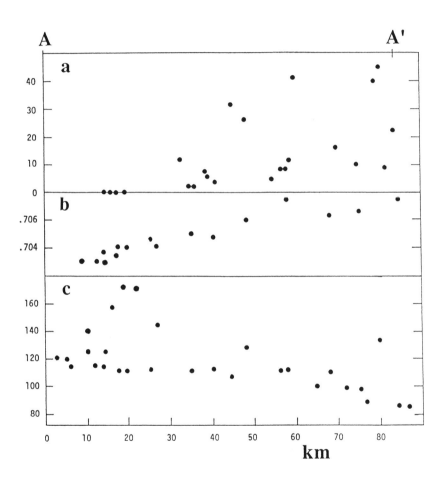

A.4. Analytical data on granitic rocks collected on or near State Route 180, the collection localities projected onto line A-A' of Fig. A.1. *a*, volume percent of potassium feldspar; *b*, initial $^{87}Sr/^{86}Sr$ ratios; *c*, U-Pb zircon ages of granitic rocks in millions of years before present. *a* and *b* point to the notion that, toward the east, the granitic rocks were formed by the melting of increasingly larger proportions of continental crust. *c* demonstrates that this melting, which produced the plutons making up the Sierra Nevada Batholith, were intruded successively eastward during the Cretaceous.

LOG OF STATE ROUTE 180 FROM FRESNO TO ROAD'S END AT CEDAR GROVE (85.5 MILES)

Miles	Signs	
0.0		Intersection of Kings Canyon Highway (State Route 180) and Clovis Boulevard, east Fresno, California. Check and set your odometer here. The highway is designated FRE180 on highway mile signs.
0.5	64.0	
12.6	76.0	
12.9		Kings River. *Historical note.* A group of Spanish explorers camped near here on January 6 (The Day of Epiphany), 1806. To commemorate this church day, they named the River "El rio de los Santos Reyes" or the river of the Holy Kings, from which is derived Kings River. Early maps spell it King's, which is not in accord with either current usage or the original name.
13.6	77.0	
16.2		The western margin of the Sierra Nevada Batholith includes Early Cretaceous mafic intrusions of gabbro, diorite, tonalite, and granodiorite. U-Pb zircon ages on gabbro and more siliceous compositions and K-Ar hornblende ages on gabbros range between 110 and 126 Ma (Sharp, 1988). Campbell Mountain and Jesse Morrow Mountain, to the south and north, respectively, consist primarily of olivine-hornblende gabbro cumulates. Note the subdued landslide topography with the upper depression and the lower hummocky lobe on the north side of Campbell Mountain.
19.8	83.0	
20.0	Stop 1	Large exfoliated boulders. Isolated outcrop of dark-colored tonalite with characteristic large sieve-textured biotite crystals. Mode: Pl, 52.4; Kf, 0; Q, 22.8; Mf, 24.8. Uranium-lead zircon age, 114 Ma (Chen and Moore, 1982).
21.1		Dark tonalite. Mode: Pl, 48.9; Kf, 0; Q, 22.2; Mf, 28.9.
21.2		Dark tonalite. Mode: Pl, 59.2; Kf, 0; Q, 19.2; Mf, 21.6.
21.7	85.0	
22.2		1,000-foot atitude.
25.7	89.0	
26.8		Squaw Valley Store.
28.2		Fine-grained, light-colored, irregularly banded tonalite. Mode: Pl, 57.4; Kf, 1.0; Q, 33.5; Mf, 8.1.
29.0		Bridge over Mill Creek.
29.8	93.0	
30.1	Stop 2	Large gravel turnout. Walk back (west) about 80 yards to see the blocks of pillow lava (Fig. 5.6) on the north side of the highway that have rolled down from Bald Mountain to the north. They are part of the Permian and Triassic Kings-Kaweah Terrane, which consists of several tectonic blocks composed of peridotite, gabbro, mafic dikes, pillow lava, rare chert, and zones of serpentinite-matrix mixed rock all metamorphosed to hornblende hornfels grade of meta-

morphism (Saleeby, 1978). The rocks are recrystallized to irregular grains of sodium plagioclase and hornblende, and are intruded by Jurassic and Cretaceous plutons. The southernmost Bald Mountain slab consists of well-preserved mafic dikes at its northern end, which grade into well-preserved pillow lava to the south.

The Kings Canyon Highway crosses over the southern end of a Jurassic plutonic complex in the vicinity of Bald Mountain. This complex consists of mafic rocks including gabbro and diorite, and tonalite with U-Pb zircon ages of both Middle and Late Jurassic (Chen and Moore, 1982; Sharp, 1988). This plutonic complex intrudes the Kings River Ophiolite along a 12-mile length and is cut by younger Cretaceous plutonic rocks. It is well exposed in roadcuts adjacent to the pillow lava.

30.9	94.0	
31.7		Clingans Junction.
32.5		2,000 feet.
32.9	96.0	
33.0		Dark granodiorite. Mode: Pl, 47.5; Kf, 11.7; Q, 25.3; Mf, 15.5.
34.4		Tonalite. Mode: Pl 55.2; Kf - 2.2; Q - 27.0; Mf -15.6.
35.4		Tonalite. Mode: Pl, 53.7; Kf, 2.3; Q, 29.3; Mf, 14.7. U-Pb zircon age, 111 Ma (Chen and Moore, 1982).
36.0		3,000 feet.
38.3		Granodiorite. Mode: Pl, 50.2; Kf, 7.2; Q, 30.4; Mf, 2.2. The irregular increase of potassium feldspar in the granitic rocks and the eastward predominance of granodiorite over quartz diorite or tonalite fixes the Quartz Diorite Line (Moore, 1959) at approximately this position.
39.5		Tonalite. Mode: Pl, 49.2; Kf, 5.1; Q, 30.9; Mf, 14.8.
39.8		4,000 feet.
40.5		Snowline Lodge. Quartz diorite. Mode: Pl, 54.0; Kf, 3.4; Q, 22.8; Mf, 19.8. U-Pb zircon age, 112 Ma (Chen and Moore, 1982).
41.2	104.0	
42.1	Stop 3	Large paved turnout. Walk west on the highway 50 yards to dark-colored late Tertiary volcanic rock (leucite-bearing basalt) in the roadcut. This is one of many small lava-flow remnants and associated vents of potassium-rich basaltic rocks scattered about on the west slope of the Sierra Nevada (Moore and Dodge, 1980). Most of the volcanic remnants differ slightly from one another in composition and represent individual small eruptions. This lava contains phenocrysts of olivine, biotite, and augite and fine-grained minerals of augite, biotite, leucite, K-feldspar, apatite, and opaque minerals. The lava contains: SiO_2, 49.3 wt.%, MgO, 11.5%, and K_2O, 5.5%. About 1.2 miles south, in Milk Ranch Canyon, a second late Cenozoic flow remnant of basalt is visible capping Stony Flat. This lava contains phenocrysts of olivine and pyroxene as well as fine-grained minerals of plagioclase, pyroxene, opaques, apatite, and minor biotite, and contains SiO_2, 55%. The whole-rock K-Ar age of the lava at Stony Flat is 3.4 Ma.

43.9		5,000 feet.
44.2	107.0	
44.8	Stop 4	Large paved turnout where a biotite-muscovite-garnet-sillimanite granite crops out east of a thin screen of metasedimentary rocks. Mode: Pl, 36.3; Kf, 31.8; Q, 31.0; Mf, 0.6; Garnet, 0.3. The rocks contain: SiO_2, 75.5 wt.%; MgO, 0.06%; Na_2O, 4.5%; and K_2O, 3.9%. The garnet-bearing rock occurs only in a marginal zone about 0.3 mile thick within the pluton, which is 10 miles long. The marginal granite has a high initial $^{87}Sr/^{86}Sr$ value of 0.7673. This rock yields a U-Pb zircon age of 106 Ma (Chen and Moore, 1982). *Historical note.* During the summer of 1864, a party of the Whitney Survey made the first scientific expedition into the High Sierra Nevada in the drainages of the Kings, Kern, and San Joaquin Rivers, adjacent to the region covered by this field trip. The party entered the mountains from Visalia, traveled through Millwood, at the site of Sequoia Lake 1.2 miles northeast of here, and entered the giant sequoia trees at General Grant Grove (Farquhar, 1965).
45.3	108.0	
47.3	110.0	6,000 feet.
48.2	111.0	
48.3		Park Entrance Station. Note that sampling or collecting rocks is not allowed in the park without a collecting permit.
48.9	Stop 5	Big Stump Picnic Area; restrooms. Light-colored granite. Mode: Pl, 37.1; Kf, 26.1; Q, 31.4; Mf, 5.4.
49.9		The Wye. Intersection of State Route 180 with the Generals Highway to Giant Forest and other parts of Sequoia National Park.
51.3		Visitors' Center, Grant Grove. *Historical note.* The 1864 Whitney Survey party traveled from Grant Grove through Big Meadows, 6 miles east of here, and climbed a peak on the Kings-Kaweah Divide, which they named Mount Silliman. From this point, they descended Sugarloaf Creek to Roaring River and ascended a high peak on the Great Western Divide, which they named Mount Brewer, after the party chief. From this vantage point, several high peaks on the main Sierra Nevada Crest, about 9 miles to the east, were mapped and named, including Mounts Whitney, Goddard, Williamson, and Tyndall. Clarence King and Richard Cotter then made a five-day backpacking trip over the Kings-Kern divide, where they climbed Mount Tyndall on the main Sierra Nevada Crest. After rejoining the main party near Mount Brewer, the group returned to Big Meadow and worked down into the great canyon of the South Fork of the Kings River to near the present site of Cedar Grove (Farquhar, 1965).
52.6	112.09	Park boundary.
54.3		Cherry Gap, 6,897 feet.
54.5	114.0	
57.3		Hume Lake turnoff; 3 miles to lake. A small area of Tertiary basalt occurs near this turnoff. Two miles east on the Hume Lake road, and on the upper road to Camp Seven, one can obtain excellent views of the canyon of the Middle Fork of the Kings River, Mount Clarence King, and Tehipite Dome.

58.6			Dark Giant Forest Granodiorite. Mode: Pl, 49.8; Kf, 4.7; Q, 21.1; Mf, 24.4.
58.6	118.0		Views into Kings Canyon.
58.8		Stop 6	Olivine basalt lava flow in stream channel (Fig. 8.2). Basalt rests on about 6 feet of sediment composed of granitic sand, basaltic ash, and lapilli, which in turn rest on weathered granodiorite. The basalt contains phenocrysts of olivine, pyroxene, and plagioclase along with fine-grained minerals of olivine, pyroxene, plagioclase, potassium feldspar, apatite, opaques, and rare biotite. The rock contains 47.2 wt.% SiO_2, 10.9 wt.% MgO, 2.45 wt.% Na_2O, and 1.2 wt.% K_2O. To the east from this point can be seen the Middle Palisade and University Peak on the main Sierra Nevada Crest and Mounts Gardiner and Cotter on King Spur. Mount Clarence King is obscured by the Monarch Divide.
59.6	119.0		
62.0		Stop 7	Junction View (Fig. 9.1). From here, one can look down into the junction of the Middle and South Forks of the Kings River, observing the V-shape of these giant river-cut canyons. Faintly visible, upstream, one can see the high walls of the glacially carved Yosemite-like canyons of these rivers separated by the Monarch Divide. Points visible, from north to south (left to right), are Deer Ridge, Spanish Mountain (10,051 feet), Tombstone Ridge, Tehipite Valley, Monarch Divide, Wren Peak (9,450 feet), and University Peak (13,632 feet) on the main Sierra Crest to the southeast. The canyon below the junction is one of the deepest in North America, dropping 8,011 feet from Spanish Mountain (10,051 feet) on the north to the river (Fig. 1.6). Much of the north wall of the canyon is carved in the granodiorite of Brush Canyon.
64.1	123.55		Bridge over Ten Mile Creek.
64.2			Barton's Resort. Mafic granodiorite and the quartz diorite of Yucca Point. Mode: Pl, 48.4; Kf, 8.2; Q, 20.2; Mf, 23.2. The quartz diorite of Yucca Point is surrounded and overlain by the younger granodiorite of Brush Canyon in Kings Canyon.
64.5			Mafic granodiorite and quartz diorite of Yucca Point. Mode: Pl, 43.2; Kf, 8.0; Q, 27.3; Mf, 21.5.
65.5			Mafic granodiorite and quartz diorite of Yucca Point. Mode: Pl, 43.5; Kf, 11.1; Q, 27.3; Mf, 18.1. U-Pb zircon age, 110 Ma (Chen and Moore, 1982).
65.6	125.0		
66.3			Fine-grained porphyritic granodiorite dike apparently related to the granodiorite of Brush Canyon. The nearly horizontal dikes of this rock are visible across the canyon cutting the quartz diorite of Yucca Point on Monarch Divide. This unit contains 70–74 wt.% SiO_2, and here on the highway the mineral content is: Pl, 28.4; Kf, 41.1; Q, 29.0; Mf, 1.5. This dike is among the youngest dated granitic rocks, with a U-Pb zircon age of 86 Ma (Chen and Moore, 1982). The road passes through only small offshoots of the Brush Canyon unit, but across the river the pluton is more than 7 miles wide, and much of the canyon wall is carved into it. The main mass of the granodiorite of Brush Canyon has yielded Ar-Ar ages of about 90 Ma (Brent Turrin, written communication, 1992).

APPENDIX: ROAD AND TRAIL GUIDES　367

67.2		This part of the highway cuts obliquely through the contacts between the western units of metasedimentary rocks in the Boyden Cave Roof Pendant, as follows: biotite-feldspar-quartz schist, calcium silicate schist, andalusite-biotite schist. Local abundant north-northwest-striking second-generation folds also occur, along with locally abundant dikes and small masses of the granodiorite of Brush Canyon and garnet-bearing aplite-pegmatite dikes.
68.5	128.0	
68.6		Redwood Creek, with a high, lacy waterfall in metamorphic rocks south of the highway.
69.3		Prisoner Camp Turnout. The anomalous flat and knob toward the river are produced by a landslide originating on the steep slope east of the highway. Convicts worked on the construction of this part of the Kings Canyon Highway during the 1930s and were housed near this site.
	Stop 8	
69.5	129.0	Horseshoe Bend. This stop is near the central part of a well-bedded, well-sorted, feldspar-rich, quartz sandstone unit of the Boyden Cave Roof Pendant within the Kings Terrane. Apparent cross-bedding and some graded bedding occur in strata that are nearly vertical. The dominant rock here is a massive gray to white quartz sandstone, with an apparent thickness of about 3,600 feet. Microscopic studies of the quartzites show that quartz generally comprises more than 90% of the rock, with as much as 15% potassium feldspar. Because these originally horizontal sedimentary beds have been rotated into a vertical position, the direction toward the top (younger part) of the sequence is in doubt. Field measurements indicate that the dominant direction toward the top of the layered quartzite unit is toward the west, but care must be taken to separate intersecting foliations from cross-bedding in this unit. 　The quartz-rich metasandstones in the western part of the Boyden Cave Roof Pendant are variously interpreted as forming in a shallow-water, tidal- and wave-influenced environment, or as the remnants of a large channel in the Kings Sequence fan system. The ultimate source of the sediment, as determined from age data from zircon grains, was presumably a Precambrian continental mass. 　Looking northwest from the turnout at Horseshoe Bend, the various units of the western part of the Boyden Cave Roof Pendant are spectacularly exposed on the southern slopes of Monarch Divide, including the bold outcrops of blue-gray marble (Fig. 5.5).
	Stop 9	
70.7	130.13	Bridge over the South Fork of the Kings River. At the Boyden Cave parking area, west of the bridge, is massive, vertically layered, blue-gray marble with an apparent thickness of about 2,000 feet (Fig. 5.5). To the southeast, the marble unit is thinned by faulting and pinches out in the upper part of Boulder Creek. Local, small- to moderate-sized caves lie within the marble unit, as at Boyden Cave. At river level, water entering small cave openings in marble are visible on the south flank of Monarch Divide.

East of the marble unit is a zone of slaty rocks that contains fragments of other rock types, including blocks of marble, quartzite, and schist in a fine-grained matrix.

Sparse fossils, mainly crinoids and ammonites of Early Jurassic age, have been recovered from this unit along the trail following the lower reaches of Boulder Creek (Jones and Moore, 1973).

71.3 — Beware of dangerous highway traffic along this part of the Kings Canyon highway!

A spectacular exposure of east-northeast-striking minor chevron folds is on the north side of the highway in a large tectonic block of interlayered calcium silicate phyllite in the slaty unit (Fig. 5.2).

Toward the east, the highway crosses alternating units of quartz-rich and biotite-rich slaty rocks that trend north-northeast and dip steeply.

Farther east, the tops of the sedimentary bedding appear to be toward the east in the quartz- and biotite-rich slaty rocks (Saleeby and others, 1978). Tops (the upward, younger, direction in a layered sequence) are determined by the grading in thin beds from a quartz-rich (originally sandy) base to a mica-rich (originally clay-rich) top, and by small-scale channels that cut into the layers below.

South of the canyon, the rocks in the central and western parts of the Boyden Cave Roof Pendant are faulted against the metavolcanic rocks of the eastern part of the pendant (Nokleberg, 1983). Along the Kings River, these important relationships are obscured by an intervening granitic pluton, the granodiorite of Tombstone Creek.

72.5   132.0

72.6 — Contact between the granodiorite of Tombstone Creek to the west and the rocks of the Mesozoic metavolcanic belt (The Goddard Terrane) to the east (Fig. 5.1). The granodiorite of Tombstone Creek has a preferred U-Pb zircon age of 99 to 102 Ma (Chen and Moore, 1982; Saleeby and others, 1990), and the metarhyolite tuff unit has a U-Pb zircon age of 106 Ma (Saleeby and others, 1990). These relations indicate a narrow time span for (1) deposition of the metavolcanic rocks at about 102 to 106 Ma; (2) steep regional tilting of the metavolcanic unit; and (3) intrusion of the granodiorite of Tombstone Creek at about 99 to 102 Ma.

74.3 — 4,000 feet.

75.6   135.0

75.9 — Grizzly Creek. Granodiorite of Lightning Creek. Mode: Pl, 46.3; Kf, 15.5; Q, 22.5; Mf, 15.7. Preferred U-Pb zircon age of 100 Ma (Saleeby and others, 1990). The granodiorite of Lightning Creek intrudes the eastern margin of metamorphosed rhyolites.

The main trunk glacier that occupied the South Fork Canyon reached to about this point during its greatest advance in the Pleistocene.

79.2 — Bridge across the South Fork of the Kings River. Granodiorite of Lookout Peak. Mode: Pl, 49.6; Kf, 9.7; Q, 21.5; Mf, 19.2; U-Pb zircon age, 97 Ma (Chen and Moore, 1982). Despite a low potassium feldspar content, this rock generally

		has large potassium feldspar phenocrysts. The granodiorite of Lookout Peak, along with the correlative 98-Ma granodiorite of Castle Creek to the south, forms the outermost members of the zoned Mitchell Intrusive Suite, which extends 36 miles south.
80.3		Cedar Grove Village turnoff. Still in the granodiorite of Lookout Peak.
81.2	Stop 10	Canyon Viewpoint (5,025 feet). Turnout on north side. Good view up U-shaped glacial valley (Figs. 2.28, 2.32).
81.7		Turnout on the north side of the highway. Granite of North Mountain. Contact between the granite of North Mountain and the granodiorite of Lookout Peak is in the saddle behind the knob to the south.
82.3		Turnout on north side of the highway. Float boulders of the granite of North Mountain. Mode: Pl, 30.4; Kf, 38.9; Q, 27.9; Mf, 2.8.
83.1	Stop 11	Roaring River turnout on east side of Roaring River across bridge. Short trail to Roaring River Falls (Fig. 9.18). The granite of North Mountain, which crops out near the falls, constitutes a small pluton about 3 miles in diameter. The top of the pluton is exposed high on the canyon walls, where it is mantled by a thin hood of metamorphic rocks. The granitic rocks of the North Mountain unit consist of a heterogeneous, fine-grained, light-colored granite with admixed dark granitic rocks and many aplite dikes. Average mode: Pl, 20.9; Kf, 44.9; Q, 33.2; Mf, 1.0.
83.6		Turnout on the north side of the highway, just across Kings River Bridge. This is the best place to see the contact of the granite of North Mountain on the south wall of the canyon.
83.8		Turnout on south side of the highway. Contact of the granite of North Mountain is visible on the south wall. This is the best place to walk to the contact on the north canyon wall and to see the metamorphic screen between the Lookout Peak and North Mountain Plutons.
84.6		Turnout on south side of the canyon. Abundant isolated boulders of the granodiorite of Lookout Peak, showing large Carlsbad-twinned potassium feldspar crystals. Mode: Pl, 57.7; Kf, 8.6; Q, 16.7; Mf, 17.0.
84.8		Grand Sentinel Viewpoint on south side of the highway (Fig. 2.35). Glacial grooves and polish can be seen high on the canyon walls, indicating that the Pleistocene glaciers were hundreds of feet thick at this point. *Historical note.* John Muir explored Kings Canyon in 1873, 1875, and 1877, and wrote of the Grand Sentinel: "Beyond the Gable Group, and separated slightly from it by the beautiful Avalanche Canyon and Cascades, stands the bold and majestic mass of the Grand Sentinel, 3300 feet high, with a split vertical front presented to the valley, as sheer and nearly as extensive as the front of the Yosemite Half Dome. Projecting out into the valley from the base of this sheer front is the Lower Sentinel, 2,400 feet high; and on either side, the West and East Sentinels, about the same height, forming altogether the boldest and most massively sculptured group in the valley" (Fig. 2.35: Muir, 1891).
85.5	Stop 12	Road's End turnout on north side of the loop. Note that sampling or collecting of rocks in the park is not allowed without a collecting permit.

From here, a distinctive turquoise copper stain can be seen at about the 6,800-foot-altitude on the northeast face of Grand Sentinel. The copper occurs in minor sulfide mineralization with epidote and quartz in a system of left-lateral joints (shears) trending northeast.

Passing through the Roads End loop is the contact with the porphyritic granodiorite of the Paradise Pluton, which extends east for about 6 miles. This pluton is characterized by "camouflaged" potassium-feldspar phenocrysts filled with concentrically arranged inclusions of biotite, hornblende, and plagioclase. Float boulders of this rock are visible here. Average mode: Pl, 43.6; Kf, 22.0; Q, 22.5; biotite, 6.7; hornblende, 3.5; accessory minerals, 1.8. U-Pb zircon age, 83 to 86 Ma.

The Paradise Pluton forms the outer shell of one of the largest and youngest nested plutonic sequences in the Sierra Nevada Batholith, with dimensions of 16 by 50 miles, and an area of 470 square miles. The pluton extends N 30° W from Olancha Peak, on the Sierra Nevada Crest, to near Granite Pass, on Monarch Divide. The core of the plutonic sequence is occupied by the younger Whitney Pluton, which is characterized by large (up to 4-inch) potassium feldspar phenocrysts. The Whitney Pluton has a U-Pb zircon age of 83 Ma and K-Ar ages of 79 to 82 Ma (Evernden and Kistler, 1970). This concentric plutonic sequence is similar in composition, age, size, structure, and tectonic setting to the Tuolumne Intrusive Suite in Yosemite National Park about 90 miles to the north-northwest.

Historical note. The 1864 Whitney Survey party passed this spot and made several attempts to explore the region to the north. They made their way up Copper Creek to Granite Basin and from the crest of the divide near Granite Pass (10,673 feet) found that their pack animals could not descend into the Middle Fork of the Kings River. From the crest, they mapped and named Mount Goddard on the San Joaquin–Kings Divide, the Palisades on the main Sierra Nevada Crest, and three nearby peaks on Kings Spur: Mounts Gardiner, Cotter, and Clarence King. Returning to the South Fork Canyon, they worked their way east of this spot up Bubbs Creek to Kearsarge Pass and down to Owens Valley (Farquhar, 1965).

State Route 198 and the Generals Highway

This part of the field trip begins in the San Joaquin Valley in eastern Visalia on California Route 198, at Country Center Drive. The highway is designated TUL198 on highway mile signs, but no signs are present in the parks. The road goes east into the Sierra Nevada Foothills, past Lake Kaweah and the town of Three Rivers, and enters Sequoia National Park at the Ash Mountain entrance 36.5 miles from Visalia. From the park entrance, the road (now designated the Generals Highway) continues up the canyon of the Middle Fork of the Kaweah River, then generally northwest, up steep switchbacks on the north wall of the canyon to Giant Forest (campground, store, gas, food). Continuing northwest, the road passes the park campground and store at Lodgepole on the Marble Fork of the Kaweah River and continues across the drainage of the North Fork of the Kaweah River to join with State Route 180 near Grant Grove, 81 miles from Visalia.

The first half of the route traverses the Kings-Kaweah Terrane, then across dark granodiorite and associated mafic plutonic rocks of the western Sierra Nevada Batholith. Near Lake Kaweah, and continuing to Three Rivers, the highway crosses two major roof pendants of metasedimentary rocks of the Kings Terrane. Thereafter the highway traverses the outer part of the 99-Ma Giant Forest Granodiorite.

LOG OF STATE ROUTE 198 AND GENERALS HIGHWAY (81.3 MILES)

Miles	Signs	
0.0	8.0	Junction of State Route 198 and Country Center Drive in Visalia. Shopping center south of the highway includes Denny's, Albertson's, and K-Mart. Reset your odometer here.
1.0	9.0	
1.4	9.4	Highway 63 exit.
10.0	18.0	
10.7		Highway 65 to Exeter and Porterville.
11.0	19.0	
11.7		Route 245 to Woodlake on left. Homers Nose is visible to the east, from here to about mile 14. The 600-foot-high nearly vertical cliff is formed by domelike exfoliation in the granite of Case Mountain, which covers a broad area in the drainages of the East and South Forks of the Kaweah River.
13.0	21.0	
13.2		High Sierra Road. The long, low Badger Hill just to the south is underlain by serpentine-bearing mixed rock containing Permian and Carboniferous ophiolitic rocks.
13.8		Yokohl Creek. The Rocky Hill granodiorite stock can be found 2 miles on the road that turns off beyond the creek. It has yielded K-Ar ages of 132 Ma (Putnam and Alfors, 1965), 114 and 115 Ma (Evernden and Kistler, 1970), and a Pb-U zircon age of 120 Ma (Saleeby and Sharp, 1980). A tonalite found in the Sierra Foothills 6 miles north yielded a questionable U-Pb zircon age of 116 Ma (Chen and Moore, 1982).
		Historical note. Just off the highway is a bronze plaque commemorating the Jordan Toll Trail: "This monument is near the beginning of the Jordan trail laid out by John Jordan in the year 1861. The trail shortened the route to the silver mines in the Coso Range near Owens Lake east of the Sierra Nevada

Mountains. The trail led up Yokohl Creek, across the south end of Blue Ridge, up Bear Creek to near Balch Park, thence to Hossack Meadow, south side of Jordon Peak, Kern Flat on Kern River, and crossed the main summit of the Sierra via Jordan Hot Springs and Monache Meadows. The trail down the east slope was by Olancha Pass, down Walker Creek and to Owen's Lake. Portions of the trail are still in use."

14.0	22.0	
15.1	23.0	The roadcut is in tonalite with abundant mafic inclusions. Biotite K-Ar age 109 Ma (Putnam and Alfors, 1965).
19.2		Lemon Cove Post Office. Wutchumna Hill, immediately northwest of town, is underlain by hornblende hypersthene gabbro.
20.1	28.0	
20.7	Stop A	The road to the left leads 1.3 miles to the spillway of Kaweah Lake Dam. Thin-bedded quartzite and calcium silicate hornfels, showing chevron folding and marble layers, is exposed in the spillway. The distinctive thin-bedded quartzite (metachert) is characteristic of the metamorphic rocks west of the park in the Kings Terrane. A 1-foot-thick mafic dike dips gently west, cutting metamorphic rocks.
21.0	29.00	
21.2		Gently dipping dark dikes cutting tonalite are well exposed in road cuts on the main highway. The majority of mafic dikes near Lake Kaweah contain 53–62 wt.% SiO_2 and 5.2–2.6 wt.% MgO ; one dike at the west end of the roadcut is more primitive, with 48 wt.% SiO_2 and 9.2 wt.% MgO.
23.5	31.50	
25.5		Horse Creek.
27.0	35.00	
27.4	Stop B	Good exposures of the garnet-bearing Tharps Peak Granodiorite (Liggett, 1990) dated at 109 Ma by the Rb-Sr method. Turn off behind the hump on the lake side—but beware of highway traffic. The Tharps Peak Granodiorite contains biotite and muscovite. Silica content of the granodiorite is 72–74 wt.%, and associated felsic dikes attain 76 wt.%. The tonalite pluton east of the Tharps Peak Granodiorite has been dated at 108 Ma (Pb-U zircon; Chen and Moore, 1982), within the age-limits error for the Tharps Peak Granodiorite.
27.6	35.50	
27.7		Slick Rock Turnout. The relatively dark granodiorite here yields a zircon Pb-U age of 108 Ma (Chen and Moore, 1982).
28.0	36.00	
28.9	37.00	
29.4		Western town limit of Three Rivers. The white flower-like pattern on the north face of Blossom Peak to the south is formed by a thin light-colored granite dike about 1 foot thick cutting vertically layered dark calcium hornfels rock. The mountain face is roughly parallel to the partly exposed dike, giving an irregular white pattern (resembling a blossom) on the gray metamorphic

rock. These metasedimentary rocks are part of the Blossom Peak Roof Pendant of Kings Sequence rocks, some 8 miles long, that crosses the Kaweah River.

30.0	38.00	
30.5		North Fork Kaweah River Drive.

Historical Note. Up the North Fork of the river 5 miles, the Kaweah Colony established the camp of Advance in the late 1880s. This was the base camp for an ambitious effort to build a road to Giant Forest for the purpose of logging. The threat of logging the giant sequoias helped to marshal the concern and the energies of those who sought to protect the big trees and to create Sequoia National Park. The road that the colony built, although now in disrepair, was the only road access to the Giant Forest region for many years.

30.9		Chevron gasoline station.
32.0	40.0	
32.3		Marble layers forming Comb Peak in the Sheep Ridge Roof Pendant are visible to the north. The line defining the position where the initial $^{87}Sr/^{86}Sr$ value of granitic rocks is 0.706 lies near here.
33.0	41.0	
34.1	42.0	
34.4		Intersection with Mineral King Road. (A road log describing the geology along the 25-mile-long road to Mineral King follows this guide.)
35.4		View of Moro Rock.
36.0	43.94	Kaweah River Bridge.
36.4		Roadcut blasted in diorite, followed by a road on the right (with a gate) that leads to a stream-gauging station on the Middle Fork of the Kaweah River. Water-washed outcrops expose heterogeneous hornblende diorite, with SiO_2 of 55–59 wt.% and MgO of 4–3 wt.%, typical of small mixed masses of dioritic rocks common near contacts between plutons and roof pendants.
36.5		Entrance to Sequoia National Park, the first national park in California.

Historical note. The park was created by act of Congress on September 25, 1890. On October 1, 1890, six days later, a second act created Yosemite National Park and enlarged Sequoia National Park to include Giant Forest and the General Grant Grove.

36.6		Sequoia National Park sign.
37.4		Park Headquarters, Visitors' Center.
38.1	Stop C	Coarse-grained marble within small metasedimentary pendant enclosed by the granite of Frys Point. Paved pullout on right.
38.8		Tunnel Rock. Tunnel has been blasted through a large block of the granite of Frys Point that has rolled downhill from the cliffs above.
39.8		2,000 feet.
40.2		Bridge over the Marble Fork of Kaweah River.
40.3		Potwisha Campground turnoff. A short walk down to the river will bring one to numerous Indian mortar holes in bedrock used for grinding acorns. Mortars

		have been milled into the dark granodiorite of Milk Ranch Peak, which contains conspicuous black hornblende crystals. U-Pb ages between 97 and 99 Ma (Chen and Moore, 1982).
40.6		The Castle Rock spires and domes are visible. They are carved from the slightly layered light-colored granite of Dome Creek.
41.8		Sign pointing out Mount Stewart (12,205 feet) just north of Kaweah Gap on the Great Western Divide. The peak, carved out of the granite of Eagle Scout Peak, is named for George W. Stewart, a newspaper publisher in Visalia and a leader in the movement to establish Sequoia National Park.
42.6	Stop D	Hospital Rock. Rest rooms, water. The parking lot is littered with large boulders, up to many feet in diameter, that were deposited during major rockfalls and avalanches off Moro Rock 1.5 miles and 4,000 feet above to the north. A large granodiorite boulder here has split after coming to rest. Indian pictographs, utilizing a red iron oxide compound, are painted on the flat joint face of this fracture. For the next 4.1 miles up the road from this point, numerous large boulders of the dark Giant Forest Granodiorite that are part of this avalanche, or series of avalanches, are visible adjacent to the road. The bedrock beneath the road is almost entirely composed of metasedimentary rocks of the Sequoia Roof Pendant. A small road from here leads about 1.5 miles to Buckeye Flat Campground on the Middle Fork of the Kaweah River. There, metasedimentary rocks of the Sequoia Roof Pendant are well exposed in waterworn outcrops (Fig. 5.3). The aluminum silicate mineral andalusite is visible in the slate and schist that are the metamorphosed equivalents of mudstone. *Historical Note.* Cattleman Hale Tharp was the first non-native visitor to this site, in 1858. At that time a year-round settlement of about 200 Indians was in residence.
43.3		3,000 feet.
44.2		Switchback on the east side of the Sequoia Roof Pendant, where the road barely enters the porphyritic marginal variety of the Giant Forest Granodiorite.
46.0		Big Fern Spring. A small trail leads up a stream that flows across vertically layered, thin-bedded quartz-biotite schist with quartzite layers. Metamorphic rocks collected here, apparently derived from aluminum-rich quartz-poor shaly rocks, contain plagioclase, biotite, potassium feldspar, minor quartz, andalusite, sillimanite, cordierite, and corundum (Clifford Hopson, written communication, July 1993). Two large granodiorite avalanche boulders here are about 20 feet in diameter.
47.4		Amphitheater Point (telephone). Good view of Moro Rock and Castle Rocks to the south, across the canyon of the Kaweah River. Mount Eisen (12,160 feet) is visible on the Great Western Divide. The Sequoia Roof Pendant can be seen crossing Paradise Ridge in a saddle 4 miles to the southeast, just east of the prominent Milk Ranch Peak. Brownish units in the roof pendant are quartzite; lighter-colored blue-gray units are marble, on which yucca plants commonly grow.

48.3		Granite Spring on the east side of the Sequoia Roof Pendant just inside the outcrop area of the Giant Forest Granodiorite. The sharp vertical contact of the granodiorite with metasedimentary rocks is visible slightly uphill in the highway roadcut.
48.5		Highest granodiorite boulders in avalanches from Moro Rock.
49.0	Stop E	Deer Ridge Lookout (telephone). Switchback Peak to the west with blue-gray marble below. Walk to the northeast side of the viewpoint to see several marble layers to the north that are nearly vertical, discontinuous, and folded in places (Fig. 5.4). These marble units form tabular bodies that can be traced for as much as 5 miles. Chamise, manzanita, buckeye, deerbrush, blue oak, and poison oak thrive on Deer Ridge and lower slopes.
49.5		5,000 feet.
49.9		Eleven Range Overlook. Note the marble layers crossing the canyon below. The highway has just entered the dark-colored (average 17% dark minerals) Giant Forest Granodiorite (U-Pb zircon age of 99 Ma; Chen and Moore, 1982), in which it will remain for the next 28 miles, to mile 78, except for short excursions into the granodiorite of Big Meadows. The Giant Forest Granodiorite is the oldest unit of the zoned Sequoia Intrusive Suite, a series of granodiorite and granite plutons in a roughly concentric arrangement centered at Shell Mountain in the Giant Forest Quadrangle. The composition and texture of the Giant Forest Granodiorite, as well as its abundance of mafic inclusions, are quite typical of the large hornblende granodiorite plutons of the Sierra Nevada Batholith. Contents of SiO_2 fall in the range 60.5–69 wt.%, but most are 62–65 wt.%; K_2O lies in the range 2.3–4.2 wt.%, but most are near 3.0 wt.%; and, MgO lies in the range 1–2 wt.%. Note the conspicuous honey-brown wedge-shaped mineral, sphene, which is not common in the tonalitic rocks of the western foothills.
50.5		Crystal Cave Road turnoff. The cave, 3.5 straight-line miles northwest, is in a marble layer within the Sequoia Roof Pendant.
53.1		Giant Forest Visitors' Center, store, and restaurant. The road to the right leads to Moro Rock in the Giant Forest Granodiorite (Fig. 11.17). A short, steep trail leads to the summit, where good views display the peaks of the Great Western Divide to the east, Castle Rocks to the south, and the western foothills and Central Valley to the west.
55.2		General Sherman Tree turnoff. The General Sherman tree, *Sequoiadendron gigantea*, aged about 2,200 years, is one of the world's largest living things (N. Stevenson, oral communication, 1998). It is 274.9 feet high and 102.6 feet in circumference at the base, with a volume of 52,500 cubic feet.
55.8		Wolverton turnoff. Slightly more than 1 mile east on this road, at Long Meadow, a small ski area has been developed. The meadow is formed behind the southern lateral moraines of the Tahoe and Tioga Glaciations deposited by major glaciers that flowed west down the U-shaped Tokopah Valley of the Marble Fork of the Kaweah River. This morainal material can be examined in road cuts leading to the ski area parking lot.

57.4		Lodgepole turnoff. The main glacier of the Tahoe Glaciation in Tokopah Valley terminated about 1 mile west of the highway at an altitude of 6,000 feet. About one-half mile east of the highway in Tokopah Valley is the east contact (concealed) of the Giant Forest Granodiorite, where it intrudes the overlying older granite of Lodgepole, a pale-orange coarse-grained granite (72–76 wt.% SiO_2) that underlies many of the neighboring domes and peaks, including Mt. Silliman (11,188 feet) and Alta Peak (11,204 feet). A questionable Pb-U age of the granite of Lodgepole suggests that it is older than 115 Ma (Chen and Moore, 1982).
57.5		Marble Fork Bridge.
58.3		Clover Creek Bridge.
59.1		7,000 feet.
61.1		Halstead Creek.
62.5		Suwanee Creek.
64.2	Stop F	Little Baldy Saddle (7,335 feet). Beware of danger from traffic. A vertical channel or pipe-like structure in granitic rocks containing orbicules cuts the Giant Forest Granodiorite about 150 yards west of the road at the saddle turnout (Fig. 6.15). The orbicules are elongate parallel to the walls of the pipe and are notably smaller at the edges than in the center of the pipe. Concentrations of mafic minerals define an internal contact between two generations of orbicule-bearing pipes. The pipe was originally about 20 feet in diameter, before it was cut and offset by an aplite dike. On the other (east) side of the highway, a trail leads up Little Baldy, a typical Sierra granitic dome in the Giant Forest Granodiorite, where good views can be had from the summit. The round-trip trail is 3.4 miles with a walking time of 2 to 3 hours. Internal contacts within the Giant Forest Granodiorite, which are defined by changes in hornblende abundance and habit, are well displayed in the blasted road cuts to the south of Little Baldy Saddle.
65.7		Dorst Creek Campground entrance.
66.4		Dorst Creek.
67.1		Cabin Creek.
68.4		Lost Grove.
69.1		Sequoia National Forest (to the northwest) and Sequoia National Park Boundary.
70.2		Stony Creek. Campground turnoff. Metamorphosed rhyolite tuffs of Cretaceous age (Pb-U zircon age 110 Ma; J. Saleeby, written communication, 1991) are well exposed about one-fourth mile down Stony Creek from the lower part of the campground (Fig. 5.7). This turnoff is within the granite of Big Meadows. Interesting contact relations of the granite of Big Meadows with the granite of Weaver Lake can be viewed by hiking northeast, upstream along Stony Creek.
70.7		Stony Creek Village.

75.4		Big Meadow road turnoff. About 3 miles east at the Big Meadows Ranger Station, the trail southeast to Shell Mountain affords an opportunity to examine the Sequoia Intrusive Suite in the northeastern part of the Giant Forest 15-minute Quadrangle. This concentric plutonic complex, about 7.5 by 26 miles, consists of the Giant Forest Granodiorite and several younger and smaller granitic masses—the granite of Big Meadows, the granodiorite of Clover Creek, the granite of Weaver Lake (youngest), and tentatively the granite of Chimney Rock. These units were emplaced sequentially, generally toward the central part of the intrusive suite, in the order listed. In general they become lighter-colored and more SiO_2-rich with time, except for the granodiorite of Clover Creek, which is more mafic than the older granite of Big Meadows. The youngest unit, the granite of Weaver Lake, forms a central funnel-shaped mass about 4.4 miles in diameter that sent out numerous upward-dipping sills intrusive into the granite of Big Meadows. Tabular masses and blocks of the granite of Big Meadows are isolated between these intrusive sheets. The units of the Sequoia Intrusive Suite were emplaced and cooled at about 99 Ma. The nested succession and chemical coherence of the units indicate that they are genetically related, and formed from a common magma through successive intrusions as temperature systematically decreased.
75.6		Big Baldy Trailhead. A ridge-crest walk of about 2 miles to Big Baldy provides good exposures of metasedimentary rocks of the Big Baldy Roof Pendant.
77.6	Stop G	Sierra View Overlook. Peaks visible to the north are, from left (west to east): Spanish Mountain (10,051 feet), Obelisk (9,970 feet), Tombstone Ridge, Mount McGee (12,604 feet, 31 air miles), Mount Reinstein (12,604 feet), Mount Goddard (13,568 feet), Kettle Dome (9,446 feet), Finger Peak (12,404 feet), Burnt Mountain (10,602 feet), Mount McDuffie (13,271 feet), and Tunemah Peak (11,894 feet, 23 air miles). None of these peaks are on the main Sierra Crest, which is 5 miles beyond Mounts Goddard and McDuffie. On the right can be seen Buck Rock Lookout, which lies on the outer margin of the granite of Big Meadows of the Sequoia Intrusive Suite (U-Pb zircon age, 99 Ma; Chen and Moore, 1982).
77.7		Quail Flat. Hume Lake road.
78.6		Park-Forest boundary. Grant Grove section, Kings Canyon National Park.
81.3		The Wye: end of Generals Highway and the intersection with State Route 180.

The Mineral King Road

The Mineral King Road extends 25 miles east from State Route 198 near Hammond, east of the town of Three Rivers, to the summer hamlet of Mineral King at an altitude of 7,900 feet. The road—more than a hundred years old—was built to serve the mining camp of Mineral King, which never produced much silver. The narrow paved road is steep and winding, is closed in the winter, and is the only road in the southern Sierra that reaches to the edge of the high country.

Most of the road is in dark granodiorite cut by large, irregularly shaped, light-colored sills and dikes sent out by the younger granite of Case Mountain.

LOG OF MINERAL KING ROAD, FROM STATE ROUTE 198 TO MINERAL KING (25 MILES)

Miles	Signs	
0.0		Beginning of Mineral King Road (designated MTN375 on road mile signs) at State Route 198, just above Hammond. The granodiorite of Three Rivers (U-Pb zircon age, 104 Ma; Chen and Moore, 1982), a relatively dark granodiorite carrying abundant mafic inclusions, occurs from the beginning of the road to Mile 3.5. The rock is cut by many dikes, including a lighter-colored granitic unit probably related to the granite of Case Mountain, exposed to the east.
1.0	1.00	
3.0	3.00	
3.5	3.50	The granite of Case Mountain extends from here to Mile 16.2, although an older granodiorite occurs within this span. The granite of Case Mountain is light-colored, relatively coarse-grained, and commonly porphyritic, and intrudes the granodioritic masses that border it on the east and west. Its margins are commonly irregular because of the many dikes and sills sent out by the main body. K-Ar ages are 83 to 90 Ma uncorrected (Evernden and Kistler, 1970). Samples of the granite of Case Mountain contain 73–75 wt.% SiO_2 and 3.4–4.3 wt.% K_2O.
4.0	4.00	
5.1		Nearly horizontal 3-foot-thick pegmatite dike in road cut.
6.2		Beginning of the dark granodiorite of Three Rivers, which contains abundant mafic inclusions. This rock unit is exposed along the road to Mile 9.5.
6.6		Bridge over East Fork of Kaweah River.
7.0	7.0	Settlement of Oak Grove (telephone).
8.5		Grunigen creek.
9.0	9.0	
9.5		Sequoia National Park boundary. Reenter the granite of Case Mountain, which extends to Mile 16.2.
9.7		Coldspring, concrete water trough at 3,200 feet.
10.0		A large pegmatite dike about 30 feet thick, which cuts the granite of Case

APPENDIX: ROAD AND TRAIL GUIDES 379

		Mountain, crops out as a white knob almost hidden behind the brush on the south side (right) of the road.
10.4		Park entrance station. Note that sample collecting is not allowed within the park without a permit.
10.9		Overhanging cliff made up of the granite of Case Mountain.
13.0	13.04	
15.7	15.76	
16.2		Redwood Creek and concrete water trough (5,706 feet). Contact of the granite of Case Mountain with the granodiorite of Castle Creek (U-Pb zircon age, 98 Ma; called Eagle Lake quartz monzodiorite by Busby-Spera and Saleeby, 1987). The granodiorite of Castle Creek is the outermost and oldest unit of the Mitchell Intrusive Suite, a north-trending series of granodiorite and granite plutons 36 miles long that are arranged in a roughly concentric pattern centered to the north in the Triple Divide Peak Quadrangle. The granodiorite of Castle Creek is lighter-colored than, but otherwise resembles, the granodiorite of Giant Forest, the oldest unit of the zoned Sequoia Intrusive Suite exposed along the Generals Highway. The granodiorite of Castle Creek has SiO_2 averaging 67 wt.% and K_2O averaging 3.4 wt.%.
17.7	17.68	Small stream.
19.2		Atwell Mill Ranger Station.
9.4		Atwell Mill Campground.
19.9		Deadwood Creek.
20.1	20.21	
20.5		Cabin Cove: "Seventh Heaven" sign over the road.
21.1		Silver City Store.
22.0		Highbridge Creek.
23.8		Mineral King Ranger Station, 7,580 feet. The western margin of the Mineral King Roof Pendant is not exposed on the road but passes close to the ranger station. From the ranger station to road's end, the road traverses metarhyolite tuff, calcium silicate rock, and slate. Metavolcanic rocks in the Mineral King Roof Pendant have yielded U-Pb zircon ages of 217 Ma (Late Triassic) and 190 Ma (Early Jurassic), and the metasedimentary rocks have yielded Late Triassic and Early Jurassic marine fossils.
24.2		The travertine terrace 100 feet north of the road is apparently a postglacial mineral spring deposit.
23.9	23.94	
24.6		Trailhead to Sawtooth Pass (parking, telephone).
24.7		Bridge over Monarch Creek.
25.0		Corral, road's end, within U-shaped glacial valley all underlain by metamorphosed volcanic rocks.

The John Muir Trail

The John Muir trail is a high-altitude wilderness trail that remains, throughout it length, slightly west of the crest of the range as it follows the backbone of the Sierra Nevada. It leads from Mount Whitney on the south to Yosemite Valley on the north, a total of 212.5 miles, and about half of its length (102.1 miles) is within the Sequoia and Kings Canyon National Parks area, where it crosses spectacularly exposed geologic features. The trail is not crossed by any road for its entire length.

In 1914 (the year of John Muir's death) the Sierra Club established a committee to investigate funding for a high trail that would extend from Yosemite to Mount Whitney and remain close to the range crest. Money was appropriated by the California State Legislature and work began on the trail in 1915. Much of the construction was done under the supervision of the U.S. Forest Service. The final, most difficult parts of the trail, both of which are in the parks area, were the Foresters Pass segment, over the Kings–Kern Divide, and the Mather Pass segment, from the South Fork of the Kings River to the headwaters of the Middle Fork of the Kings River. The Foresters Pass segment was completed in 1932 by the National Park Service on the Kern side and by the Forest Service on the Kings side. The Mather Pass segment was completed in 1938 by the Forest Service. Theodore Solomons had conceived the trail in 1892; it was finally completed some 46 years later.

In this trail guide, distances are given from south to north for each of four trail segments: (1) Mount Whitney to Foresters Pass (25.0 miles), (2) Foresters Pass to Mather Pass (36.7 miles), (3) Mather Pass to Muir Pass (22.0 miles), and (4) Muir Pass to Piute Creek (18.4 miles). The trail has no mile markers, but trail junctions are generally well marked by signs giving the mileages to points of interest, such as trail heads and passes. The relevant published geologic maps are indicated for each segment.

LOG OF THE JOHN MUIR TRAIL, SEGMENT FROM MOUNT WHITNEY TO FORESTERS PASS (25 MILES)

Mount Whitney Geologic Quadrangle Map, GQ-1545

Miles north of Mount Whitney

0.0	From the summit of Mount Whitney (14,495 feet) the sweeping view reveals a large part of the southern Sierra Nevada (Figs. 1.7, 2.18, 2.30, 3.11, 3.28). It is the highest point of the eastern or main crest of the range, which reaches north as far as the eye can see in rank on rank of snowy peaks. Southward the range

retains its rugged character only to Mount Langley, beyond which it rapidly decreases in height and ruggedness. To the west the Great Western Divide dominates the landscape, and is highest in the length from Mount Brewer on the north to the Kaweah Peaks on the south. In the huge glacial canyon between these parallel ranges flows the Kern River, heading in a series of basins draining the south side of the Kings–Kern divide. To the east, and 2 miles down, is Owens Valley and the dry bed of Owens Lake. Eastward, range after desert range are visible deep into Nevada.

The summit of Mount Whitney and the trail along the entire segment from Mount Whitney to Foresters Pass is underlain by granitic rocks of the Mount Whitney Intrusive Suite. This includes of a series of nested gigantic granitic bodies that were intruded as molten material during the Cretaceous Period (Fig. 6.7). They are among the youngest and largest intrusive sequences in the Sierra Nevada. The composite intrusion is oval in plan, 52 miles long, up to 14 miles wide, and 430 square miles in area, its long dimension extending northwest.

The nested sequence probably represents a single melting event, and was initially emplaced as a single intrusion about 85 Ma. After emplacement, the mass began cooling and solidifying from the walls inward. Before complete solidification, the partly molten and somewhat more differentiated SiO_2-rich core of the intrusive mass surged upward and intruded its own solidified walls. Later, as this inner mass cooled and partly solidified, the same upward intrusion of the center occurred again. The resulting sequence consists of an outer shell composed of the granodiorite of Sugarloaf on the west and the granodiorite of Lone Pine Creek on the east, an intermediate shell composed of the Paradise Granodiorite, and an innermost mass composed of the Whitney Granodiorite (average mode: Pl, 44.9; Kf, 23.5; Q, 24.9; Mf, 6.7), which is quite uniform over a broad area. Progressing from the outside inward, each rock mass is younger and composed of lower-temperature melting materials. After solidification of the upper part of the Whitney Granodiorite, porphyritic granite dikes and sills originating from deeper still-molten parts of the pluton were emplaced in fractures. These can be observed near Mount Muir, on Pinnacle Ridge, on the trail to the summit of Mount Whitney, and on the summit itself.

Because of the giant crystals (phenocrysts) of potassium feldspar that are scattered throughout, the Whitney Granodiorite is a stunning rock. These well-formed crystals, 1 to 2½ inches long, generally etched into relief by weathering, are among the largest crystals in any major intrusive unit in the Sierra Nevada. During the 1864 Brewer expedition, Clarence King noted that these crystals provided good handholds for climbing, but produced an uncomfortable sleeping platform. As a general matter, the size of the crystals within the rock of a pluton increases with the size of the pluton, inasmuch as a larger pluton cools more slowly than a smaller one and allows more crystallization time. It is thus no accident that the largest crystals occur in the largest pluton.

2.5 Trail Crest (13,480 feet), at the pass on the Sierra Crest and at the junction with the trail connecting Owens Valley with points west. From this pass, rock

5.0	glaciers are visible to the east on both sides of the switchbacking trail, and to the west on the steep cirque walls south and east of Hitchcock Lakes.
	After descending 2,000 feet from Trail Crest and passing the Hitchcock Lakes, we come into an area of extensive glaciated surfaces on the Whitney Granodiorite. Well exposed on these outcrops are steeply dipping joints and small faults trending nearly east. Where these faults offset other structures such as aplite dikes, the sense of offset is apparent and is generally left-lateral—when observed from one side of the fault, the other side is offset in a horizontal sense to the left.
9.0	Crabtree Meadows. North of here, the trail continues for 10 miles along an irregular bench at 10,000 to 11,000 feet, between the Sierra Crest and the brink of Kern Canyon to the west. Along this bench the trail crosses a series of lateral moraines generated long ago by glaciers that headed in crestal snowfields and flowed west to connect with the main Kern Canyon trunk glacier. Along this entire segment, the bedrock is consistently the Whitney Granodiorite, a remarkably uniform rock with its large characteristic potassium feldspar phenocrysts. Just north of Crabtree Meadow, we climb moraines flanking the north side of the canyon of Whitney Creek and cross Sandy Meadow.
11.4	Divide between Whitney and Wallace Creeks (10,964 feet). A strenuous 1-mile hike from this divide east up the west ridge of Mount Young to an altitude of 11,900 feet brings us to a ragged ridge outcrop of a pipe filled with orbicular granite. The nucleus of each orb is generally a large potassium feldspar phenocryst.
13.0	East end of the High Sierra Trail, on the north bank of Wallace Creek, at 10,300 feet. Down to Wallace Creek and on up to the Bighorn Plateau on the divide between Wallace Creek and Tyndall Creek, we cross about a dozen lateral moraine crests generated by ice lobes once occupying tributary drainages of Wallace Creek.
19.0	Tyndall Creek. Another series of moraines is crossed, both as we descend from the Bighorn Plateau into the gorge of Tyndall Creek and as we climb out on the northwest side on the trail toward Foresters Pass.
23.0	Contact between the Whitney Granodiorite and the Paradise Granodiorite, opposite the center of a small lake east of the trail at 12,500 feet. The rocks are similar, but the Paradise is somewhat darker (average mode: Pl, 46.2; Kf, 20.9; Q, 23.0; Mf, 9.9), and contains phenocrysts of potassium feldspar that are markedly smaller than those of the Whitney Granodiorite. Also, the phenocrysts are darker because they contain tiny inclusions of biotite and hornblende. The contact is quite sharp and dips 30 to 40° under the Paradise Granodiorite. Contact relations indicate that the Whitney Granodiorite is younger and intrusive into the Paradise. The Paradise (86 Ma) remains as the bedrock to the summit of Foresters Pass on the principal divide separating the Kings River and Kern River drainages.
25.0	Foresters Pass (13,200 feet), in the Paradise Granodiorite.

APPENDIX: ROAD AND TRAIL GUIDES

LOG OF THE JOHN MUIR TRAIL, SEGMENT FROM FORESTERS PASS TO MATHER PASS (36.7 MILES)

Mount Whitney Geologic Quadrangle Map, GQ-1545   *Mount Pinchot geologic map, Bulletin 1130*
Big Pine geologic map, Professional Paper 470

The segment of the John Muir Trail from Foresters Pass to Mather Pass passes through a geologic region more complex than those north or south. Included in this area is the Oak Creek Roof Pendant, which contains Jurassic metamorphic rocks primarily of volcanic origin, as well as several Jurassic granitic plutons associated with the metamorphic rocks, which are themselves somewhat metamorphosed. Many of these pre-Cretaceous rocks are cut by a swarm of dark-colored Late Jurassic dikes, the Independence Dike Swarm.

Miles north of Foresters Pass

1.5	The trail enters a mass of dark dioritic rock apparently caught up in the contact zone between the Paradise Granodiorite Pluton and the next major intrusive mass to the north, the Bullfrog Pluton. The contact is offset in this vicinity by two faults with total right-lateral offset of a few hundred yards. They lie north of the northern mapped termination of the right-lateral Kern Canyon Fault, which displays offsets of up to 8 miles. Perhaps these faults, and other nearby small right-lateral faults, account for the northward decrease in displacement of the Kern Canyon Fault. For the next mile, after crossing the contact between the Paradise Granodiorite and the diorite, the trail is approximately parallel with the diorite-Bullfrog Pluton contact, and passes through these rocks of contrasting character to eventually cross into the granite of the Bullfrog Pluton. Also in this contact zone, the trail is crossed by a major northeast-trending granitic dike called the Golden Bear Dike, after Golden Bear Lake in Center Basin, lying to the east. The dike is composed of porphyritic granite with well-formed crystals of quartz and potassium feldspar set in an especially fine-grained matrix. This vertical dike is about 33 feet thick here and has been mapped for a distance of 10 miles, from 3 miles southwest of the trail, crossing east over the Sierra Crest between University Peak and Mount Bradley and down the east range front to its base.
2.7	We now enter the main mass of the granite of the Bullfrog Pluton. The rock is an extremely light-colored granite (average mode: Pl, 33.0; Kf, 39.6; Q, 24.9; Mf, 2.5). The Bullfrog Pluton, which trends northwest, is 15 miles long and up to 6 miles wide, and covers 55 square miles. It was intruded about 100 Ma and hence is considerably older than the Paradise Granodiorite. For the next 7 miles, to within one-half mile of Glen Pass, the trail remains in the Bullfrog Granite.
3.5	Center Basin Trail. Somewhat more than a mile up this trail takes us to Golden Bear Lake in Center Basin. Excellent exposures of the Golden Bear Dike crop out near the upper end of the lake.

7.0	Junction with Bubbs Creek Trail at an altitude of 9,700 feet on the canyon floor, 3,500 feet down from Foresters Pass. The trail beyond climbs up to Glen Pass on the divide between Woods Creek and Bubbs Creek, both in the upper drainage of the South Fork of the Kings River.
8.2	Junction with trail to Kearsarge Pass (Fig. 2.21).
9.9	One-half mile on the south side of Glen Pass the trail leaves the Bullfrog Granite and passes into a dark, somewhat heterogeneous diorite.
10.4	Glen Pass, 11,978 feet. North of Glen Pass, the trail passes from the diorite through a mass of Bullfrog Granite and back into the dark diorite. During the descent from the pass, looking toward the east, we have excellent views of Painted Lady and Dragon Peak, which are carved from the dark Onion Valley Mafic Complex. The Painted Lady owes its splendor to massive horizontal dikes of aplite, which cut this dark rock in striking fashion.
11.4	Opposite Upper Rae Lake, the diorite (and the Bullfrog Pluton) is in fault contact with the Dragon Pluton, which is composed of a quartz-poor medium-colored granodiorite (average mode: Pl, 54.6; Kf, 28.0; Q, 8.3; Mf, 9.0).
15.1	On the north side of a small lake where the trail to Baxter Pass joins the John Muir Trail, the Dragon Pluton (commonly containing massive inclusions and dikes of other granitic rocks) is in contact with the White Fork Pluton. Composed of an older granodiorite, the White Fork Pluton is extensively sheared and altered, and cut by numerous northwest-trending dark-colored dikes of the Independence Dike Swarm. The fact that it is cut by the dikes and also is intimately associated with septa of metamorphosed sedimentary rocks indicates that the White Fork Pluton is one of the older granitic masses, probably Jurassic in age.
17.2	We enter a mixed rock dominated by dark-colored diorite and cut by dark diorite dikes.
18.5	We encounter the side of a mass of Bullfrog Granite projecting north from the main mass, which we had crossed to the south in upper Bubbs Creek.
19.1	The John Muir Trail reaches the junction of the main and South Forks of Woods Creek, and also the junction with the trail leading down to the canyon of the South Fork of the Kings River. In the middle of this stretch of Bullfrog Granite, we can see, high on the canyon wall north of the trail junction, the white spires of the Castle Domes; they are carved out of the white, coarse-grained, little-jointed Bullfrog Granite. North of the Bullfrog Granite we pass through the same diorite mass cut by mafic dikes that we encountered south of the trail junction.
20.5	We traverse again across the White Fork Granodiorite Pluton, which is cut by numerous northwest-trending dark-colored dikes.
21.3	The first of several thin screens (septa) of metamorphosed sedimentary rocks is encountered. These rocks were originally flat-lying, clay-rich, and limy mudstone. Their long history of crumpling, shearing, heating, and recrystallization has converted them into steeply dipping schist and hornfels.

21.5	Beyond the metamorphic rocks is a sequence of mixed rocks dominated by the light-colored granitic rock of the Twin Lakes Pluton. It is mixed with darker-colored granitic rocks and contains large inclusions of limy metamorphosed sediments, the whole sequence cut by dark diorite dikes.
22.6	Junction with the trail across Woods Lake Basin to Sawmill Pass. This trail crosses directly into the northern part of the Woods Lake mass of the Tinemaha Granodiorite, a northwest-trending pluton 11.5 miles long that is intricately cut by mafic dikes. The granodiorite (average mode: Pl, 41.8; Kf, 21.8; Q, 23.6; Mf, 12.0) is generally slightly porphyritic with small phenocrysts of potassium feldspar. The features of these mafic dikes are beautifully exposed in glaciated outcrops on Mount Cedric Wright and near Woods Lake, a few miles east on this trail. The same pluton, with its array of dikes, is accessible by road in the canyon of Independence Creek on the east flank of the range, west of the town of Independence, but the ragged cliff outcrops along Independence Creek do not expose the dikes as well as the fresh, smooth, glaciated outcrops near the John Muir Trail in this region.
23.1	At lower Twin Lake, we leave the granite of the Twin Lakes Pluton and enter a second screen of metamorphosed mudstone and shale that is only about 0.3 mile thick. A third, even narrower screen lies from 23.9 to 24.0 miles on the trail. Between the last two screens is a narrow mass of dark granodiorite of the McDoogle Pluton.
24.0	Beyond the third metamorphic screen we enter the Lamarck Pluton, composed of a slightly porphyritic lighter-colored granodiorite (average mode: Pl, 47.6; Kf, 20.3; Q, 24.8; Mf, 7.4). The Lamarck Pluton, emplaced about 97 Ma, is 47 miles long and extends north well out of the map area.
25.2	The trail reenters the McDoogle Pluton as we climb toward Pinchot Pass. The Lamarck Pluton is younger than the McDoogle and clearly has intruded it from below. Here we can examine the roof of the Lamarck Pluton, well exposed on the walls of the cirque on the southwest side of Mount Wynne. Flat-dipping granite dikes fed from the Lamarck Pluton are conspicuous where they cut the dark rocks of Striped Mountain and Mount Wynne, east of the pluton.
26.1	Pinchot Pass (12,050 feet), on the divide between the South Fork of the Kings River and Woods Creek.
26.2	North of Pinchot Pass 100 yards, we again enter the Lamarck Granodiorite, which seemingly occurs in separate bodies north and south of Pinchot Pass, but the tops of both masses dip toward one another under the pass, indicating that they belong to a single intrusion. The Lamarck Granodiorite remains the dominant rock for 10.6 miles along the trail to Mather Pass, but the trail is near the west side of the Lamarck Pluton and in several places it crosses into the Cartridge Pass Pluton on the west (average mode: Pl, 47.8; Kf, 18.3; Q, 23.4; Mf, 9.7). From Lake Marjorie one can look into a cirque to the south now occupied by a rock glacier (Fig. 9.7). At Lake Marjorie the trail loops into the Cartridge Pass Pluton (Fig. 6.6) for one-fourth mile, on the north margin of the lake.

30.5	The John Muir trail crosses the South Fork of the Kings River near a trail junction. The trail to the west leads down the river, and that to the east crosses over the main Sierra Crest at Taboose Pass. A side trip of about 2 miles to Taboose Pass takes one into the Red Mountain Creek pluton, composed of a very light-colored granite (Fig. 6.27). This older granite is cut by numerous dark dikes of the Independence swarm. The dikes are easily studied because of the stark contrast where they cut white granite.
32.3 to 32.8	The trail crosses a mass of dark diorite commonly mixed with granite, caught up in the Lamarck Granodiorite.
32.8	For the next 2.5 miles the trail follows closely along the magnificently exposed contact of the younger Cartridge Pass Pluton, on the west, with the Lamarck Pluton on the east. The granodiorite of the Cartridge Pass Pluton (especially in its marginal zone) is darker and finer-grained than the Lamarck Granodiorite. The rock of the Cartridge Pass Pluton has a layering that lies parallel to the contact, and the contact transects a layering in the Lamarck Granodiorite (Fig. 6.17). In addition, thin dikes of the Cartridge Pass Pluton cut the Lamarck Granodiorite. These features indicate that the Cartridge Pass Pluton is the younger.
36.7	Mather Pass (12,050 feet). To this point the trail has remained in the Lamarck Granodiorite and its associated dark diorite masses.

LOG OF THE JOHN MUIR TRAIL, SEGMENT FROM MATHER PASS TO MUIR PASS (22.0 MILES)

Big Pine geologic map, Professional Paper 470   *Mount Goddard Geologic Quadrangle Map, GQ-428*

Miles north of Mather Pass

0.0	Mather Pass, on the divide between the Middle and South Forks of the Kings River, is underlain by a rather complex contact zone between medium-colored granodiorite belonging to the Lamarck Granodiorite and a small sliver of Cartridge Pass Granodiorite.
0.5	North of the pass one-half mile, the trail crosses into the light-colored granite of Evolution Basin, a distinctive unit that makes up a northwest-trending pluton some 20 miles long and 4 miles wide. This rock, among the most SiO_2-rich rocks in the Sierra, is extremely light colored (average mode: Pl, 26.1; Kf, 38.1; Q, 33.8; Mf, 2.0), and consequently is dazzlingly white and reflective. K-Ar biotite ages of 85, 80, and 79 Ma were reported by Evernden and Kistler (1970).
1.5	The trail passes into marginal outcrops of the older Lamarck Granodiorite, a much darker rock that makes up an even larger elongate pluton adjacent to the north of the Evolution Basin Pluton. This thin, lenticular, northwest-trending pluton, 37 miles long and about 6 miles wide in the middle, was intruded about 90 Ma. The Lamarck Granodiorite is moderately dark-colored with about

16 percent dark minerals, but near the contact there are many heterogeneous, darker rock types in irregular masses up to several hundred feet wide.

The Lamarck Granodiorite contains abundant inclusions of darker rock richer in the dark minerals. These inclusions, commonly about 6 inches to 1 foot long, are flattened, and thus define a planar structure through the rock. The structure defined by these parallel disk-shaped inclusions is generally steep and parallel to steep contacts with other rocks, but in the southern part of the pluton it dips gently or is nearly horizontal. This structure apparently reflects the shape of the roof of the pluton that must have overlain this part of the Lamarck, not far above the exposed level.

4.0	Lower Palisade Lake.
4.5	The trail reenters a mass of mixed dioritic rock and then a 1-mile-broad area of the white granite of Evolution Basin, which is a northern projection of the main mass of the pluton.
5.5	We pass into a south-projecting lobe of the Lamarck Granodiorite.
7.0	Lower Deer Meadows.
8.0	The trail moves back into the light-colored rock of Evolution Basin, but a band of dark diorite lies between the Lamarck Granodiorite and the white granite. The contact between the dark granodiorite and its associated darker diorite with the light-colored granite is magnificently displayed on both walls of the canyon of Palisade Creek below Deer Meadow.
	Except for a narrow septum of metamorphic rock, the trail remains, for the 14 miles to Muir Pass, in the light-colored granite rock of Evolution Basin. On the slopes to the right (northeast) are prominent dark cliffs underlain by the Lamarck Granodiorite. Likewise, up on the left (southwest) are the dark metamorphosed volcanic rocks of the Goddard Roof Pendant. These rocks underlie the 12,000- to 13,000-foot crags of the Black Divide, including Mount Woodworth, Wheel Mountain, Mount McDuffie, and Black Giant. These metamorphosed volcanic rocks seem to be more resistant to glacial erosion than are the neighboring granitic rocks. Small glaciers, occupying high cirques on the northeast slope of Black Giant, are visible.
15.2	Little Pete Meadow.
19.7	We pass into a spur of metavolcanic rock connected to the main Goddard Roof Pendant, which nearly divides the granite of Evolution Basin into two masses. On the north part of the spur, and extending onto the shore of Helen Lake, the trail passes through a giant plutonic breccia composed of blocks of metamorphic rock caught up in a matrix of light-colored granitic rocks.
21.0	At the lower end of Helen Lake, we pass from the intrusive breccia back into the light-colored granite of Evolution Basin.
22.0	Muir Pass (12,059 feet), on the divide between the South Fork of the San Joaquin and the Middle Fork of the Kings Rivers.

LOG OF THE JOHN MUIR TRAIL, SEGMENT FROM MUIR PASS TO PIUTE CREEK (18.4 MILES)

Mount Goddard Geologic Quadrangle Map, GQ-428   *Blackcap Mountain Geologic Quadrangle Map, GQ-429*

Miles north of Muir Pass

	Descending north from Muir Pass the white granite of Evolution Basin is magnificently exposed along the shores of Wanda Lake and Sapphire Lake, where the lack of dark inclusions and the scarcity of dark minerals makes it nearly impossible to define a directional structure in the rock.
5.3	About one-fourth mile before reaching the upper end of Evolution Lake, we cross a sharp contact between this light-colored granite of Evolution Basin and the dark Lamarck Granodiorite. The granite is younger and dips under the granodiorite at 50–65°. Interestingly, this contact passes nearly over the summit of several major peaks, including Mounts Wallace, Haeckel, and Spencer on the right (east) of the trail and The Hermit on the left (west) of the trail. Perhaps the juxtaposition of these two rock types produces a joint-free rock more resistant to frost action and glacial erosion.
9.8	Colby Meadow.
12.8	Evolution Meadow at the lower end of the glacial trough of Evolution Valley. Some 1,200 feet above the meadow on a bench on the south side of the canyon, at 10,400 feet, is a small erosional remnant of young basalt. This basalt is probably related to the scattered erosional remnants to the south on and near Blackcap Mountain, some of which have been dated at about 3.5 Ma.
14.4	We pass from the Lamarck Granodiorite into the main northern part of the Mount Goddard Roof Pendant, which is composed of metamorphosed volcanic rocks. A metamorphosed rhyolite tuff, strongly lineated in places, crops out beside the trail.
14.7	We pass into a metamorphosed tuffaceous sediment.
14.8	Trail to Goddard Canyon.
15.8	The trail partly parallels a septum of light-colored granitic rocks that are cut by fine-grained granitic dikes.
17.0	The trail is underlain by a darker sheared granodiorite cut by light-colored dikes sent out from the septum of granitic rocks.
18.4	Paiute Creek and Trail (7,900 feet), the end of the segment of the John Muir Trail covered in this guide. The John Muir Trail continues north for 110.4 miles (total length, 212.5 miles) and terminates in Yosemite Valley.

The High Sierra Trail

The 49.1-mile-long High Sierra Trail was constructed in 1928–32 to furnish a link between Giant Forest and the John Muir Trail, by crossing the Great Western Divide and the Kern Canyon (Fig. A.2). The trail, accessed by a 3-mile road from Giant Forest Village, begins at Crescent Meadow, contours along the north wall of the great canyon of the Middle Fork of the Kaweah, crosses the Great Western Divide at Kaweah Gap, drops down into the Kern Canyon, and follows the Kern River north to Wallace Creek, where it climbs out of the Canyon and joins the John Muir Trail.

The trail provides good exposures of three major granitic intrusive complexes, the three becoming progressively younger to the east. They are the Sequoia Intrusive Suite (97–102 Ma), the Mitchell Intrusive Suite (91–98 Ma), and the Whitney Intrusive Suite (83–86 Ma). The trail crosses only a single tiny screen of metamorphic rock.

LOG OF THE HIGH SIERRA TRAIL, SEGMENT FROM CRESCENT MEADOW TO JOHN MUIR TRAIL (49.1 MILES)

Triple Divide Peak Geologic Quadrangle Map, GQ-1636   *Mount Whitney Geologic Quadrangle Map, GQ-1545*
Kern Peak Geologic Quadrangle Map, GQ-1584

Points along the trail are all designated as miles east of the trailhead at Crescent Meadow. Trail signs showing mileage are found at passes and trail junctions.

Miles north of Crescent Meadow

0	At Crescent Meadow the trail begins in the Giant Forest Granodiorite, a dark granitic rock belonging to the Sequoia Intrusive Suite, which contains prominent dark hornblende crystals and about 19 percent dark-colored minerals. Dark, fine-grained inclusions commonly 2 to 5 inches in size are abundant, usually 2 to 6 per square yard. A zone about one-half mile wide around the outer margin of the pluton contains phenocrysts of potassium feldspar.
3.2	The Giant Forest Granodiorite is in sharp contact with an older and much lighter-colored pluton made of the granite of Lodgepole. This rock is relatively coarse-grained, light-colored, and commonly stained orange (average mode: Pl, 30.0; Kf, 36.1; Q, 30.3; Mf, 4.5), and has a SiO_2 content of about 75 wt.%. Lead-uranium ages indicate that the rock crystallized more than 115 million years ago.
	The trail contours along the north canyon wall of the Middle Fork of the Kaweah River, some 3,000 feet below.
4.8	Contact between the granite of Lodgepole and the granodiorite of Castle Creek, a dark-colored medium-grained granodiorite (average mode: Pl, 47.5; Kf, 15.1; Q, 18.7; Mf, 18.8) with a SiO_2 content of about 68 wt.% and an age of about 98 Ma.
7.5	We pass from the granodiorite of Castle Creek into the Mitchell Peak Granodiorite, which was intruded 91 Ma. The rock is also relatively dark-colored (average mode: Pl, 49.5; Kf, 14.7; Q, 18.3; Mf, 17.5), but distinctly finer-grained. It contains phenocrysts of potassium feldspar, hornblende, biotite, and

	plagioclase. Vertical joints are commonly absent in the Mitchell Peak Granodiorite, and it is therefore a dome-former. As we pass through the pluton, one can catch glimpses of Little Blue Dome and Sugarbowl Dome to the east, decorating the north wall of the canyon of the Middle Fork of the Kaweah River. Ball Dome (Fig. 11.18), also in the granodiorite of Mitchell Peak, lies 10 miles north at the headwaters of Sugarloaf Creek.
11.0	Bearpaw Meadow.
13.1 to 14.8	Just a short distance beyond the junction with the trail to Elizabeth Pass, the trail passes back into a narrow shell of the granodiorite of Castle Creek, and then into a narrow band of the granite of Eagle Scout Peak. This light-colored rock (5 percent dark minerals) has areas that contain small (one-half inch to an inch) phenocrysts of potassium feldspar.
15.0	We pass into the very light-colored granitic pluton made up of the granite of Tamarack Lake (average mode: Pl, 29.7; Kf, 36.2; Q, 29.7; Mf, 3.9). This small pluton, which seems to be only shallowly unroofed, is less than 2 miles in diameter, and its contacts dip gently outward. As we walk up the trail, the sheer wall to the north exposes these light-colored granites, cut by dark dikes, and the enclosing masses of the dark diorite of Hamilton Lake. Also in this vicinity is a nice example of an avalanche chute cutting the granite wall north of the trail (Fig. 9.13).
15.8	Just east of the large Hamilton Lake, we climb out of the granite of Tamarack Lake into a mass of very dark diorite (average mode: Pl, 59.3; Kf, 1.2; Q, 6.8; Mf, 32.8).
18.0	We pass from the diorite back into the granite of Eagle Scout Peak.
19.7	Kaweah Gap on the crest of the Great Western Divide.
22.3	We pass through a narrow screen of metamorphic rock that separates the east margin of the Eagle Scout Peak Pluton from the next pluton, the granodiorite of Chagoopa. Where exposed near the trail, the metamorphic rock is a calcium silicate schist and hornfels produced by the metamorphism of a limy mudstone. Coarse garnet crystals speckle some rock layers near the trail.
22.5	The granodiorite of Chagoopa, bordering the screen on the east, is relatively dark (average mode: Pl, 45.3; Kf, 26.1; Q, 17.2; Mf, 11.3), but is somewhat lighter-colored in the interior of the pluton and contains small phenocrysts of potassium feldspar. The trail now climbs out of the upper part of the Big Arroyo onto the Chagoopa Plateau, an old erosional surface, which extends for the next several miles to the brink of the Kern Canyon.
27.5 to 32.8	The Chagoopa Plateau is underlain primarily by the granodiorite of Chagoopa, but it is extensively mantled by glacial moraine. These moraines were deposited by glaciers flowing south from the south slopes of the Kaweah Peaks Ridge.
33.0	As we reach the brink of the Kern Canyon, bedrock reappears from under the glacial cover, and we find that we have crossed the contact out of the granodiorite of Chagoopa into a dark dioritic rock. Here, the glacial canyon is about 2,000 feet deep and remarkably straight (Figs. 9.16, 11.7). The trail follows a narrow joint slot in diorite down to the floor of Kern Canyon.

35.3	We reach the floor of the great south-trending Kern Canyon at Upper Funston Meadow. The glacial trough follows a major strike-slip fault. On the west wall the dark diorite we encountered on the Chagoopa Plateau crops out. On the east wall is the near-white granite of Upper Funston Meadow. The nearly vertical contacts between various granitic plutons are offset across the fault, and the right-lateral strike-slip offset increases to the south. Here, the offset is about 4.4 miles, and 18 miles south it is 6.8 miles (Moore and du Bray, 1978).

Generally, the main fault trace is buried by river gravel and talus, which mantle much of the canyon bottom, but nearly parallel branches of the main fault appear on the canyon walls. The timing of offset along the fault is not well known. The fault offsets granitic rock contacts as young as 83 Ma and, to the south, does not offset lava flows 3.5 Ma, nor are earthquake epicenters clustered on the fault. Recrystallization of relatively high-temperature minerals is extensive among the crushed rocks of the fault zone, and the mineral tourmaline is abundantly distributed, which suggests that the fault zone is ancient and has been deeply eroded. This, and the fact that some rocks along related faults are plastically deformed, suggests that offset followed closely the period of granite emplacement and cooling about 80 Ma. |
| 37.3 | Kern Hot Spring, on the floor of the valley, is 2 trail miles north of Upper Funston Meadow. Along the way we view the spectacular Chagoopa Falls, cascading down the west canyon wall. The hot spring is small, but a soak is a welcome relief for trail-weary muscles. Eight miles south along the fault trace near the mouth of Golden Trout Creek, another mineral spring occurs, a CO_2-rich soda spring. These features indicate that the fault zone still acts as a passage for various hydrothermal fluids.

West of the hot spring, we pass the contact of the diorite on the west wall with the granodiorite of Chagoopa, a variable rock that is porphyritic in places. On the canyon wall east of the hot spring, we still observe the granite of Upper Funston Meadow, which is cut by dark-colored dikes. However, 1 mile north of the hot spring the Paradise Granodiorite descends to the level of the canyon floor, displacing the light-colored granite on the east wall. The Paradise Granodiorite is the main unit in the nested Whitney Intrusive Suite (average mode: Pl, 46.2; Kf, 20.9; Q, 23.0; Mf, 9.9), and is 86 Ma. The rock is distinctly porphyritic, with abundant phenocrysts of potassium feldspar. |
39.3	The main trace of the Kern Canyon Fault passes from the axis of the valley onto a series of bluffs on the west valley wall. In this segment the Paradise Granodiorite, east of the fault, is in contact with a sliver of the granite of Upper Funston Meadow, west of the fault.
40.5	The sliver of granite ends, and the Paradise Granodiorite crops out on both sides of the fault. Continuing north up the canyon, the Paradise Granodiorite remains on both sides of the fault and canyon, and the trace of the fault is increasingly difficult to follow.
44.3	Junction Meadow occupies the floor of the canyon at the point where Wallace Creek enters from the east and the Kern-Kaweah River enters from the west. About 1 mile north of Junction Meadow, the High Sierra Trail leaves the Kern

Canyon and climbs the east wall on the north side of Wallace Creek, which it follows to the John Muir Trail.

47.1 About 1.8 miles east of the junction, the trail crosses the main contact between the Paradise Granodiorite on the west and the Whitney Granodiorite on the east, in the Wallace Creek gorge at 9,400 feet. The contact is sharp, and the Whitney dips west at about 35° beneath the Paradise. The Whitney Granodiorite is slightly lighter-colored (average mode: Pl, 44.9; Kf, 23.5; Q, 24.9; Mf, 6.7), clearly younger, and intrusive into the Paradise. Both rocks are coarse-grained granite-granodiorites with large potassium feldspar phenocrysts and very few small dark inclusions. The primary difference between the two rock units—and one that is quite obvious—is that the Whitney Granodiorite contains larger and cleaner potassium feldspar phenocrysts, commonly 1 to 2 inches in size.

49.1 Junction of the High Sierra Trail with the John Muir Trail, and the end of the High Sierra Trail. At the junction, we remain in the Whitney Granodiorite, about 83 Ma.

REFERENCE MATTER

Glossary

accretionary lapillus (pl. **lapilli**)   A small spherical mass of fine-grained, concentrically structured volcanic ash, commonly ¼ to ½ inch in size. Lapilli form in the air in an ash-charged volcanic cloud, by accretion of ash around an ash particle due to the condensation of moisture on the core, and then fall to the ground as mud balls.

alidade   A straightedge rule, equipped with simple or telescopic sights, for use on a *plane table* to sight along, and draw in (on a map positioned on the plane table), the direction to distant objects. See *azimuth*.

alluvium   Clastic sedimentary material, such as clay, sand, and/or gravel, deposited by a stream or river along its course.

amphibole   Any of a group of common rock-forming silicate minerals, rich in iron and magnesium, that form elongate crystals. One of the most common is hornblende.

andalusite   An aluminum silicate mineral, Al_2SiO_5, that commonly forms in metamorphosed shales at relatively low pressure and temperature. It is one of three minerals with the same composition (the others are sillimanite and kyanite), the three forming under different conditions of pressure and temperature.

andesite   A fine-grained grayish volcanic rock intermediate in composition between basalt and rhyolite. It commonly contains about 55 to 60% silica and is the dominant lava type in most large continental volcanoes.

apatite   Any of a group of calcium phosphate minerals, $Ca_5(PO_4)_3(F,OH,Cl)$, that generally occur in small amounts in all igneous rocks.

aplite   A light-colored, fine-grained igneous rock, generally granitic in composition. Aplite commonly intrudes cracks, forming the light-colored dikes that so commonly intrude granitic masses and their wall rocks.

arête   A sharp, rugged mountain ridge situated typically between two glaciated valleys.

ash   Fine volcanic fragments blown into the air during an eruption. Ash of this sort that has consolidated into rock is called *tuff*.

asthenosphere	The somewhat plastic layer of the Earth's mantle lying beneath the lithosphere, on which the lithospheric plates move. It lies from 60 to 200 miles below the Earth's surface. See *core, crust, lithosphere, mantle*.
atmospheric pressure	The force per unit area in any part of the Earth's atmosphere. At sea level, atmospheric pressure is about 14.66 pounds per square inch, which is equivalent to the weight of a column of mercury 29.92 inches high.
atomic number	The number, unique to each chemical element, that is the number of protons or electrons within each atom of that element.
atomic weight	The relative weight of the atom of an element as compared to that of a standard element. Oxygen, with an atomic weight of 16, is the standard element.
augite	A common dark-colored, rock-forming mineral, $Ca,Fe,Mg(SiO_3)_2$, that is a member of the pyroxene group. Augite and sodium calcium feldspar are the most abundant constituents of basalt.
auriferous	Gold-bearing, as various ores.
azimuth	The angle measured in the horizontal plane, from the point of an observer, between a fixed direction (commonly north) and an object at some distance from the observer.
barometer	An instrument for measuring atmospheric pressure and hence for determining both altitude and changes in the weather. A *cistern* or *mercury* barometer utilizes a column of mercury to balance the air pressure; an *aneroid* barometer measures change in air pressure by monitoring the compression of a flexible box within which there is almost no air.
basalt	A common, fine-grained volcanic rock composed primarily of sodium-calcium feldspar and augite.
batholith	A large mass of plutonic rock, by definition at least 40 square miles in extent, composed of coarse-grained rock that originally crystallized below the surface. The Sierra Nevada Batholith is made up of many smaller *plutons*, intrusive masses largely composed of granitic rock.
bed	A single defined layer in a sequence of layered sedimentary rocks. Bedding is the arrangement of beds or layers of various thickness or character within a sedimentary sequence.
bedrock	Solid rock that underlies soil or other unconsolidated or broken surficial material.
bench mark	A relatively permanent mark implanted in an enduring natural or artificial object to mark an important survey point for which position and/or elevation have been measured. Bench marks are commonly metal disks cemented in bedrock.
bergschrund	A major crack, or *crevasse*, near the head of a glacier where moving ice has pulled away from stationary ice rooted on rock.
biotite	A dark to black mica mineral, $K(Mg,Fe)_3AlSi_3O_{10}(OH)_2$. It is one of the most common dark minerals in granitic rocks.
bornite	A coppery red to iridescent purple copper sulfide mineral, Cu_5FeS_4, a valuable ore of copper.
breccia	A coarse-grained rock made up of angular, broken-rock fragments cemented to-

	gether by mineral material. The fragments are larger than one-tenth inch, and 1-inch fragments are common.
calcite	The principal calcium carbonate mineral, $CaCO_3$. It is the primary constituent of limestone and marble, and is weakly soluble in water.
carbonate	Any of various minerals containing the carbonate ion, CO_3. Some sedimentary rocks, such as limestone, are made up almost entirely of calcium or magnesium carbonate minerals.
cave	A naturally formed chamber or tunnel, large enough for a human to enter, that formed beneath the Earth's surface. Most caves form from the dissolving of limestone by ground water.
cementation	The process whereby fragmental sediments become consolidated into hard, compact rock by the deposition of minerals in the spaces between the individual grains.
Cenozoic Era	A major division of geologic time. It is the youngest era, extending from the present back to 65 million years ago. (See Fig. 4.1.)
chalcopyrite	A bright brass-yellow copper iron sulfide mineral, $CuFeS_2$. Commonly called copper pyrite, it is the most valuable ore of copper.
chert	A dense, hard, sedimentary rock made up of extremely fine-grained silica.
cinder cone	A conical hill formed by the accumulation of fragmental volcanic material thrown out from a volcanic vent, over days or years. The ejecta are generally of *andesitic* or *basaltic* composition.
cirque	The rounded head of a steep-walled glacial valley.
clastic rock	Fragmental material formed by the breaking of preexisting rock.
clay	Any of a group of soft, water-rich, aluminum silicate minerals that result from the surface weathering of rock. Most clay minerals form as extremely fine grains, and clay, therefore, is also defined as a sediment consisting of particles that are extremely small, less than one six-thousandth inch, far below the lower limit of visibility with the unaided eye.
cleavage	One of a set of parallel planes along which various rocks or minerals will tend to split when broken.
columnar joint	One of the rock fractures that form when magma cools and contracts. The fractures bound prismatic rock masses that are commonly hexagonal (six-sided) in cross section. Columnar joints, found in lava flows and in small intrusive masses such as dikes and sills, form perpendicular to the cooling surfaces. An individual postlike fragment of basaltic lava bounded by columnar joints is termed a *basalt column*.
comb layering	A type of igneous layering in which embedded elongated mineral crystals are arranged perpendicularly to the plane of the layers containing the crystals.
conglomerate	A clastic sedimentary rock formed of rounded pebbles cemented together by mineral material.
contact metamorphism	The recrystallization of rocks that have been heated where they have come into contact with a hot intrusive mass of magma.

cordierite	A generally blue iron, magnesium, aluminum, silicate metamorphic mineral, $(Fe,Mg)_2Al_4Si_5O_{18}$, commonly formed from shaly sediments subjected to low-pressure metamorphism.
core	The central part of the Earth, beginning about 1,800 miles below the surface and composed largely of iron. See *asthenosphere, crust, lithosphere, mantle*.
corundum	An aluminum oxide mineral, Al_2O_3, that forms from the metamorphism of aluminum-rich sedimentary rocks. It is extremely tough and makes up the gem varieties ruby and sapphire.
country rock	Preexisting rock into which plutonic rocks, dikes, sills, or ore deposits have been emplaced.
crevasse	A deep fissure or crack in a glacier formed by the stresses of glacial movement. See *bergschrund*.
cross-bedding	The original layering of sedimentary rocks in which, as originally formed, the beds are inclined, commonly at a moderate angle to the horizontal.
crust	The upper layer of the Earth's lithosphere, thus the outermost part of the Earth. The continental crust is about 20 to 30 miles thick; the oceanic crust, about 4 miles thick. See *asthenosphere, core, lithosphere, mantle*.
crystalline	Having the nature of a crystal, in which atoms are arranged in a regular three-dimensional lattice. Said of a rock composed wholly of crystals.
dacite	A fine-grained volcanic rock intermediate in composition between rhyolite and andesite.
deformation	The process of compression, extension, folding, faulting, and shearing of rock units as a result of Earth forces.
differentiation	In a cooling and solidifying magma, the process that causes the stages of remaining liquid to change composition. Generally, the high-temperature, early-crystallizing minerals are rich in iron and magnesium and poor in silicon. The crystallization and removal of these minerals from the liquid causes the remaining differentiated liquid to become poorer in iron and magnesium, and richer in silicon.
dike	A thin tabular mass of igneous rock formed by the solidification of magma that has intruded into a crack and cuts across, perhaps vertically, the bedding of the rocks it intrudes. See *sill*.
diopside	A white to green silicate mineral, $CaMg(SiO_3)_2$, common as a metamorphic mineral associated with crystalline limestone.
diorite	An igneous rock with a composition midway between gabbro and granite, generally containing sodium-calcium feldspar, hornblende, and pyroxene, with little quartz.
dip	The angle between an inclined plane (as in rock layers) and the horizontal plane. The dip angle is measured in a vertical plane from the horizontal to the inclined plane.
dolomite	A calcium-magnesium carbonate mineral, $CaMg(CO_3)_2$.
dome	In granitic terrain, a dome-shaped landform commonly formed by the exfoliation of curved shells of relatively structureless granitic rock. See *exfoliation, foliation*.

element	In chemistry, any one of a limited number (about 100) of distinct varieties of matter that, whether singly or in combination, compose substances of all kinds.
enclave	See *mafic inclusion*.
epicenter	That point on the Earth's surface directly above the site of origin of an earthquake.
epidote	A yellow-green water-bearing silicate mineral, $Ca_2(Al,Fe)_3Si_3O_{12}(OH)$, common in metamorphosed limy sedimentary rocks and in chemically altered zones in granitic rocks.
epoch	A short geologic time unit; a subdivision of a period. (See Fig. 4.1.)
era	Any of several long geologic time units, each divided into periods. (See Fig. 4.1.)
exfoliation	The process of fracturing or spalling of a rock mass into curved shells that resemble the layers of an onion. See *dome*, *foliation*.
fault	A fracture in rock along which movement has occurred. See *normal fault* (gravity fault), *thrust fault* (reverse fault).
feldspar	Any of various forms of a mineral group containing silicon and aluminum and including either potassium (potassium feldspar and orthoclase) or both sodium and calcium (plagioclase). The feldspars are the most abundant minerals in the Earth's crust.
felsic rock	Any of various igneous rocks, high in silica, that contain a large proportion of feldspar and quartz. They are generally light-colored and contrast with dark-colored mafic rocks. The term *felsic* is also applied to minerals, notably feldspars and quartz.
foliation	The layered texture of metamorphic rocks caused by the parallel arrangement of platy minerals such as mica. Foliated rocks commonly split and break most easily along a plane parallel with the foliation. See *exfoliation*.
fossil	The remains or traces of ancient life, embedded or imprinted in rock.
gabbro	A dark-colored, commonly coarse-grained plutonic rock made up primarily of the minerals sodium-calcium feldspar and pyroxene, and sometimes also olivine; the intrusive equivalent of basalt.
galena	A heavy mineral, lead sulfide, PbS, having a silvery metallic luster and perfect cubic cleavage. It is the primary ore mineral for lead, and commonly contains silver.
garnet	Any of a group of silicate minerals of widely varying composition commonly containing calcium, magnesium, iron, and aluminum. It is common in both metamorphic rocks and igneous rocks.
geologic column	A diagram showing the divisions of geologic time, with the oldest at the bottom and the youngest at the top. (See Fig. 4.1.)
glaciation	The process of formation, advance, and retreat of glaciers, or ice sheets.
glacier	A large mass of ice, formed by the compaction of snow over time, that moves slowly downhill in response to its own weight.
graben	A trenchlike structure created when a tract of ground has dropped down between two faults. See *horst*.

graded bedding	A type of bedding in sedimentary rock in which the size of the sedimentary particles grades upward in each layer from coarse to fine.
granite	A light-colored, quartz-rich plutonic rock, commonly igneous, that contains more than 20 percent quartz, as well as sodium-calcium feldspar, potassium feldspar, and some dark minerals. More than one-third of the feldspar is potassium feldspar. See *granodiorite* and Fig. 4.3.
granitic rock	Any of various light-colored quartz-bearing plutonic rocks, commonly igneous, that also contain sodium-calcium feldspar and potassium feldspar. A general term that includes granite (as defined above), granodiorite, and tonalite.
granitization	A metamorphic process by which solid rock is converted to a granitic rock by recrystallization and the action of permeating fluids, without passing through a magmatic (liquid) stage. See *magmatism*.
granodiorite	A light-colored quartz-rich plutonic rock, commonly igneous, that contains more than 20 percent quartz, sodium-calcium feldspar, potassium feldspar, and some dark minerals. From one-tenth to one-third of the feldspar is potassium feldspar. Granodiorite is the most abundant rock type in the Sierra Nevada Batholith. See *granite*.
gravel	An unconsolidated sediment composed primarily of more or less rounded rock fragments larger than sand (diameter greater than $1/12$ inch).
gravity fault	See *normal fault*.
hachure	One of a series or array of short lines used on a map to portray the differing elevations and slopes of mountains and valleys.
half-life	The time necessary for a sample of a radioactive isotope to decay to one-half of its original amount. See *radioactivity*.
Holocene Epoch	That part of the Quaternary Period lasting from 10 thousand years ago to the present. (See Fig. 4.1.)
hornblende	A dark, rock-forming silicate mineral, $(Ca,Na)_2(Mg,Fe,Al)_3 Si_8O_{22}(OH)_2$, commonly occurring as elongated crystals in granitic rocks.
hornfels	A dense metamorphic rock showing no mineral alignment, and commonly derived from a shaley rock.
horst	An elongate, elevated block created when a tract of ground bounded by two faults has been uplifted above the surrounding terrain. See *graben*.
hypersthene	A common rock-forming iron magnesium silicate mineral, $Fe,MgSiO_3$, in the pyroxene group. It is gray to black in color and is an important constituent of many igneous rocks.
idocrase	A complex calcium silicate mineral containing water, $Ca_{10}Mg_2Al_4(SiO_4)_5(Si_2O_7)_2(OH)_4$, common in contact-metamorphosed limestone. It is also called *vesuvianite*.
igneous rock	Any of various rock types that form when hot, molten rock (magma) has cooled and solidified either below or above the Earth's surface.
ilmenite	A black iron titanium oxide mineral, $FeTiO_3$, a common minor mineral in igneous rocks.

GLOSSARY

interglacial Falling in the interval between two successive glacial periods, hence in a period of warmer climate.

intrusive rock A mass of igneous rock, generally medium to coarse-grained, that was molten when it came in contact with (and commonly penetrated into) its wall rock, and which cooled and solidified below the surface. See *volcanic rock, plutonic rock*.

isotope Any of the several forms, or nuclides, that an atom of a given element can assume. The various isotopes of an element all contain the same number of electrons and protons and hence have the same atomic number, but they contain different numbers of neutrons in their nucleus and therefore have different atomic weights.

joint A rock fracture along which no movement has occurred.

kyanite An aluminum silicate metamorphic mineral, Al_2SO_5, that commonly forms in metamorphosed shales at relatively high pressure. It is one of three minerals with the same composition (the others are sillimanite and andalusite), the three forming under different conditions of pressure and temperature.

lapillus (pl. lapilli) A fragment of volcanic rock ranging from $\frac{1}{10}$ inch to 2½ inches in size.

lava The molten rock that erupts from a volcano, or the rock that solidifies from that molten rock. See *magma, volcanic rock*.

leucite A potassium aluminum silicate mineral, $KAlSi_2O_6$, common in igneous rocks particularly poor in silica and enriched in potassium.

leveling A surveying procedure for determining the relative elevations of selected points on the Earth's surface. It utilizes a horizontally looking telescopic instrument and a graduated rod about 10 feet long.

limestone A sedimentary rock made up largely of calcium carbonate ($CaCO_3$). Most limestone is formed by the accumulation of organic remains (coral and shells) or by direct precipitation from seawater.

limy Containing a significant amount of lime or limestone.

lithosphere The rigid outer layer of the Earth made up of the crust and the uppermost layer of the mantle. It is from 40 to 90 miles thick, thicker where it comprises the continents, and thinner where it comprises the ocean basins.

lithospheric plates Blocks or segments of the lithosphere, about nine of which are large, a dozen smaller, each unique in its size and shape. The plates are underlain by, and move over, the more plastic asthenosphere at a rate of about 1 to 2 inches per year. Mountains, deep-sea trenches, and the other great features of the Earth's surface are formed as the plates move about and create great forces where they meet.

lode A mineral deposit found in a system of veins in bedrock, as opposed to a deposit in unconsolidated sediment.

Ma One million years ago.

mafic inclusion One of the hand-sized fragments of dark plutonic rock, commonly ovoid in shape, that are commonly distributed through a mass of granitic rock. Also called *enclave*.

mafic rock Any of various igneous rocks or minerals that contain a large proportion of iron and magnesium and are low in silica. They are generally dark-colored and contrast with light-colored felsic rocks and minerals. Common mafic rocks are basalt and gabbro, and common mafic minerals are hornblende, pyroxene, biotite, and olivine.

magma   Naturally occurring molten rock beneath the Earth's surface, called *lava* when it erupts onto the surface. Magma commonly contains a variety of dispersed crystals and gases. See *lava, volcanic rock*.

magmatism   The formation of a body of rock by the cooling and solidification of magma. See *granitization*.

magnetite   A black, strongly magnetic iron oxide mineral, Fe_3O_4.

mantle   That part of the interior of the Earth lying beneath the crust and above the core. The mantle has several zones that vary greatly in rigidity and density. See *asthenosphere, core, crust, lithosphere*.

marble   A fine- to coarse-grained metamorphic rock recrystallized from limestone. It is composed primarily of calcite.

meridian   Any line passing through the Earth's poles and connecting all points of equal longitude. A north-south line. See *parallel*.

Mesozoic Era   The major division of geologic time between the Cenozoic and Paleozoic Eras, lasting from 245 to 65 million years ago. (See Fig. 4.1.)

meta-   A prefix indicating that a rock type has been metamorphosed, for example metachert, metavolcanic, metarhyolitic, metasedimentary.

metamorphic grade   The rank or extent of metamorphism in a rock as measured by the assemblage of metamorphic minerals and the extent to which recrystallization has changed the original rock.

metamorphic rock   Rock whose original features, and mineral content, have undergone solid-state changes, such as recrystallization and deformation, as a result of prolonged high pressure and temperature as well as the introduction of new chemical constituents. See *igneous rock, sedimentary rock*.

mica   Any of a group of aluminum silicate minerals characterized by a good, platy cleavage. The most common rock-forming mica minerals are muscovite and biotite.

mineral   An element or chemical compound that is crystalline in form and has formed as a result of geologic processes.

mode   The actual mineral composition of a rock, usually expressed as a set of volume percentages.

Mohorovicic discontinuity   The boundary between the Earth's crust and its mantle, marked by a sharp change in the velocity of seismic waves. Also called *Moho* and *M-discontinuity*.

molybdenite   A soft, greenish, lead-gray molydenum sulfide mineral, MoS_2, the principal ore of molybdenum.

moraine   A mound or ridge of rock debris that has been transported and deposited directly by glacial ice and has not yet been sorted by running water. A mound deposited at the farthest extent of the ice is a *terminal moraine*, mounds deposited at successive positions as the ice front retreats are *recessional moraines*, a mound formed at the side of the glacier is a *lateral moraine*, and that central mound formed where two

	lateral moraines are juxtaposed because of the joining of two branches of a glacier is a *medial moraine*. See *till*.
muscovite	A pale to translucent mica mineral, $KAl_2(AlSi_3)O_{10}(OH)_2$, that is common in metamorphic rocks and pegmatites.
normal fault	An inclined fracture in rock in which the overlying block has moved downward. Also called a *gravity fault*. See *thrust fault*.
obsidian	A volcanic glass formed by the rapid quenching of lava. It is usually black, yet rich in silica.
oceanic spreading ridge	A mountain range on the ocean floor formed by upwelling magma, marking the site of divergence of two tectonic plates.
olivine	A glassy green, rock-forming iron and magnesium silicate mineral, $(Fe,Mg)_2SiO_4$, common in basalt and gabbro and more mafic rocks. See *serpentine*.
ophiolite	Any of a group of mafic volcanic and plutonic rocks and associated sedimentary rocks that occur in oceanic crust and are distinctive of such a setting. Pillow lava, peridotite (which chemically alters to the mineral serpentine), and chert are characteristic.
orbicular granite	A granitic rock containing nearly spherical, tightly packed, concentrically layered granitic masses, called *orbicules*, that are commonly 1 to 6 inches in diameter.
orthoclase	An abundant rock-forming potassium feldspar mineral, $KAlSi_3O_8$, commonly forming crystals up to several inches across in coarse-grained granitic rocks.
Paleozoic Era	The major division of geologic time between the Precambrian and Mesozoic eras, lasting from 570 to 245 million years ago. (See Fig. 4.1.)
parallel	Any line on the surface of the Earth drawn parallel to the Equator connecting all points of equal latitude. An east-west line. See *meridian*.
pegmatite	An exceptionally coarse-grained igneous rock, commonly granitic in composition, that occurs in small bodies such as dikes or pods.
peridotite	A coarse-grained plutonic rock containing primarily olivine, with or without other mafic minerals such as pyroxene and amphibole. Peridotite alters chemically to serpentine.
period	The primary unit of the geologic time scale, made up of epochs; a subdivision of an era. (See Fig. 4.1.)
phenocryst	A crystal that is distinctly larger than most other crystals in an igneous rock. Rock containing phenocrysts is said to be *porphyritic* in texture.
phyllite	A well-foliated metamorphic rock in which the mineral grains are small and just visible. Intermediate in grain size (and metamorphic grade) between slate and schist.
pillow lava	A lava flow characterized by rounded masses resembling, in the shape and size of its lobes, a pile of pillows. Such lava flows (commonly of basalt) result from cooling of the lava under water.
plagioclase	A sodium-calcium feldspar mineral, the most abundant mineral in the crust of the Earth. It forms as a mixture of relatively varying amounts of two end members: $NaAlSi_3O_8$ and $CaAl_2Si_2O_8$.

plane table	A surveying instrument consisting of a drawing board mounted on a tripod. It is used with a sighting ruler (*alidade*) to plot graphically (on an affixed map) the lines of sight to distant targets, during a survey made directly from field observations.
plate tectonics	An Earth model based on the concept that a number of lithospheric plates (composed of continental and/or oceanic crust and uppermost mantle) move collectively over the surface of the Earth. Much of the Earth's dynamic geologic processes occur where the plates meet and move relative to one another.
Pleistocene	A relatively recent epoch of the Cenozoic Era in the geologic time scale. It lasted from about 2 million to 10 thousand years ago and is also called the Ice Age, because it includes several periods of extensive glaciation. (See Fig. 4.1.)
pluton	A single, massive body of plutonic rock, commonly consisting of intrusive igneous rock and of mappable scale. An assemblage of plutons is a *batholith*, and an individual pluton may be large enough (greater than 40 square miles) to be considered a batholith.
plutonic rock	Rock formed at considerable depth and usually medium- to coarse-grained. It may form by crystallization of magma or by metamorphic crystallization. See *volcanic rock*, *intrusive rock*.
porphyritic rock	Igneous rock in which large, generally well-formed crystals (*phenocrysts*) are set in a matrix of finer crystals. Generally formed by a two-stage cooling history.
potassium feldspar	An abundant rock-forming potassium aluminum silicate mineral, $KAlSi_3O_8$. A common variety is *orthoclase*.
Precambrian Era	All that part of geologic time before the Paleozoic Era, lasting from about 4,500 million years ago (the approximate origin of the Earth) to 570 million years ago. In its younger parts, it contains extremely simple life forms. (See Fig. 4.1.)
pumice	A frothy, porous volcanic glass, commonly of rhyolitic composition.
pyrite	A common, worthless, brassy-yellow iron sulfide mineral, FeS_2, "fool's gold." It is commonly associated with valuable minerals in quartz veins.
pyroclastic rock	Rock made up of fragments of rock erupted during explosive volcanic eruptions. See *ash*, *tuff*.
pyroxene	Any of a group of dark, rock-forming silicate minerals. Most are rich in iron and magnesium.
quartz	An extremely common, light-colored, hard, resistant mineral formed of silica, SiO_2. It has no cleavage, and because of its resistance to abrasion and weathering, is very common in sands and sandstone.
quartz diorite	A quartz-poor granitic rock, commonly igneous, that contains less than 20% quartz, as well as sodium-calcium feldspar and some dark minerals. Less than one-tenth of the feldspar is potassium feldspar (Fig. 4.3). In an early classification it included *tonalite*.
quartz diorite line	A line traceable through the western part of North America that divides Mesozoic and Cenozoic granitic rocks into two regional zones. West of the line the dominant granitic rock is tonalite (formerly classified as quartz diorite), and east of it, granodiorite (Fig. 6.2).

quartzite	A metamorphosed sandstone composed dominantly of quartz grains.
quartz monzodiorite	A quartz-poor granitic rock, commonly igneous, that contains less than 20% quartz, as well as sodium-calcium feldspar, potassium feldspar, and some dark minerals. From one-tenth to one-third of the feldspar is potassium feldspar (Fig. 4.3).
quartz monzonite	A quartz-poor granitic rock, commonly igneous, that contains less than 20% quartz, as well as sodium-calcium feldspar, potassium feldspar, and some dark minerals. From one-third to two-thirds of the feldspar is potassium feldspar (Fig. 4.3).
Quaternary Period	The youngest period of Earth history, the period that includes the Pleistocene and Holocene Epochs and extends to the present. It began about 1.6 million years ago. (See Fig. 4.1.)
radioactivity	A spontaneous, regular process whereby an isotope of one element emits rays or particles as its nuclei disintegrate steadily and produce different isotopes of the same or different elements. See *half-life*.
radiometric dating	The process of determining the age of geologic material by analysis of the proportions of naturally occurring radioactive elements (*isotopes*) and their decay products (daughter isotopes) within the material.
recrystallization	The development of new crystalline mineral grains in a rock at the expense of previous mineral grains, as a result of changing conditions of heat and pressure. The new grains may be of similar or different mineral and chemical composition. See *metamorphic rock*.
reverse fault	See *thrust fault*.
rhyolite	A light-colored, fine-grained or glassy volcanic rock with a composition similar to that of granite. Rhyolite is an extrusive rock, granite an intrusive rock.
Richter scale	A scale based on the intensity of seismic (earthquake) waves as recorded on a seismograph, which indicates the amount of energy released by an earthquake.
rock glacier	A tongue-shaped mass of rock debris mixed with ice that moves slowly downslope in a fashion similar to that of an ice glacier.
roof pendant	A downward projection, into an igneous intrusion, of the country rocks forming the roof over the intrusion.
sand	An unconsolidated sediment consisting of small rock or mineral particles having a diameter ranging from $1/400$ to $1/12$ inch (a size range between that of the lower limit of visibility with the unaided eye and that of the head of a wooden match).
sandstone	Rock formed by the accumulation and cementation of primarily sand particles.
scheelite	A yellowish-white calcium tungstate mineral, $CaWO$, the most common ore of tungsten. It is commonly found near the region of contact between granitic intrusions and limy sediments.
schist	A medium-grained, well-foliated metamorphic rock, distinctly coarser than phyllite, in which flat mineral grains lie parallel with the layering. Grain size commonly ranges from about $1/25$ to $1/5$ inch.
sedimentary rock	Rock resulting from the consolidation of loose sediment that has accumulated in layers.
seismic velocity	The speed at which the various types of earthquake vibrations or waves propagate through Earth materials, changing notably at the Mohorovicic discontinuity.

seismograph	An instrument used to record the time and strength of the vibrations or waves produced by an earthquake.
serpentine	Any of a group of soft, water-rich minerals formed by low-temperature alteration of magnesium-rich silicates such as olivine and pyroxene. See *peridotite*.
sextant	A handheld instrument used for measuring the angular distance between two objects or between an object and the horizon. Used for surveying and navigation.
shale	A fine-grained laminated rock formed of clay and silt.
silica	Silicon dioxide, SiO_2, commonly occurring as the mineral quartz.
silicate	Any of numerous compounds or minerals containing silicon and oxygen and one or more other elements. The minerals are further grouped by the most important element(s) combined with the silicon and oxygen, for example, calcium aluminum silicate. Most of the rock making up the Earth is composed of silicate minerals.
sill	A tabular sheet of intrusive igneous rock that lies parallel to, and has split apart, the layering in the country rock. See *dike*.
sillimanite	An aluminum silicate metamorphic mineral, Al_2SiO_5, that commonly forms in metamorphosed shales at relatively high temperature. It is one of three minerals with the same composition (the others are andalusite and kyanite), the three forming under different conditions of pressure and temperature.
silt	An unconsolidated assemblage of sedimentary rock fragments or particles each larger than coarse clay and smaller than very fine sand, with a particle diameter ranging from $1/6000$ to $1/400$ inch.
siltstone	A rock formed by the accumulation and cementation of primarily silt particles.
slate	A well-foliated, fine-grained metamorphic rock that splits readily into flat sheets. Grain size is so fine that individual crystals cannot be seen with the unaided eye. See *phyllite, schist*.
sodium-calcium feldspar	An extremely abundant rock-forming aluminum silicate mineral termed *plagioclase*. It forms as a mixture, in any proportions, of a sodium end member, $NaAlSi_3O_8$, and a calcium end member, $CaAl_2Si_2O_8$.
specific gravity	The ratio of the weight of a material relative to the weight of an equal volume of water, thus a measure of density. The specific gravity of water is 1, and of, for example, mercury, 13.5.
sphene	A honey-brown calcium titanium silicate mineral, $CaTiSiO_5$, common as a minor constituent in granitic rocks.
spirit level	A sensitive device consisting of a glass tube partly filled with a low-viscosity liquid (such as alcohol), used to determine a horizontal line, as is done with a carpenter's level. The position of the bubble in the tube is used to establish the level position of various instruments, for example a transit.
stalactite	An accumulation, usually of calcium carbonate minerals precipitated by dripping water, that hangs like icicles from the ceilings of caves. See *travertine, stalagmite*.
stalagmite	An accumulation of carbonate minerals, precipitated by dripping water, that builds up from the floor of caves. See *travertine, stalactite*.

steatite	A fine-grained, commonly greenish metamorphic rock rich in the soft mineral talc, $Mg_3Si_4O_{10}(OH)_2$. Steatite, also called soapstone, can be easily carved into ornamental and useful objects.
stoping	The process whereby intruding magma detaches and engulfs pieces of the wall rock, which then sink down, or are otherwise assimilated within the growing magma chamber.
striation	A linear groove in a rock or mineral. Glacial striations are scratches and furrows inscribed on a bedrock surface that are cut by sharp rocks embedded in the ice moving over the rock surface.
strike	The line of intersection of an inclined plane (such as that of a sedimentary bed or a fault) with an imaginary horizontal plane.
strike slip	Fault movement (slip) in a horizontal direction, parallel to the strike of the fault. A right-lateral strike-slip fault is one in which the side of the fault opposite that on which an observer stands has moved to the right. (If the observer then moves to the far side and looks back, the offset is the same.)
subduction	The process by which one lithospheric tectonic plate sinks beneath another as the two plates press against one another.
sulfide	A mineral in which sulfur is combined with one or more metallic elements. Pyrite (iron sulfide, FeS_2) is a sulfide mineral.
superposition	A geologic law stating that in a sequence of layered sedimentary rocks, the layers on top are younger than those below.
tactite	A rock, of varied mineral composition, formed by contact metamorphism of limy rocks. It is usually rich in the mineral garnet.
talc	A soft, greenish magnesium silicate metamorphic mineral, $Mg_3Si_4O_{10}(OH)_2$. See *steatite*.
talus	A pile or fan of loose, angular rock fragments that have been shed from, and accumulate at the base of, a cliff or other steep rock mass.
tectonics	That part of the science of geology concerned with the broad architecture and dynamics of the upper part of the Earth's lithosphere, including the crust.
terrane	A fault-bounded geologic entity, commonly of regional extent, accreted to a continent, possessing a distinct assemblage of rock units, structure, and geologic history, an assemblage differing from that of adjacent rock masses. The terrane can be visualized as having once been all or part of a tectonic plate that may have traveled some distance before being accreted to the continent at its present location.
Tertiary Period	That part of the Cenozoic Era lasting from 65 to 1.6 million years ago. (See Fig. 4.1.)
thrust fault	An inclined fracture in rock in which the overlying block has moved upward. Also called a *reverse fault*. See *normal fault*.
till	Unsorted and unstratified sediment deposited from glacier ice and not reworked by running water. See *moraine*.
tonalite	A light-colored plutonic rock, commonly igneous, that contains more than 20% quartz, sodium-calcium feldspar, and some dark minerals. Less than one-tenth of

the feldspar is potassium feldspar (Fig. 4.3). Tonalite is an abundant rock type in the western part of the Sierra Nevada Batholith. (An earlier classification called this composition *quartz diorite*).

topography — The general configuration of the land surface, including its relief (variation in surface elevation) and the positions of its natural and manmade features. On *topographic* maps, relief is generally shown by lines of equal elevation (contour lines); such maps differ from *planimetric* maps, which do not depict elevation or relief in measurable form.

tourmaline — Any of a group of aluminum silicate minerals of complex composition that contain boron and water as well as several other constituents.

transform fault — A strike-slip fault that offsets spreading oceanic ridges.

transit — A surveying instrument mounted on adjustable wooden legs and generally equipped with a telescope for measuring horizontal and vertical angles. The telescope of the transit can be lifted and turned end-for-end around a horizontal axis.

travertine — A cave or spring deposit composed of calcium carbonate, $CaCO_3$. It is typically precipitated from surface or groundwater, and may also be precipitated by calcareous algae.

triangulation — A system of surveying in which stations on the ground (usually mountain peaks) are considered as the vertices of a network of triangles. The angles of the triangles are measured instrumentally, and when the length of the sides of some of the triangles are then determined, the size of the entire network of triangles can be computed. This network is then the skeleton of a map covering the area of interest.

tuff — A pyroclastic rock made up of cemented volcanic ash fragments that originated as ash-fall and ash-flow deposits.

ultramafic rock — Dark-colored igneous rock that contains exceptionally large proportions of magnesium and iron. See *mafic rock*.

vernier scale — A short scale, mounted so as to slide along the primary scale on a measuring device, that is used to measure fractional parts of the smallest divisions of the primary scale.

volatile component — Any of several constituents of magmas or solid rock that have low vaporization temperatures and are likely to bubble out of magmas when pressure is reduced, or to lower the melting point of solid rock. The most common volatile components are water and carbon dioxide.

volcanic rock — Fine-grained rock formed above the surface of the Earth by the cooling and crystallization of molten material erupted from a volcanic vent. Such molten material is termed *magma* below the surface and *lava* above. See *intrusive rock*, *ophiolite*, *plutonic rock*.

water table — That level below the ground surface at which water stands in wells. Below it the rock or soil is saturated with groundwater.

zircon — A hard zirconium silicate mineral, $ZrSiO_4$, the principal ore of zirconium. It is a common minor accessory mineral in granitic rocks.

References Cited

CHAPTER I

Chalfant, W. A., 1933, *The Story of Inyo*: Chalfant Press, Bishop, Calif., 430 pp.

Clewlow, C. W., Jr., 1978, Prehistoric rock art, *in* W. C. Sturtevant, ed., *Handbook of North American Indians*, v. 8, California: Smithsonian Institution, Washington, D.C., pp. 619–625.

Clyde, N., 1962, *Close Ups of the High Sierra*: La Siesta Press, Glendale, California, 79 pp.

Elliot, W. W., ed., 1881, 1884, *History of Fresno County, California*: W. W. Elliot and Co., Publishers, San Francisco, 246 pp.

Elsasser, A. B., 1972, *Indians of Sequoia and Kings Canyon National Parks*: Sequoia Natural History Association, Three Rivers, Calif., 56 pp.

Matthes, F., 1960, Glacial reconnaissance of Sequoia National Park, California, U.S. Geological Survey Professional Paper 504-A, 58 pp. [Prepared posthumously by F. Fryxell.]

Muir, J., 1887, *Picturesque California: The Rocky Mtns. and the Pacific Slope*: J. Dewing Publishing Co., New York and San Francisco, 508 pp.

Norris, R. M., and R. W. Webb, 1990, *Geology of California*: John Wiley & Sons, New York, 541 pp.

Porcella, S. F., and C. M. Burns, 1998, *Climbing California's Fourteeners*: The Mountaineers, Seattle, 269 pp.

Starr, W. A., Jr., 1934, *Guide to the John Muir Trail and the High Sierra Region*: Sierra Club, San Francisco, 4th Edition, 130 pp.

Steward, J. H., 1933, Ethnography of the Owens Valley Paiute: *University of California Publications in American Archaeology and Ethnology*, v. 33, pp. 233–350.

Thelin, G. P., and R. J. Pike, 1991, Landforms of the conterminous United States—A digital shaded-relief portrayal: U.S. Geological Survey Map I-2206.

Thompson, T. H., 1892, *Official Historical Atlas Map of Tulare County*: Tulare, Calif., 147 pp.

CHAPTER 2

Alsup, W., 1987, *Such a Landscape! Reports of William Henry Brewer*: Yosemite Association and Sequoia Natural History Association, 120 pp.

Bade, W. F., 1924, *The Life and Letters of John Muir*: Houghton Mifflin Company, Boston, 2 vols., 399 pp. and 454 pp.

Blake, W. P., 1858, *Report of a Geological Reconnaissance in California, Made in Connection with the Expedition to Survey Routes in California, to Connect with the Surveys of Routes for a Railroad from the Mississippi River to the Pacific Ocean under the Command of Lieut. R. S. Williamson, Corps Top. Eng'rs, in 1853*: H. Bailliere, New York, 336 pp.

Block, R. H., 1982, *The Whitney Survey of California, 1860–74: A Study of Environmental Science and Exploration*: University of California, Los Angeles, Ph.D. dissertation, 480 pp.

Brewer, W. H., 1930, *Up and Down California in 1860–1864: The Journal of William H. Brewer, Professor of Agriculture in the Sheffield Scientific School from 1864 to 1903*, ed. F. P. Farquhar: Yale University Press, New Haven, 601 pp.

Brewster, E. T., 1909, *Life and Letters of Josiah Dwight Whitney*: Houghton Mifflin Co., Boston and New York, 411 pp. [Inside front cover is bookplate with etching of Mount Whitney.]

Chalfant, W. A., 1933, *The Story of Inyo*: Chalfant Press, Bishop, Calif., 430 pp.

Clyde, N., 1962, *Close Ups of the High Sierra*: La Siesta Press, Glendale, Calif., 79 pp.

Colby, W. E., 1918, Notes and correspondence: *Sierra Club Bulletin*, v. 10, n. 3.

Dilsaver, L. M., and W. C. Tweed, 1990, *Challenge of the Big Trees*: Sequoia Natural History Association, Three Rivers, Calif., 379 pp.

Farquhar, F. P., 1965, *History of the Sierra Nevada*: University of California Press, Berkeley, 262 pp.

Fremont, J. C., 1887, *Memoirs of My Life*: Belford, Clarke & Company, Chicago, 653 pp.

Goddard, G. H., 1855, Report of a survey of a portion of the eastern boundary of California, and of a reconnaissance of the old Carson and Johnson immigrant roads over the Sierra Nevada, *in Annual Report of the Surveyor-General of the State of California*: James Allan, State Printer, Session of 1856, pp. 91–186.

———, 1857, Map of the State of California: Britton and Rey, San Francisco. [Compiled from the U.S. land and coast surveys, the several military, scientific, and rail road explorations, the state and county boundary surveys, made under the order of the Surveyor General of California, and from private surveys.]

Goodyear, W. A., 1888, Inyo County, *in* W. Irelan, Jr., *Eighth Annual Report of the State Mineralogist*: California State Mining Bureau, Sacramento, pp. 224–309.

Gray, A., 1880, *Botany*: John Wilson and Son, University Press, Cambridge, Mass., Second (Revised) Edition, 628 pp. [Gold insignia of California Geological Survey embossed on back cover.]

Hoffmann, C. F., 1873, Topographical map of central California together with a part of Nevada: J. D. Whitney, State Geologist, Geological Survey of California. [Four sheets, scale 1:375,000.]

Irelan, W., Jr., 1891, Preliminary mineralogical and geological map of the state of

California: California State Mining Bureau. [In color; scale 12 miles to the inch.]

King, C., 1876. Geological and topographical atlas accompanying the report of the Geological Exploration of the Fortieth Parallel: Chief Engineer, War Department, Washington, D.C.

———, 1935 [1872], *Mountaineering in the Sierra Nevada*, ed. F. P. Farquhar: W. W. Norton & Co., New York, 320 pp.

Merriam, C. H., 1923, First crossing of the Sierra Nevada: Jedediah Smith's trip from California to Salt Lake in 1827, *Sierra Club Bulletin*, v. 11, pp. 375–379.

Middleton, W. E. K., 1969, *Catalog of Meteorological Instruments in the Museum of History and Technology*: Smithsonian Institution Press, Washington, D.C.

Muir, J., 1873, On actual glaciers in California: *American Journal of Science*, v. 5, pp. 69–71.

———, 1874, Mountain sculpture, origin of Yosemite valleys: *Overland Monthly*, v. 12, pp. 393–403.

———, 1887, *Picturesque California: The Rocky Mtns. and the Pacific Slope*: J. Dewing Publishing Co., New York and San Francisco, 508 pp.

———, 1891, A rival of the Yosemite: The cañon of the South Fork of Kings River, California: *Century Magazine*, v. 43, Nov., pp. 77–97.

———, 1938, *John of the Mountains*, ed. L. M. Wolfe: Houghton Mifflin Company, Cambridge, Mass., 459 pp.

———, 1968, *South of Yosemite*, ed. F. R. Gunsky: Wilderness Press, Berkeley, Calif., 1,030 pp.

Raymond, R. W., 1911, Memoir of William Phipps Blake: *Geological Society of America Bulletin*, v. 22, pp. 36–47.

Taylor, Z., 1850, Message from the President of the United States, transmitting information in answer to a resolution of the House of the 31st of December, 1849, on the subject of California and New Mexico, 31st Congress, 1st Session, Ex. Doc. No. 17, Washington, D.C., 976 pp.

Whitney, J. D., 1865, *Geology: Report of Progress and Synopsis of the Field Work, from 1860 to 1864*: Sacramento, Geological Survey of California, v. 1, 498 pp. [Gold insignia of Survey embossed on back cover.]

———, 1874, *The Yosemite Guide-Book*: Sacramento, Geological Survey of California, 186 pp. [Gold insignia of Survey embossed on back cover.]

Wright, J. W. A., 1883, Guide to the scenery of the Sierra Nevada, *in History of Fresno County, California*: W. W. Elliot and Co., San Francisco, 246 pp.

CHAPTER 3

Adams, A., 1944, The photography of Joseph N. LeConte: *Sierra Club Bulletin*, v. 25, n. 5, pp. 41–46.

Browning, P., 1986, *Place Names of the Sierra Nevada*: Wilderness Press, Berkeley, Calif., 253 pp.

Clyde, N., 1962, *Close Ups of the High Sierra*: La Siesta Press, Glendale, Calif., 79 pp.

Dawdy, D. O., 1993, *George Montague Wheeler—The Man and the Myth*: Swallow Press/Ohio University Press, Athens, 122 pp.

Durrell, C., 1940, Metamorphism in the southern Sierra Nevada northeast of Visalia, California: University of California Department of Geology, Bulletin, v. 24, n. 1, pp. 1–117.

Dyer, H., 1893, The Mt. Whitney trail: *Sierra Club Bulletin*, v. 1, n. 1, pp. 1–8. [Includes an attached map by "J.N.L." of the Kings Canyon–Mount Whitney region.]

Farquhar, F. P., 1965, *History of the Sierra Nevada*: University of California Press, Berkeley, 262 pp.

Gannett, S., and D. H. Baldwin, 1908, Spirit leveling in California, 1896–1907, inclusive: U.S. Geological Survey Bulletin 342, 172 pp.

Irelan, W., Jr., 1891, Preliminary mineralogical and geological map of the state of California: California State Mining Bureau. [In color; scale 12 miles to the inch.]

King, C., 1935 (1872), *Mountaineering in the Sierra Nevada*, ed. F. P. Farquhar: W. W. Norton & Co., New York, 320 pp.

———, 1876. Geological and topographical atlas accompanying the report of the Geological Exploration of the Fortieth Parallel: Chief Engineer, War Department, Washington, D.C.

Knopf, A., and P. Thelen, 1905, Sketch map of the geology of Mineral King, California: University of California Department of Geology, Bulletin, v. 4, pp. 227–262.

Krauskopf, K., 1953, Tungsten deposits of Madera, Fresno, and Tulare Counties, California: California Division of Mines Special Report 35, 83 pp.

Lawson, A. C., 1903, The geomorphogeny of the upper Kern Basin: University of California Department of Geology, Bulletin, v. 3, pp. 291–376.

LeConte, J. N., 1893, Maps of the Sierra: Sierra Club, San Francisco. [Two maps, one of the Yosemite region and one of the Kings River region.]

———, 1896, Map of the central portion of the Sierra Nevada Mountains and of the Yosemite Valley: Sierra Club, San Francisco, publication no. 12.

———, 1899, A revised map of the High Sierra: *Sierra Club Bulletin*, v. 2, n. 5, p. 285 and plate 38.

———, 1905, The Evolution group of peaks: *Sierra Club Bulletin*, v. 5, n. 3, pp. 229–237.

Matthes, F., 1960, Glacial reconnaissance of Sequoia National Park, California, U.S. Geological Survey Professional Paper 504-A, 58 pp. [Prepared posthumously by F. Fryxell.]

Muir, J., 1891, A rival of the Yosemite: *Century Magazine*, v. 43, Nov., pp. 77–97.

Ross, D. C., 1958, Igneous and metamorphic rocks of parts of Sequoia and Kings Canyon National Parks, California: California Division of Mines and Geology Special Report 53, 24 pp.

Sargent, S., 1989, *Solomons of the Sierra*: Flying Spur Press, Yosemite, California, 132 pp.

Sobel, D., 1995, *Longitude*: Walker and Co., New York, 184 pp.

Solomons, T. S., 1895, A search for a high mountain route from the Yosemite to the King's River Cañon: *Sierra Club Bulletin*, pp. 221–237.

———, 1940, The beginnings of the John Muir Trail: *Sierra Club Bulletin*, v. 25, n. 1, pp. 28–40.

U.S. Coast and Geodetic Survey, 1882, Report of the Superintendent on the progress of work during the fiscal year ending with June, 1880: Government Printing Office, Washington, D.C., 419 pp.

Uzes, F. D., 1977, *Chaining the Land*: Landmark Enterprises, Sacramento, Calif., 315 pp.

Wentworth, G. A., 1903, *Plane and Spherical Trigonometry, Surveying, and Tables*, 2nd Edition: Ginn and Company, New York, 255 pp.

Wheeler, G. M., 1872, Preliminary report concerning explorations and surveys principally in Nevada and Arizona during 1871: Government Printing Office, Washington, D.C. [Includes map: scale 24 miles to the inch.]

———, 1874, Geographical Surveys West of the 100th Meridian, Topographical Atlas sheet 65, Expedition of 1871. [Scale 8 miles to the inch.]

———, 1889, Report upon United States Geographical Surveys West of the 100th Meridian, Vol. I—Geographical report: The Chief of Engineers, U.S. Army, Government Printing Office, Washington, D.C., 780 pp.

Whitney, J. D., 1865, *Geology: Report of Progress and Synopsis of the Field Work, from 1860 to 1864*: Sacramento, Geological Survey of California, v. 1, 498 pp.

Wilkins, T., 1958, *Clarence King, a Biography:* Macmillan Co., New York, 441 pp.

Wright, J. W. A., 1883, Guide to the scenery of the Sierra Nevada, *in* W. W. Elliot, ed., *History of Fresno County, California*: W. W. Elliot and Co., San Francisco, 246 pp.

CHAPTER 4

Dalrymple, G. B., 1991, *The Age of the Earth*: Stanford University Press, Stanford, Calif., 474 pp.

Hamilton, W., 1978, Mesozoic tectonics of the western United States, *in* D. G. Howell and K. A. McDougall, eds., *Mesozoic Paleogeography of the Western United States*: Society of Economic Paleontologists and Mineralogists, Pacific Section, pp. 33–70.

Kious, W. J., and R. I. Tilling, 1996, *This dynamic Earth: The story of plate tectonics*: Government Printing Office, Washington, D.C., 77 pp.

Locke, A., P. Billingsly, and E. B. Mayo, 1940, Sierra Nevada tectonic pattern: *Geological Society of America Bulletin*, v. 51, pp. 1513–1540.

Saleeby, J. B., and C. Busby, 1993, Paleogeographic and tectonic setting of axial and western metamorphic framework rocks of the southern Sierra Nevada, California, *in* G. Dunn and K. McDougall, eds., *Mesozoic Paleogeography of the Western United States*, II, Society of Economic Paleontologists and Mineralogists, Pacific Section, book 71, pp. 197–226.

Streckeisen, A. L., 1973, Plutonic rocks, classification and nomenclature recommended by the International Union of Geological Sciences, Subcommission on the systematics of igneous rocks: *Geotimes*, v. 18, no. 10, pp. 26–30.

CHAPTER 5

Dickinson, W. R., C. A. Hopson, and J. B. Saleeby, 1996, Alternate origins of the Coast Range Ophiolite (California): Introduction and implications: *GSA Today*, v. 6, no. 2, pp. 1–10.

Hamilton, W., 1978, Mesozoic tectonics of the western United States, *in* D. G.

Howell and K. A. McDougall, eds., *Mesozoic Paleogeography of the Western United States*: Society of Economic Paleontologists and Mineralogists, Pacific Section, pp. 33–70.

Jones, D. L., and J. G. Moore, 1973, Lower Jurassic ammonite from the south-central Sierra Nevada, California: *U.S. Geological Survey Journal of Research*, v. 1, pp. 453–458.

Moore, J. G., 1975, Mechanism of formation of pillow lava: *American Scientist*, v. 63, pp. 269–277.

Moore, J. G., and D. L. Peck, 1962, Accretionary lapilli in volcanic rocks of the western continental United States: *Journal of Geology*, v. 70, pp. 182–193.

Moore, J. N., and C. T. Foster, Jr., 1980, Lower Paleozoic metasedimentary rocks in the east-central Sierra Nevada, California: Correlation with Great Basin formations: *Geological Society of America Bulletin*, pt. 1, v. 91, pp. 37–43.

Nokleberg, W. J., 1983, Wall rocks of the central Sierra Nevada Batholith, California—A collage of accreted tectonostratigraphic terranes: U.S. Geological Survey Professional Paper 1255, 28 pp.

Palmer, A. N., 1991, Origin and morphology of limestone caves: *Geological Society of America Bulletin*, v. 103, pp. 1–21.

Saleeby, J. B., 1978, Kings River ophiolite, southwest Sierra Nevada foothills, California, *Geological Society of America Bulletin*, v. 89, pp. 617–636.

Saleeby, J. B., and C. Busby, 1993, Paleogeographic and tectonic setting of axial and western metamorphic framework rocks of the southern Sierra Nevada, California, *in* G. Dunn and K. McDougall, eds., *Mesozoic Paleogeography of the Western United States*, II, Society of Economic Paleontologists and Mineralogists, Pacific Section, book 71, pp. 197–226.

Saleeby, J. B., S. E. Goodin, W. D. Sharp, and C. J. Busby, 1978, Early Mesozoic paleotectonic-paleogeographic reconstruction of the southern Sierra Nevada region, *in* D. G. Howell and K. A. McDougall, eds., *Mesozoic Paleogeography of the Western United States*: Society of Economic Paleontologists and Mineralogists, Pacific Coast Section, Paleogeography Symposium 2, pp. 311–336.

Schweickert, R. A., and M. M. Lahren, 1991, Age and tectonic significance of metamorphic rocks along the axis of the Sierra Nevada Batholith, A critical reappraisal, *in* J. D. Cooper and C. H. Stevens, eds., *Paleozoic Paleogeography of the Western United States* II, Society of Economic Paleontologists and Mineralogists, Pacific Section, v. 67, pp. 653–676.

Tinsley, J. C., D. J. DesMarais, G. McCoy, B. W. Rogers, and S. R. Ulfeldt, 1981, Lilburn Cave's contributions to the natural history of Sequoia and Kings Canyon National Parks, California, USA., ed. B. F. Beck, *Proceedings of the International Congress of Speleology* (8th), Bowling Green, Kentucky, v. 1, pp. 287–290.

CHAPTER 6

Bateman, P. C., 1992, Plutonism in the central part of the Sierra Nevada Batholith, California: U.S. Geological Survey Professional Paper 1483, 186 pp.

Carl, B. S., A. F. Glazner, J. M. Bartley, D. A. Dinter, and D. S. Coleman, 1998, Independence Dikes and mafic rocks of the eastern Sierra: Guidebook to Field Trip No. 4, Annual Meeting, Cordilleran Section of the Geological Society of America, pp. 4-1 to 4-26.

Chen, J. H., and J. G. Moore, 1979, Late Jurassic Independence Dike Swarm in eastern California: *Geology*, v. 7, pp. 129–133.

———, 1982, Uranium-lead isotopic ages from the Sierra Nevada Batholith, California: *Journal of Geophysical Research*, v. 87, pp. 4761–4784.

Evernden, J. F., and R. W. Kistler, 1970, Chronology of emplacement of Mesozoic batholithic complexes in California and Western Nevada: U.S. Geological Survey Professional Paper 623, 42 pp.

Hamilton, W. B., 1995, Subduction systems and magmatism, *in* J. L. Smellie, ed., Volcanism associated with extension at consuming plate margins: Geological Society of America Special Publications n. 81, pp. 3–28.

Kistler, R. W., 1990, Two different lithospheric types in the Sierra Nevada, California, *in* J. L. Anderson, ed., The nature and origin of Cordilleran magmatism: Geological Society of America Memoir 174, pp. 271–281.

Moore, J. G., 1959, The quartz diorite boundary line in the western United States: *Journal of Geology*, v. 67, pp. 198–210.

———, 1963, Geology of the Mount Pinchot Quadrangle, southern Sierra Nevada, California: U.S. Geological Survey Bulletin 1130, 152 pp.

Moore, J. G., and C. A. Hopson, 1961, The Independence Dike Swarm in eastern California: *American Journal of Science*, v. 259, pp. 241–259.

Moore, J. G., and J. P. Lockwood, 1973, Origin of comb layering and orbicular structure, Sierra Nevada Batholith, California: *Geological Society of America Bulletin*, v. 84, pp. 1–20.

Oliver, H. W., J. G. Moore, and R. F. Sicora, 1993, Internal structure of the Sierra Nevada batholith based on specific gravity and gravity measurements: *Geophysical Research Letters*, v. 20, pp. 2179–2182.

Ross, D. C., 1989, The metamorphic and plutonic rocks of the southernmost Sierra Nevada, California, and their tectonic framework: U.S. Geological Survey Professional Paper 1381, 159 pp.

Ross, M. I., and C. R. Scotese, 1997, PaleoGeographic Information System/Mac: Earth in Motion Technologies, Houston, version 4.0.0.

Sisson, T. W., T. L. Grove, and D. S. Coleman, 1996, Hornblende gabbro complex at Onion Valley, Californa, and a mixing origin for the Sierra Nevada Batholith: *Contributions to Mineralogy and Petrology*, v. 126, pp. 81–108.

Smith, G. I., 1962, Large lateral displacement on Garlock Fault, California, as measured from offset dike swarm: *Bulletin of the American Association of Petroleum Geologists*, v. 46, pp. 85–104.

Tobisch, O. T., J. B. Saleeby, R. R. Renne, B. McNulty, and W. Tong, 1995, Variations in deformation fields during development of a large-volume magmatic arc, central Sierra Nevada, California: *Geological Society of America Bulletin*, v. 107, pp. 148–166.

CHAPTER 7

Colby, W. E., 1918, A story by Thomas Keough: *Sierra Club Bulletin*, v. 10, n. 3.

Goodyear, W. A., 1888, Minerals of Inyo County, *in Eighth Annual Report of the State Mineralogist*: California State Printing Office, Sacramento, pp. 224–309.

Irelan, W., Jr., 1891, Preliminary mineralogical and geological map of the state of

California: California State Mining Bureau. [In color; scale 12 miles to the inch.]

Krauskopf, K. B., 1953, Tungsten deposits of Madera, Fresno, and Tulare Counties, California: California State Division of Mines Special Report 35, 83 pp.

LeConte, J. N., 1890, *A Summer of Travel in the High Sierra*: Lewis Osborne, Ashland, Oregon (1972), 144 pp.

Porter, S. T., 1965, The silver rush at Mineral King, California, 1873–1882: Unpublished typescript, 157 pp.

Wheeler, G. M., 1889, *Report upon United States Geographical Surveys West of the 100th Meridian*, vol. 1, Geographical Report: The Chief of Engineers, U.S. Army, Government Printing Office, Washington, D.C., 780 pp.

CHAPTER 8

Bacon, C. R., and W. A. Duffield, 1981, Late Cenozoic rhyolites from the Kern Plateau, southern Sierra Nevada, California: *American Journal of Science*, v. 281, pp. 1–34.

Hoffmann, C. F., 1873, Topographical map of central California together with a part of Nevada: Geological Survey of California, J. D. Whitney, State Geologist. [Four sheets; scale 1: 375,000.]

Moore, J. G., and F. C. W. Dodge, 1980, Late Cenozoic volcanic rocks of the southern Sierra Nevada, California: I. Geology and petrology: Geological Society of America Bulletin, Part I, v. 91, pp. 515–518; Part II, v. 91, pp. 1995–2038.

Muir, J., 1902, *Tulare County Times*, August 28, Visalia.

Ormerod, D. S., N. W. Rogers, and C. J. Hawkesworth, 1991, Melting in the lithospheric mantle: Inverse modeling of alkali-olivine basalts from the Big Pine Volcanic Field, California: *Contributions to Mineralogy and Petrology*, v. 108, pp. 305–317.

Roper Wickstrom, C. K., 1993, Spatial and temporal characteristics of high altitude site patterning in the southern Sierra Nevada: Center for Archaeological Research at Davis, Publication No. 11, pp. 285–301.

CHAPTER 9

Bade, W. F., 1924, *The Life and Letters of John Muir*: Houghton Mifflin Company, Boston, 2 vols., 399 pp. and 454 pp.

Bischoff, J. L., K. M. Menking, J. P. Fitts, and J. A. Fitzpatrick, 1997, Climatic oscillations 10,000–155,000 yrs B.P. at Owens Lake, California, reflected in glacial rock flour abundance and lake salinity in core OL-92: *Quaternary Research*, v. 48, pp. 313–325.

Clark, D. H., E. J. Steig, N. Potter, J. Fitzpatrick, A. B. Updike, and G. M. Clark, 1996, Old ice in rock glaciers may provide long-term climate records: *Eos*, v. 77, pp. 217–222.

Davis, W. M., 1906, The sculpture of mountains by glaciers: *Scottish Geographical Magazine*, v. 22, pp. 76–89.

Grove, J. M., 1988, *The Little Ice Age*: Methuen, London, 498 pp.

King, P. B., 1962, The physical geography of William Morris Davis: Unpublished lecture notes (1927–1929) and collected drawings, 237 pp. [Assembled by P. B. King.]

Martinson, D. G., N. G. Pisias, J. D. Hays, J. Imbrie, T. C. Moore, Jr., and N. J. Shackleton, 1987, Age dating and the orbital theory of the ice ages: Development of a high-resolution 0 to 300,000-year chronostratigraphy: *Quaternary Research*, v. 27, pp. 1–29.

Matthes, F. R., 1930, Geologic history of the Yosemite Valley: U.S. Geological Survey Professional Paper 160, 137 pp.

Muir, J., 1887, *Picturesque California: The Rocky Mtns. and the Pacific Slope*: J. Dewing Publishing Co., New York and San Francisco, 508 pp.

———, 1894, *The Mountains of California*: Century, New York, 160 pp.

Ruddiman, W. F., M. E. Raymo, W. L. Prell, and J. E. Kutzbach, 1997, The uplift-climate connection: A synthesis, *in* Tectonic uplift and climate change, ed. W. F. Ruddiman: Plenum Press, New York, pp. 471–515.

Smith, G. I., J. L. Bischoff, and J. P. Bradbury, 1997, Synthesis of the paleoclimatic record from Owens Lake core OL-92, *in* An 800,000-year paleoclimatic record from Core OL-92, Owens Lake, southeast California, ed. G. I. Smith and J. L. Bischoff: Geological Society of America Special Paper 317, pp. 143–160.

Wahrhaftig, C., 1984, Geomorphology and glacial geology—Wolverton and Crescent Meadow areas and vicinity, Sequoia National Park, California: U.S. Geological Survey Open-File report 84-400, 52 pp.

Whitney, J. D., 1865, *Geology: Report of Progress and Synopsis of the Field Work, from 1860 to 1864*: Sacramento, Geological Survey of California, v. 1, 498 pp.

CHAPTER 10

Fry, W., 1931, The great Sequoia avalanche: Historical Series, Sequoia Nature Guide Service, Bulletin 8.

Lawson, A. C., 1904, Geomorphogeny of the upper Kern Basin, California: University of California Department of Geological Science, Bulletin, v. 3, pp. 291–386.

CHAPTER 11

Bateman, P. C., 1992, Plutonism in the central part of the Sierra Nevada Batholith, California: U.S. Geological Survey Professional Paper 1485, 186 pp.

Beanland, S., and M. Clark, 1994, The Owens Valley Fault Zone, eastern California, and surface faulting associated with the 1872 earthquake: U.S. Geological Survey Bulletin 1982, 29 pp.

Gilbert, G. K., 1904, Domes and dome structure of the High Sierra: *Geological Society of America Bulletin*, v. 15, pp. 29–36.

Goter, S. K., D. H. Oppenheimer, J. J. Mori, M. K. Savage, and R. P. Masse, 1994, Earthquakes in California and Nevada: U.S. Geological Survey Open File Report 94-647. [Map, scale 1:1,000,000.]

Hollett, K. J., W. R. Danskin, W. F. McCaffrey, and C. L. Walti, 1991, Geology and water resources of Owens Valley, California: U.S. Geological Survey Water-Supply Paper 2370, Chapter B, 77 pp.

Huber, N. K., 1981, Amount and timing of late Cenozoic uplift and tilt of the central Sierra Nevada, California—Evidence from the upper San Joaquin River Basin: U.S. Geological Survey Professional Paper 1197, 28 pp.

Kistler, R. W., 1990, Two different lithosphere types in the Sierra Nevada, California, *in* Anderson, J. L., ed., *The nature and origin of Cordilleran magmatism*: Geological Society of America Memoir 174, pp. 271–281.

Lockwood, J. P., and J. G. Moore, 1979, Regional deformation of the Sierra Nevada, California, on conjugate microfault sets: *Journal of Geophysical Research*, v. 84, pp. 6041–6049.

Muir, J., 1891, A rival of the Yosemite: *Century Magazine*, v. 43, Nov., pp. 77–97.

———, 1912, *The Yosemite*: The Century Co., New York, 37 pp.

Nokleberg, W. J., 1983, Wall rocks of the central Sierra Nevada Batholith, California: A collage of accreted tectono-stratigraphic terranes: U.S. Geological Survey Professional Paper 1255, 28 pp.

Wernicke, B., and others, 1996, Origin of high mountains in the continents: The southern Sierra Nevada: *Science*, v. 271, pp. 190–193.

Whitney, J. D., 1865, *Geology: Synopsis of the Field Work from 1860 to 1864*: Sacramento, Geological Survey of California, 498 pp.

Wolfe, J. A., H. E. Schorn, C. E. Forest, and P. Molnar, 1997, Paleobotanical evidence for high altitudes in Nevada during the Miocene: *Science*, v. 276, pp. 1672–1675.

APPENDIX

Alpha, T. R., 1977, Oblique map of Sequoia and Kings Canyon National Parks: Sequoia Natural History Association.

Busby-Spera, C. J., and J. Saleeby, 1987, Geologic guide to the Mineral King area, Sequoia National Park, California: Society of Economic Paleontologists and Mineralogists, v. 56, 44 pp.

Chen, J. H., and J. G. Moore, 1982, Uranium-lead isotopic ages from the Sierra Nevada Batholith, California: *Journal of Geophysical Research*, v. 87, pp. 4761–4784.

Evernden, J. F., and R. W. Kistler, 1970, Chronology of emplacement of Mesozoic batholithic complexes in California and western Nevada: U.S. Geological Survey Professional Paper 623, 42 pp.

Farquhar, F. P., 1965, *History of the Sierra Nevada*: Berkeley, University of California Press, 262 pp.

Jones, D. L., and J. G. Moore, 1973, Lower Jurassic ammonite from the south-central Sierra Nevada, California: *U.S. Geological Survey Journal of Research*, v. 1, pp. 453–458.

Liggett, D. L., 1990, Geochemistry of the garnet-bearing Tharps Peak granodiorite and its relation to other members of the Lake Kaweah Intrusive Suite, southwestern Sierra Nevada, California, *in* J. L. Anderson, ed., *The Nature and Origin of Cordilleran Magmatism*: Geological Society of America Memoir 174, pp. 225–236.

Moore, J. G., 1959, The quartz diorite boundary line in the western United States: *Journal of Geology*, v. 67, pp. 198–210.

Moore, J. G., and F. C. W. Dodge, 1980, Late Cenozoic volcanic rocks of the southern Sierra Nevada, California: *Geological Society of America Bulletin*, Part I, v. 91, pp. 515–518.

Moore, J. G., and E. du Bray, 1978, Mapped offset on the right-lateral Kern Canyon Fault, southern Sierra Nevada, California: *Geology*, v. 6, pp. 205–208.

Moore, J. G., W. J. Nokleberg, and T. W. Sisson, 1994, Geologic road guide to Kings Canyon and Sequoia National Parks, central Sierra Nevada, California: U.S. Geological Survey Open-File Report 94-650, 61 pp.

Muir, J., 1891, A rival of the Yosemite: The cañon of the South Fork of Kings River, California: *Century Magazine*, v. 43, Nov., pp. 77–97.

Nokleberg, W. J., 1983, Wall rocks of the central Sierra Nevada Batholith, California: A collage of accreted tectono-stratigraphic terranes: U.S. Geological Survey Professional Paper 1255, 28 pp.

Putman, G. W., and J. T. Alfors, 1965, Depth of intrusion and age of the Rocky Hill Stock, Tulare County, California: *Geological Society of America Bulletin*, v. 76, pp. 357–364.

Saleeby, J. B., 1978, Kings River Ophiolite, southwest Sierra Nevada foothills, California: *Geological Society of America Bulletin*, v. 89, pp. 617–636.

Saleeby, J. B., S. E. Goodin, W. D. Sharp, and C. J. Busby, 1978, Early Mesozoic paleotectonic-paleogeographic reconstruction of the southern Sierra Nevada region, *in* D. G. Howell and K. A. McDougall, eds., *Mesozoic Paleogeography of the Western United States*: Society of Economic Paleontologists and Mineralogists, Pacific Coast Section, Paleogeography Symposium 2, pp. 311–336.

Saleeby, J. B., R. W. Kistler, S. Longiaru, J. G. Moore, and W. J. Nokleberg, 1990, Middle Cretaceous silicic metavolcanic rocks in the Kings Canyon area, central Sierra Nevada, California, *in* J. L. Anderson, ed., *The Nature and Origin of Cordilleran Magmatism*: Geological Society of America Memoir 174, pp. 251–270.

Saleeby, J. B., and W. D. Sharp, 1980, Chronology of the structural and petrologic development of the southwest Sierra Nevada foothills, California: Summary: *Geological Society of America Bulletin*, Part I, v. 91, pp. 317–320.

Sharp, W. D., 1988. Pre-Cretaceous crustal evolution in the Sierra Nevada region, California, *in* W. G. Ernst, ed., *The Geotectonic Development of California*: Prentice-Hall, Englewood Cliffs, N.J., pp. 823–864.

Index

In this index an "f" after a number indicates a separate reference on the next page, and an "ff" indicates separate references on the next two pages. A continuous discussion over two or more pages is indicated by a span of page numbers, e.g., "57–59." *Passim* is used for a cluster of references in close but not consecutive sequence. Relevant figures and tables are referenced within the page spans in which they appear.

Accretionary lapilli, 196–97 (Fig. 5.7), 198
Alidade, 150–51 (Fig. 3.25), 155 (Fig. 3.28)
Arête, 300–301 (Fig. 9.11), 302
Asthenosphere, 184, 186 (Figs. 4.5–4.6), 319, 321–22 (Fig. 11.1), 326
Astronomical observation, for latitude and longitude, 104–11 (Figs. 3.3–3.5, Table 3.1), 145
Avalanche chute, 54–55 (Fig. 2.17), 302–3 (Fig. 9.13), 390

Ball Dome, 342, 345 (Fig. 11.18), 390
Barometer for altitude, 36, 47, 48–49 (Fig. 2.14), 66 (Fig. 2.23), 119, 122 (Fig. 3.12)
Basalt, 275–76 (Fig. 8.2), 277. *See also* Big Pine Volcanic Field; Cenozoic volcanic rocks; Golden Trout Creek Volcanic Field
Baseline measurement, 111, 122f
Basin and Range Province, 3 (Fig. 1.2), 322, 345
Batholith, 180, 209–12 *passim* (Fig. 6.2). *See also* Sierra Nevada Batholith

Bedrock mortars, 16–17 (Figs. 1.10–11)
Bergschrund, 64 (Fig. 2.22), 289–90, 290–91 (Fig. 9.4), 299; Muir on, 292
Big Pine Roof Pendant, 202
Big Pine Volcanic Field, 273–75, 277ff (Figs. 8.3–8.4), 326
Bishop Creek Roof Pendant, 191, 202, 271
Bishop Tuff, 277
Black Mountain Glacier, Muir on, 292
Blake, William, 30–34 (Fig. 2.7), 165, 351
Boyden Cave, 194–95 (Fig. 5.5), 202
Boyden Cave Roof Pendant, 191, 194–95 (Fig. 5.5), 198, 199–201 (Fig. 5.8), 202, 367f
Brewer, Mount, 51f (Fig. 2.16), 62, 365
Brewer, William, 40, 42, 49–51, 59, 62, 351f; and 1864 expedition of California Geological Survey, 5, 44–60, 65–66 (Fig. 2.23), 339, 365
Bullion, Mount, 45

California Division of Mines, 68

California Geological Survey, 21, 39, 41 (Fig. 2.10), 43 (Fig. 2.12), 50 (Fig. 2.15), 58, 67f; 1864 expedition of, 5, 42 (Fig. 2.11), 44–60, 65–66 (Fig. 2.23), 360
California–Nevada boundary, 104, 124–25
Cartridge Pass Pluton, 217 (Fig. 6.6), 233 (Fig. 6.17), 385f
Cave formation, 190–91 (Fig. 5.1), 202–6 (Figs. 5.11–5.12). *See also individual caves by name*
Cenozoic volcanic rocks, 273–83 (Fig. 8.5); maps of, 273, 274–75 (Fig. 8.1)
Central Valley (California), sediments of, 262, 343, 346
Chemical analyses, of granitic rock, 239, 244–49 (Tables 6.1–6.2)
Cirque, 292, 294 (Fig. 9.6), 300–301 (Fig. 9.11)
Cirque Peak, 56 (Fig. 2.18)
Clarence King, Mount, 60f (Fig. 2.20)
Cliff Mine, 63
Clyde, Norman (1885–1972), 11, 140–42 (Fig. 3.21)
Coast Range (California), 262, 315
Colby, William E., 140
Comb layering, 228–29 (Fig. 6.13), 232 (Fig. 6.16)
Copper and molybdenum deposits, 269–70
Corcoran, Mount, *see* Langley, Mount
Coso Volcanic Field, 275
Cotter, Mount, 60
Cotter, Richard, 45, 50 (Fig. 2.15), 51–55 (Fig. 2.17), 60, 99 (Fig. 3.1)
Crabtree, James, 267
Crystal Cave, 202, 205 (Fig. 5.12)
Crystallization of granitic rocks, 237–38, 246–49 (Fig. 6.21)

Dana, Mount, 40, 44, 140
Darwin, Mount, 74–75 (Fig. 2.26)
Davis, George R., 140, 154f (Fig. 3.28)
Day Needle Peak, 10 (Fig. 1.7), 335 (Fig. 11.8)
Differentiation, *see* Crystallization of granitic rocks
Dikes, 224–25 (Fig. 6.10), 233–34 (Fig. 6.17), 340 (Fig. 11.12). *See also* Independence Dike Swarm

Emerald Peak, 135
Empire Mine, 268–69 (Fig. 7.3)
Erosion, 288, 315, 343–47
Estimation of dark minerals, 239–41 (Figs. 6.19–6.20)
Exploration and mapping, history of and relationship between, 3–5, 47, 145, 349–53

Farallon Plate, 212–13 (Fig. 6.3), 259, 260–61 (Fig. 6.29)
Faults, 322, 326 (Fig. 11.4), 334. *See also individual faults by name*
Flower Lake, 117 (Fig. 3.10)
Fortieth Parallel, *see* U.S. Geological Exploration
Fort Independence, 20, 37
Fossils, 199–202 (Fig. 5.8), 201 (Fig. 5.9)
Fowler, Thomas, and Empire Mine, 268–69 (Fig. 7.3)
Fremont, John C.: biography, 23f (Fig. 2.1), 29, 351ff; expeditions, 23–27 (Fig. 2.2), 29; 1850 map by, 27–29 (Fig. 2.3), 104; and California Republic, 27

Gardiner, James, 44f, 60f (Fig. 2.20), 98f (Fig. 3.1), 351
Gardiner, Mount, 60f (Fig. 2.20)
Gardner, James, *see* Gardiner, James
Garlock Fault, 257
Garnet Dike Mine, 271
General Grant Grove, 93, 365
General Grant National Park, 92–93 (Fig. 2.39), 95
General Land Office, 20, 90, 122
General Sherman tree, 375
Geologic column, *see* Geologic time scale
Geologic history of Sierra Nevada, 6–7, 8 (Fig. 1.5)
Geologic investigation, future of, 353–55
Geologic mapping, 165–69
Geologic time scale, 171–75 (Fig. 4.1)
Giant Forest, 20, 82, 86f
Giant Forest Granodiorite, 229, 231 (Fig. 6.15), 374ff, 377, 389
Gilbert, Grove K., 113 (Fig. 3.7), 139, 330, 340
Glaciation and glaciers: 71, 165, 288f, 300–301 (Fig. 9.11), 304f (Fig. 9.14), 307f (Figs. 9.17–9.18); John Muir on,

285–88, 289; during ice ages, 286–87 (Fig. 9.2), 289f (Fig. 9.3); timing and causes of, 310–14. *See also* Ice Ages; Moraines; *and individual glaciers by name*
Global positioning system (GPS), 353–54
Goddard, George H.: biography, 34–37 (Fig. 2.8), 51
Goddard, Mount, 51, 63–65 (Fig. 2.22), 135, 365
Goddard Terrane, 190–91 (Fig. 5.1), 193, 198–99, 201–2, 207, 237, 251
Gold, 263, 265–66
Golden Trout Creek (formerly Volcano Creek), 90–91 (Fig. 2.38), 165, 273, 282, 287, 283
Golden Trout Creek Volcanic Field, 280–82 (Figs. 8.6–8.7)
Goodyear, Watson A. (geologist), 11, 43 (Fig. 2.12), 67, 118–19
Grand Sentinel, Kings Canyon, 78–79 (Fig. 2.29), 86 (Fig. 2.35), 127, 270, 369
Granite, 66, 181, 242–44, 246–49
Granitic rocks, 209–62; classification of, 178 (Fig. 4.2), 180–84 (Fig. 4.3); minerals in, 178–79; layering in, 221–24 (Figs. 6.8–6.9); dikes in, 224–25 (Fig. 6.10); mafic inclusions in, 225–27 (Figs. 6.11–6.12); comb layering in, 228–29 (Fig. 6.13), 232 (Fig. 6.16); orbicules, 229–32 (Figs. 6.14–16); analyses of, 239–249; changes in composition across Sierra, 249–54
Granitization, 66, 220
Granodiorite, 183f, 210–11 (Fig. 6.2), 249, 254. *See also* Granitic rocks
Gravity maps, 238–39
Great Western Divide, 8, 45f (Fig. 2.13), 51, 65, 292–93 (Fig. 9.5)
Groundhog Cone, 281f (Fig. 8.7)

Hanging valleys, 300–301 (Fig. 9.11), 305
High Sierra Terrane, 190–91 (Fig. 5.1), 193, 198–99, 202, 207–8, 237, 251
High Sierra Trail, 157; introduction to, 388–89; log of, 389–92
Hockett Trail, 26–27 (Fig. 2.2), 38, 58

Hoffmann, Charles F., 40, 43 (Fig. 2.12), 50 (Fig. 2.15), 67, 73, 100
Hopson, Cliff, and Independence Dike Swarm, 256
Horns, 300–301 (Fig. 9.11), 302
Hospital Rock, Kaweah River, 17, 316, 374
Hutchings, James, 82–83, 84
Hutchings, Mount, 84

Ice ages, 8, 289, 311
Independence Dike Swarm, 234, 256–59 (Figs. 6.26–6.28); trail guide to, 383f
Independence Fault, 1 (Fig. 1.1), 324–25 (Fig. 11.2), 326
Index fossils, 173. *See also* Fossils
Indian reservations, 12–13 (Fig. 1.8), 20
Indians, *see* Native Americans
Intrusion of granitic melts, 7, 220–21
Inyo–White Mountains Fault Zone, 326, 331
Irelan, William (mineralogist), 1891 map by, 88–89 (Fig. 2.37), 93, 101, 125, 127, 165

Jenkins, Olaf P., 68
John Muir Trail, 137, 139, 157, 180; guide to, 380–88
Joints, 322, 334–44 (Figs. 11.8–11.17)
Jordan, Mount, 53
Jordan Toll Trail, 38, 371–72

Kaweah, Mount, 67
Kaweah–Kern Divide, 58
Kaweah River, 8, 37f
Kearsarge District, 63, 115, 265–67 (Fig. 7.2)
Kearsarge Pass, 37–38, 60–63 (Fig. 2.21)
Keeler Needle Peak, 10 (Fig. 1.7), 335 (Fig. 11.8)
Keough, Thomas (prospector), 60–63
Kern Canyon Fault, 331–33; glacial canyon along, 301, 306–7 (Fig. 9.16), 332–33 (Fig. 11.7); in parks area, 323, 324–25 (Fig. 11.2), 331–33
Kern Lake, 317 (Fig. 10.1)
Kern River, 37, 51, 53 (Fig. 2.8)
Kern River Drainage, 53
King, Clarence (1842–1901), 44f, 47–49, 51, 53–57, 102, 143f (Fig. 3.22), 351;

exploration of 40th Parallel, 99
(Fig. 3.1), 102–3 (Fig. 3.2), 104;
Mount Whitney attempts, 100–101,
118–20
King, Mount, see Clarence King, Mount
Kings Canyon, 9 (Fig. 1.6), 59–60
(Fig. 2.19), 78–79 (Fig. 2.29), 85
Kings Canyon Highway (SR180): guide
to, 358–62 (Figs. A.1–A.4); log of,
363–69
Kings Canyon National Park, 3, 95,
263; maps of, 4–5 (Fig. 1.3), 92–93
(Fig. 2.39), 214–15 (Fig. 6.4); and
Muir, 71, 72–73 (Fig. 2.25)
Kings–Kaweah Divide, 45, 365
Kings–Kaweah Terrane, 190–91 (Fig. 5.1),
193–95, 199, 206–7, 237, 251, 363
Kings–Kern Divide, 53, 57
Kings River, 8, 22, 29, 51, 286 (Fig. 9.1),
288, 363
Kings Terrane, 190–91 (Fig. 5.1), 193–96,
199–201, 202, 237, 251

Lake Kaweah Roof Pendant, 190–91
(Fig. 5.1)
Lamark Granodiorite Pluton, 233
(Fig. 6.17), 385
Landslides, 315–18 (Fig. 10.1)
Langley, Mount, 11 (Table 1.1), 43
(Fig. 2.12), 56 (Fig. 2.18), 67, 118–19
Langley, Samuel, 88, 90
Latitude measurement, and zenith telescope, 104–7 (Fig. 3.3)
Lawson, Andrew, 165
Layering, 221–24 (Fig. 6.9), 228–29
(Fig. 6.13)
LeConte, Joseph N. (1870–1950), 128–30,
131 (Fig. 3.16), 139, 269–70, 351ff;
exploration by, 6, 129 (Fig. 3.15),
136–37 (Fig. 3.19), 156–57 (Fig. 3.29);
maps of, 122, 131f (Fig. 3.17), 136–39
(Figs. 3.19–3.20)
Leveling, 145, 146–47 (Fig. 3.24),
148–50
Lilburn Cave, 203f (Figs. 5.10–5.11)
Lithosphere, 184, 187 (Figs. 4.5–4.6),
319–20, 321 (Fig. 11.1)
Little Ice Age, 290–91 (Fig. 9.4), 311
Little Whitney Cone, 280
Lone Pine earthquake, 329–31

Lone Pine Fault, 322f, 324–25 (Fig. 11.2),
329
Longitude measurement, 105f (Figs. 3.3–
3.4), 107–11 (Table 3.1, Fig. 3.5)
Long Valley Volcanic Field, 273

Mafic complex, Onion Valley, 254f
(Fig. 6.25), 384
Mafic inclusions, 225–27 (Figs. 6.11–6.12)
Mapping, see Exploration and mapping;
Geologic mapping, Topographic mapping
Maps: relief of parts of California, 2–3
(Fig. 1.2); California Geological
Survey, 5; U.S. Army (Wheeler), 6,
111–12, 114–15 (Fig. 3.8), 116–17
(Fig. 3.9); Sierra Club, 6; U.S.
Geological Survey, 6, 55, 144–61
(Figs. 3.29–3.31); Fremont 1850
map, 27–29 (Fig. 2.3), 104; Blake
1855 Pacific Railroad Survey, 32–33
(Fig. 2.7), 34, 165; Goddard map,
34–35 (Fig. 2.8), 36–37; Hoffmann
1873 map, 50 (Fig. 2.15), 67; Jenkins
1938 map, 68; Muir 1891 map, 78–79
(Fig. 2.29), 84, 126–27 (Fig. 3.13),
LeConte 1896 map, 84, 136–37
(Fig. 3.19), 156–57 (Fig. 3.29); Irelan
1891 map, 88–89 (Fig. 2.37), 93,
127, 165; Wright 1883 map, 90–91
(Fig. 2.38), 127; King 1876 atlas
of 40th Parallel Survey, 99, 102–3
(Fig. 3.2), 104
Marshall, Robert B. (topographer), 154
(Fig. 3.27)
Mather, Stephen (director, National Park
Service), 94
Matthes, Francois, 166, 311
Mendel, Mount, 73, 74–75 (Fig. 2.26)
Metamorphic rocks, 63, 65, 189–208
(Fig. 5.1), 210 (Fig. 6.1)
Metamorphism, 189
Middle Palisade, 11
Milankovitch theory, 311–14
Mineral deposits, 263–72
Mineral King (District), 38, 94, 265, 267f
Mineral King Road Guide, 377–78; log
of, 378–79
Mineral King Roof Pendant, 190–91
(Fig. 5.1), 191, 198, 201, 267

Mineral King Wagon and Toll Road
    Company, 268
Minerals, 176–79 (Table 4.2, Fig. 4.2),
    181–84
Mines, 38, 265. *See also* Mineral King
    (District); *and individual mines by name*
Missionaries, 21f
Mitchell Intrusive Suite, 389
Modal analyses, 239, 242–43
Mohorovicic Discontinuity (Moho),
    320–21
Monache tribal group, 13; 15–17
    (Figs. 1.10–11)
Moore, James G., 166–67 (Fig. 3.33),
    168–69 (Table 3.3), 210–11
    (Fig. 6–2), 256, 261
Moraines, 47–49, 64 (Fig. 2.22), 74–75
    (Fig. 2.26), 294–99 *passim* (Figs.
    9.8–9.10), 302 (Fig. 9.12), 305–10
    (Figs. 9.19–20)
Moro Rock, 342, 344 (Fig. 11.17), 373f
Mount Goddard Roof Pendant, 190–91
    (Fig. 5.1), 198; trail guide to, 387f
Mount Whitney Military Reservation, 88,
    89–90 (Fig. 2.37), 125
Muir, John (1838–1914), 1, 37, 71–73, 77,
    88, 343; as founder of Sierra Club, 6,
    69 (Fig. 2.24), 128 (Fig. 3.14); and
    exploration of Sierra, 21, 75–77, 81
    (Fig. 2.31), 82 (Fig. 2.32), 86–87, 283,
    288; biography of, 68–70, 351f; on
    glaciation, 70–71, 77, 81 (Fig. 2.31),
    84–85, 285–88, 289; on vertical scale,
    76–78 (Figs. 2.27–2.28), 83–87 *passim*
    (Figs. 2.33–2.34, 2.36); on Kings
    River, 78–79 (Fig. 2.29), 285–88; 1891
    map by, 84, 125, 126–27 (Fig. 3.13)
Muir, Mount, 11

Na$_2$O in Owens Lake Basin, 313–14
    (Fig. 9.21)
Naming of places, 40–42, 161–65
    (Fig. 3.32)
Native Americans, 12–13 (Fig. 1.8), 17–20
    (Fig. 1.12)
New England Tunnel and Smelting
    Company, 268
North American Continental Plate, 187f,
    212–13 (Fig. 6.3), 259–60, 260–61
    (Fig. 6.29)

Oak Creek Roof Pendant, 198, 266, 383
O'Farrell, John (Harry Parole), 38, 267
Olancha Peak, 5 (Fig. 1.4), 56 (Fig. 2.18),
    90–91 (Fig. 2.38)
Onion Valley, *see* Mafic complex, Onion
    Valley
Ophiolite, 195f (Fig. 5.6)
Orbicules, 229–32 (Figs. 6.14–6.16)
Owens Lake, 8, 45
Owens River, 8, 25
Owens Valley Fault Zone, 323, 325
    (Fig. 11.3), 327 (Fig. 11.5), 329–31
Owens Valley Paiute Indians, *see* Paiute
    tribal group
Oxygen isotope analysis, 312–14

Pacific Railroad Surveys, 30–34
    (Figs. 2.5–2.7)
Paiute tribal group, 13–14 (Fig. 1.9), 20
Palisades Glacier, 290–91 (Fig. 9.4), 292
Palmer, Joseph, 318
Paradise Granodiorite, 218–19 (Fig. 6.7),
    381ff, 391f
Paradise Peak, 87 (Fig. 2.36)
Passes in Sierra Nevada, 3, 5 (Fig. 1.4);
    Kearsarge, 13, 62 (Fig. 2.21), 37,
    60–63; Walker, 22–23, 25, 67; Carson,
    25, 36; Donner, 25, 34, 84; Tehachapi,
    25, 34; Tejon, 34; Sonora, 36f; Olancha,
    38, 372; Longley, 53, 58; Millys Foot,
    58; Coyote, 58; Little Pine, 63; Hell-
    for-Sure, 65, 258 (Fig. 6.28); Silver,
    135; Muir, 135, 139f, 155 (Fig. 3.28);
    Granite, 140; Taboose, 257 (Fig. 6.27),
    328–29 (Fig. 11.6), 386; Sawmill,
    328–29 (Fig. 11.6); Foresters, 380ff;
    Mather, 380, 385f; Pinchot, 385
Peaks, *see* Sierra peaks; *and individual peaks
    by name*
Petrochemistry, 245
Petroglyphs, Owens Valley, 17–18
Pictographs, Sierra Nevada, 17f (Fig. 1.12)
Pine Creek Pendant, 199
Place names, *see* Naming of places
Plane table, 150–51 (Fig. 3.25)
Plate tectonics, 184–88 (Figs. 4.4–4.6),
    212–13 (Fig. 6.3), 259–62 (Fig. 6.29),
    319–22 (Fig. 11.1); subduction of Far-
    allon Plate, 212–13 (Fig. 6.3), 259–60,
    260–61 (Fig. 6.29)

Plutons, 180, 209–19 (Fig. 6.5), 232–34, 237–39
Point-counting, 183–84. *See also* Modal analyses
Principal meridian measurement, 122

Quartz diorite line, 210–11 (Fig. 6.2), 261, 364

Rabe, Carl, 43 (Fig. 2.12), 119, 122 (Fig. 3.12)
Radiometric dating, 174–76 (Table 4.1), 234–37 (Fig. 6.18)
Redwood Mountain Roof Pendant, 196–97 (Fig. 5.7)
Rhyolite, 277, 282
River of the Holy Kings, *see* Kings River
River systems of the Sierra, 8
Rock, 179–80. *See also* Granitic rock
Rock glacier, 54–55 (Fig. 2.17), 290 (Fig. 9.3), 292–94 (Fig. 9.5–9.7)
Roof pendants, 190–91 (Fig. 5.1), 193 (Fig. 5.3), 263. *See also individual pendants by name*
Russell, Mount, 11

San Joaquin River Drainage Basin, 347
Scheelite, 270f
Sediments, 6–7
Seismic experiment, Sierra (1995), 320–22
Sequoia Intrusive Suite, 377, 389
Sequoia National Park, 3, 8, 20, 26–27 (Fig. 2.2), 38, 263; maps of, 4–5 (Fig. 1.3), 92–93 (Fig. 2.39), 126–27 (Fig. 3.13), 214–15 (Fig. 6.4); and Muir, 71, 72–73 (Fig. 2.25); creation and enlargement of, 90–95 (Fig. 2.39), 125, 373
Sequoia Roof Pendant, 191, 193 (Fig. 5.3), 194–95 (Fig. 5.4), 205 (Fig. 5.12), 374
Sheep Mountain, *see* Langley, Mount
Sheep Ridge Roof Pendant, 191
Sierra Club, 6, 128–29 (Fig. 3.14), 131
Sierra Nevada, 1, 2–3 (Fig. 1.2), 6–7, 97f, 209, 288
Sierra Nevada Batholith, 180, 184, 188, 208, 210 (Fig. 6.1), 219–21; and other batholiths, 209–12 (Fig. 6.2); radiometric age of, 234–37 (Fig. 6.18); depth of, 237–39; variation within, 238, 249–53 *passim* (Figs. 6.22–6.24); strontium isotope ratios in, 250–51 (Fig. 6.22), 251–54. *See also* Plate tectonics; Uplift of Sierra Nevada (Block)
Sierra Nevada Block, 322
Sierra Nevada Crest, 5 (Fig. 1.4), 8f, 45, 321, 321–22 (Fig. 11.1)
Sierra Nevada Frontal Fault (Zone), 316, 323, 324–29 (Figs. 11.2, 11.5–11.6)
Sierra peaks, 5 (Fig. 1.4); greater than 14,000 ft., 3, 4–5 (Fig. 1.3), 9–13 (Table 1.1)
Silliman, Benjaman, Jr., 45
Silliman, Mount, 46 (Fig. 2.13), 365
Smith, Jedediah (fur trader and explorer), 22
Solomons, Theodore S., 133–37 (Fig. 3.18)
South Fork Cone, 280
South Guard Peak, 53
Specific gravity, 217 (Fig. 6.6), 238–39, 241–42 (Fig. 6.20)
Starr, Walter, Jr.: on Palisade group, 13
State Route 198 and Generals Highway, 370–71; log of, 371–77
Steno, Nicolaus (1638–1686), 171, 173
Stewart, George W., 374
Stewart, Mount, 374
Stony Flat Lava, 275, 364
Stoping, 220
Strontium isotope ratios, in Sierra Nevada, 250–54 (Fig. 6.22)
Structural geology, 319
Subduction, *see* Plate tectonics; Sierra Nevada Batholith
Surveying, 122–24

Tahoe, Lake, 124
Tahoe and Tioga glaciations, 295–99, 313–14 (Fig. 9.21); moraines of, 280, 296–99 (Figs. 9.9–9.10), 375
Tehipite Dome, Middle Fork, Kings River, 84, 137, 343, 346 (Fig. 11.19)
Tehipite Valley, 18 (Fig. 1.12), 73, 76–77 (Fig. 2.27), 84, 366
Terranes, 192–93, 206–8. *See also individual terranes by name*
Tharp, Hale, 19–20, 82, 374
Third Needle Peak, 10 (Fig. 1.7), 335 (Fig. 11.8)

Till, 294
Tilt, *see* Uplift of Sierra Nevada (Block)
Tioga Glaciation, 286–87 (Fig. 9.2), 290 (Fig. 9.3). *See also* Tahoe and Tioga glaciations
Tonalite (formerly quartz diorite), 210–11 (Fig. 6.2), 249
Topographic mapping, 144–53 (Fig. 3.25, Table 3.2)
Travertine hot spring deposit, 282–83
Triangulation, 138–39 (Fig. 3.20), 145–48 (Fig. 3.23)
Tungsten mines, 199, 264–65 (Fig. 7.1), 270–72
Tyndall, Mount, iiff (Frontis.), 51, 53, 54–55 (Fig. 2.17), 365

Uplift of Sierra Nevada (Block), 7f (Fig. 1.5), 322, 326, 343–47
U.S. Army, 6, 21, 93–94, 98, 122; Corps of Engineers, 23, 30
U.S. Board of Geographic Names, 162
U.S. General Land Office, *see* General Land Office
U.S. Geological Exploration of 40th Parallel, 99, 101–4 (Fig. 3.2)
U.S. Geological Survey, 6, 143; topographic maps of, 144–53 (Fig. 3.26); 30 minute maps of, 145, 153–59 (Table 3.2, Fig. 3.27, Fig. 3.30); 15 minute maps of, 157–59, 160–61 (Fig. 3.31); 7.5 minute maps of, 159–61 (Fig. 3.31); and geologic mapping, 165–69

Volcanic rocks, *see* Cenozoic volcanic rocks
Volcanoes, 187–88

Von Schmidt, Allexey W. (surveyor), 124f

Walker, Joseph (fur trader and scout), 22–23, 25
Werner, Abraham Gottlob (1749–1817), 171
Wheeler, George, 6, 111–15 (Figs. 3.6, 3.8), 116–17 (Fig. 3.9), 120–22
Whitney, Josiah Dwight, 39–42 (Fig. 2.9), 44, 58, 60, 71, 285, 330, 351f
Whitney, Mount: named, ii (Frontis.), 43 (Fig. 2.12), 50f (Fig. 2.15), 365; identification of, 1 (Fig. 1.1), 6, 9, 43 (Fig. 2.12), 56 (Fig. 2.18), 67, 100, 118–19; views of and from, 10 (Fig. 1.7), 80–81 (Fig. 2.30), 335 (Fig. 11.8), 380f; expeditions to, 56 (Fig. 2.18), 58–59, 88, 116–17 (Fig. 3.9), 118–20; surveying of, 120–22 (Fig. 3.11), 149–50; altitude of, 122 (Fig. 3.12), 160–61 (Fig. 3.31)
Whitney Granodiorite, 218–19 (Fig. 6.7), 381, 392
Whitney Intrusive Suite, 57, 218–19 (Fig. 6.7), 381, 389, 391
Whitney Survey (1864), 365, 370
Williamson, Mount, 56, 365
Williamson, Robert S., 30–34 (Fig. 2.5), 56
Wim-mel-che tribal group, 22
Woodville, California, 18, 35 (Fig. 2.8)
Wright, J. W. A., 90–91 (Fig. 2.38), 127, 165, 273

Yokohl Valley Pendant, 190–91 (Fig. 5.1), 199
Yosemite National Park, 93, 98, 125, 373
Yosemite Valley, 23, 71, 285–88

Library of Congress Cataloging-in-Publication Data

Moore, James G.
    Exploring the highest sierra / James G. Moore.
       p.   cm.
    Includes bibliographical references (p.) and index.
    ISBN 0-8047-3647-2 (alk. paper). — ISBN 0-8047-3703-7 (pbk. : alk. paper)
    1. Sierra Nevada (Calif. and Nev.)—Discovery and exploration.
2. Sierra Nevada (Calif. and Nev.) Surveys. 3. Cartography—Sierra Nevada
(Calif. and Nev.) 4. Geology—Sierra Nevada (Calif. and Nev.) I. Title.
F868.S5M66 2000
917.94'4—dc21   99-16229

∞ This book is printed on acid-free, recycled paper.

Original printing 2000

Last figure below indicates year of this printing:
09 08 07 06 05 04 03 02 01 00

This book was designed by Janet Wood and
typeset in 11/15 Bembo by James P. Brommer